The Biological and Clinical Aspects of Merkel Cell Carcinoma

The Biological and Clinical Aspects of Merkel Cell Carcinoma

Editors

Virve Koljonen
Weng-Onn Lui
Jürgen C. Becker

MDPI • Basel • Beijing • Wuhan • Barcelona • Belgrade • Manchester • Tokyo • Cluj • Tianjin

Editors
Virve Koljonen
University of Helsinki and
Helsinki University Hospital
Finland

Weng-Onn Lui
Karolinska University
Hospital
Sweden

Jürgen C. Becker
University of Duisburg-Essen
Germany

Editorial Office
MDPI
St. Alban-Anlage 66
4052 Basel, Switzerland

This is a reprint of articles from the Special Issue published online in the open access journal *Cancers* (ISSN 2072-6694) (available at: https://www.mdpi.com/journal/cancers/special_issues/MCC_Cancer).

For citation purposes, cite each article independently as indicated on the article page online and as indicated below:

LastName, A.A.; LastName, B.B.; LastName, C.C. Article Title. *Journal Name* **Year**, *Volume Number*, Page Range.

ISBN 978-3-0365-7446-2 (Hbk)
ISBN 978-3-0365-7447-9 (PDF)

Contents

cancers

Editorial

New Insights into the Biological and Clinical Aspects of Merkel Cell Carcinoma

Virve Koljonen [1,*], Weng-Onn Lui [2,*] and Jürgen C. Becker [3,4,*]

1 Department of Plastic Surgery, University of Helsinki and Helsinki University Hospital, 00029 Helsinki, Finland
2 Department of Oncology-Pathology, Karolinska Institutet, BioClinicum, Karolinska University Hospital, 171 64 Solna, Sweden
3 Translational Skin Cancer Research, German Cancer Consortium (DKTK), 45151 Essen, Germany
4 German Cancer Research Center (DKFZ), 69120 Heidelberg, Germany
* Correspondence: virve.koljonen@hus.fi (V.K.); weng-onn.lui@ki.se (W.-O.L.); j.becker@dkfz-heidelberg.de (J.C.B.)

The Special Issue in *Cancers*, "The Biological and Clinical Aspects of Merkel Cell Carcinoma", walks the avid reader through the interesting and sometimes even mysterious facets of Merkel cell carcinoma (MCC), starting at its carcinogenesis to also cover innovative treatment options.

The groundworks for MCC and its causative agent Merkel cell polyomavirus (MCPyV) are laid in an exhaustive review by Pietropaolo et al. [1]. They provide a comprehensive review of the current knowledge and spell out the undisputed role of MCPyV in oncogenesis in viral-associated MCC. Further and current evidence for the MCPyV oncogenic functions is provided by Spurgeon et al. [2]. With murine skin cancer model they show that MCPyV T antigens function in tumor promotion but not in initiation. Leaving one of the most enigmatic questions in MCC, open and thus vacant for future research.

To date, the effective treatment options for advanced MCC are still limited. In this issue, an interesting article by Sarma et al. [3] tested the effect of artesunate, an anti-malaria compound listed in the World Health Organization essential medicines [4], on MCC. They show that artesunate represses MCPyV T-antigen expression and inhibits cell growth in vitro and in vivo, suggesting its potential treatment for MCC. Fan et al. [5] concede that miR-375 is unlikely an intracellular oncogene in MCC cells and thus may rather serve for intercellular communication; indeed they subsequently published that miR-375 is functional in polarizing cancer associated fibroblasts [6]. Kervarrec et al. [7] take on the complexity of cell of origin in MCC, in which they conclude that MCPyV T antigens contribute to the acquisition of Merkel cell-like phenotype in epithelial cells.

Turning to clinical patient care, Sahi et al. [8], portray a grim picture on real life experience on the treatment of MCC patients. Although limited to Finland, it is presumed that similar situation is a common and worldwide problem, not only with MCC patients, but rather in all patients with rare cancers. Björn Andtback et al. [9] review on their past experience on adjuvant radiation therapy in MCC, strengthening the previous notion that female MCC patients, regardless of MCPyV status, actually do better compared with their male counterparts. The third clinical paper by Naseri et al. [10] summarize the consensus treatment recommendations by the Danish MCC expert group. A second paper by Naseri et al. [11] described prognostic markers which hold the potential to stratify MCC patients for different treatment regimens.

Rare cancers pose a major challenge to the medical and scientific community [12]. Due to low patient numbers and thus limited market potentials, development and testing innovative therapeutic intervention is not prioritized by the pharmacological industry. Indeed, less common cancer subtypes and rare cancers are frequently only included in basket trials among several different entities, which may leave less attention to differing

Citation: Koljonen, V.; Lui, W.-O.; Becker, J.C. New Insights into the Biological and Clinical Aspects of Merkel Cell Carcinoma. *Cancers* **2021**, *13*, 2259. https://doi.org/10.3390/cancers13092259

Received: 22 April 2021
Accepted: 30 April 2021
Published: 8 May 2021

responses. Consequently, the lack of good and well-established treatments and clinical practices produces varying treatments and varying results. Furthermore, patients with rare cancers are worse off than other cancer patients [12], because these are often not diagnosed in a timely or correct manner. False diagnoses are more often than other patients or the correct diagnosis is delayed; in either case allowing the disease to progress before adequate therapy is initiated; thus, response to treatment is not as good as it could be. Getting peer support is often overwhelming.

It is often thought that the rarity of a specific cancer, such as MCC, causes patients to being "under-diagnosed" and to receive "under-treatment", which is both unfortunately true. Published data on rare cancer are frequently based on a few patient cases or minor series with inadequate reporting [13], results that are not generalizable and it is difficult to establish a cause-and-effect relationship [14]. For example, due to the reporting bias for "successfully treated cases", chemotherapy for MCC may have been advocated longer that it was reasonable [15]. Both clinicians and journal editors should keep this notions in mind [14].

Note added in Proof:

After preparation of this editorial, two additional manuscripts were accepted. Horny et al. [16] revealed mutational landscape of virus-positive and –negative MCC cell lines that is comparable to tumor samples, suggesting their utility as preclinical models for functional studies. Hill et al. [17] suggested three subgroups of MCC based on genomic copy number variants.

Funding: This research received no external funding.

Conflicts of Interest: J.C.B. is receiving speaker's bureau honoraria from Amgen, Pfizer, Merck-Serono, Recordati and Sanofi, is a paid consultant/advisory board member/DSMB member for Almirall, Boehringer Ingelheim, InProTher, ICON, MerckSerono, Pfizer, 4SC, and Sanofi/Regeneron. His group receives research grants from Bristol-Myers Squibb, Merck Serono, HTG, IQVIA, and Alcedis. The other authors declare no conflict of interest.

References

1. Pietropaolo, V.; Prezioso, C.; Moens, U. Merkel cell polyomavirus and Merkel cell carcinoma. *Cancers* **2020**, *12*, 1774. [CrossRef] [PubMed]
2. Spurgeon, M.E.; Liem, A.; Buehler, D.; Cheng, J.; DeCaprio, J.A.; Lambert, P.F. The Merkel Cell Polyomavirus T Antigens Function as Tumor Promoters in Murine Skin. *Cancers* **2021**, *13*, 222. [CrossRef] [PubMed]
3. Sarma, B.; Willmes, C.; Angerer, L.; Adam, C.; Becker, J.C.; Kervarrec, T.; Schrama, D.; Houben, R. Artesunate Affects T Antigen expression and survival of virus-positive Merkel cell carcinoma. *Cancers* **2020**, *12*, 919. [CrossRef]
4. WHO. World Health Organization Model List of Essential Medicines: 21st List 2019; World Health Organization. 2019. Available online: https://apps.who.int/iris/handle/10665/325771 (accessed on 1 February 2021).
5. Fan, K.; Zebisch, A.; Horny, K.; Schrama, D.; Becker, J.C. Highly expressed miR-375 is not an intracellular oncogene in Merkel cell polyomavirus-associated Merkel cell carcinoma. *Cancers* **2020**, *12*, 529. [CrossRef]
6. Fan, K.; Spassova, I.; Gravemeyer, J.; Ritter, C.; Horny, K.; Lange, A.; Gambichler, T.; Odum, N.; Schrama, D.; Schadendorf, D.; et al. Merkel cell carcinoma-derived exosome-shuttle miR-375 induces fibroblast polarization by inhibition of RBPJ and p53. *Oncogene* **2020**, *40*, 980–996. [CrossRef]
7. Kervarrec, T.; Samimi, M.; Hesbacher, S.; Berthon, P.; Wobser, M.; Sallot, A.; Sarma, B.; Schweinitzer, S.; Gandon, T.; Destrieux, C. Merkel cell polyomavirus T antigens induce Merkel cell-like differentiation in GLI1-expressing epithelial cells. *Cancers* **2020**, *12*, 1989. [CrossRef] [PubMed]
8. Sahi, H.; Their, J.; Gissler, M.; Koljonen, V. Merkel Cell Carcinoma Treatment in Finland in 1986–2016-A Real-World Data Study. *Cancers* **2020**, *12*, 1224. [CrossRef]
9. Björn Andtback, H.; Björnhagen-Säfwenberg, V.; Shi, H.; Lui, W.-O.; Masucci, G.V.; Villabona, L. Sex Differences in Overall Survival and the Effect of Radiotherapy in Merkel Cell Carcinoma—A Retrospective Analysis of a Swedish Cohort. *Cancers* **2021**, *13*, 265. [CrossRef] [PubMed]
10. Naseri, S.; Steiniche, T.; Ladekarl, M.; Bønnelykke-Behrndtz, M.L.; Hölmich, L.R.; Langer, S.W.; Venzo, A.; Tabaksblat, E.; Klausen, S.; Skaarup Larsen, M. Management Recommendations for Merkel Cell Carcinoma—A Danish Perspective. *Cancers* **2020**, *12*, 554. [CrossRef] [PubMed]

11. Naseri, S.; Steiniche, T.; Georgsen, J.B.; Thomsen, R.; Ladekarl, M.; Heje, M.; Damsgaard, T.E.; Bonnelykke-Behrndtz, M.L. Tumor Ulceration, Reduced Infiltration of CD8-Lymphocytes, High Neutrophil-to-CD8-Lymphocyte Ratio and Absence of MC Virus are Negative Prognostic Markers for Patients with Merkel Cell Carcinoma. *Cancers* **2020**, *12*, 888. [CrossRef] [PubMed]
12. Gatta, G.; Capocaccia, R.; Botta, L.; Mallone, S.; De Angelis, R.; Ardanaz, E.; Comber, H.; Dimitrova, N.; Leinonen, M.K.; Siesling, S.; et al. Burden and centralised treatment in Europe of rare tumours: Results of RARECAREnet-a population-based study. *Lancet Oncol.* **2017**, *18*, 1022–1039. [CrossRef]
13. Kaszkin-Bettag, M.; Hildebrandt, W. Case reports on cancer therapies: The urgent need to improve the reporting quality. *Glob. Adv. Health Med.* **2012**, *1*, 8–10. [CrossRef] [PubMed]
14. Nissen, T.; Wynn, R. The clinical case report: A review of its merits and limitations. *BMC Res. Notes* **2014**, *7*, 264. [CrossRef]
15. Nghiem, P.; Kaufman, H.L.; Bharmal, M.; Mahnke, L.; Phatak, H.; Becker, J.C. Systematic literature review of efficacy, safety and tolerability outcomes of chemotherapy regimens in patients with metastatic Merkel cell carcinoma. *Future Oncol.* **2017**, *13*, 1263–1279. [CrossRef] [PubMed]
16. Horny, K.; Gerhardt, P.; Hebel-Cherouny, A.; Wülbeck, C.; Utikal, J.; Becker, J.C. Mutational Landscape of Virus- and UV-Associated Merkel Cell Carcinoma Cell Lines Is Comparable to Tumor Tissue. *Cancers* **2021**, *13*, 649. [CrossRef] [PubMed]
17. Hill, N.T.; Kim, D.; Busam, K.J.; Chu, E.Y.; Green, C.; Brownell, I. Distinct Signatures of Genomic Copy Number Variants Define Subgroups of Merkel Cell Carcinoma Tumors. *Cancers* **2021**, *13*, 1134. [CrossRef] [PubMed]

cancers

MDPI

Article

Distinct Signatures of Genomic Copy Number Variants Define Subgroups of Merkel Cell Carcinoma Tumors

Natasha T. Hill [1], David Kim [2], Klaus J. Busam [3], Emily Y. Chu [4], Clayton Green [2] and Isaac Brownell [1,*]

1 Dermatology Branch, National Cancer Institute, NIH, Bethesda, MD 20892, USA; natasha.hill@nih.gov
2 Marshfield Center, Marshfield, WI 02050, USA; kim.seung@marshfieldclinic.org (D.K.); clayton_green@urmc.rochester.edu (C.G.)
3 Department of Pathology, Memorial Sloan Kettering Cancer Center, New York, NY 10065, USA; busamk@MSKCC.ORG
4 Deprtment of Dermatology, Perelman School of Medicine at the University of Pennsylvania, Philadelphia, PA 19104, USA; Emily.Chu@uphs.upenn.edu
* Correspondence: isaac.brownell@nih.gov

Simple Summary: Cancer results from genetic changes in cells. These changes are often mutations that alter the DNA sequence of critical genes. However, duplications and deletions in cancer-related genes can also contribute to malignant transformation. In this study we use Nanostring technology to assess DNA copy number changes in samples of Merkel cell carcinoma (MCC), a rare and aggressive neuroendocrine skin tumor. We were able to identify recurrent amplifications and deletions in cancer-related genes. We also found that MCC tumors grouped into three distinct copy number variant profiles. The first group consisted of tumors with multiple deletions. The second group contained tumors with low levels of genomic structural alterations. The last group comprised tumors containing multiple amplifications. Our study suggests that most MCC tumors are associated with deletions in cancer-related genes or are lacking in copy number changes, whereas a small percentage of tumors are associated with genomic amplifications.

Citation: Hill, N.T.; Kim, D.; Busam, K.J.; Chu, E.Y.; Green, C.; Brownell, I. Distinct Signatures of Genomic Copy Number Variants Define Subgroups of Merkel Cell Carcinoma Tumors. *Cancers* **2021**, *13*, 1134. https://doi.org/10.3390/cancers13051134

Academic Editors: Virve Koljonen; Weng-Onn Lui and Jürgen C. Becker

Received: 25 January 2021
Accepted: 3 March 2021
Published: 6 March 2021

Abstract: Merkel cell carcinoma (MCC) is a rare, aggressive neuroendocrine skin cancer. Most MCC tumors contain integrated Merkel cell polyomavirus DNA (virus-positive MCC, VP-MCC) and carry a low somatic mutation burden whereas virus-negative MCC (VN-MCC) possess numerous ultraviolet-signature mutations. In contrast to viral oncogenes and sequence mutations, little is known about genomic structural variants in MCC. To identify copy number variants in commonly altered genes, we analyzed genomic DNA from 31 tumor samples using the Nanostring nCounter copy number cancer panel. Unsupervised clustering revealed three tumor groups with distinct genomic structural variant signatures. The first cluster was characterized by multiple recurrent deletions in genes such as *RB1* and *WT1*. The second cluster contained eight VP-MCC and displayed very few structural variations. The final cluster contained one VP-MCC and four VN-MCC with predominantly genomic amplifications in genes like *MDM4*, *SKP2*, and *KIT* and deletions in *TP53*. Overall, VN-MCC contained more structure variation than VP-MCC but did not cluster separately from VP-MCC. The observation that most MCC tumors demonstrate a deletion-dominated structural group signature, independent of virus status, suggests a shared pathophysiology among most VP-MCC and VN-MCC tumors.

Keywords: Merkel cell carcinoma; virus positive Merkel cell carcinoma; virus negative Merkel cell carcinoma and copy number variant

1. Introduction

Merkel cell carcinoma (MCC) is a rare neuroendocrine skin cancer associated with advanced age, UV-damage, and immunosuppression [1–3]. MCC is an aggressive cancer, with a lethality rate of over one-third, and thus is more deadly than malignant melanoma [1,3–5].

The incidence of MCC has increased in the past several decades in part due to improved diagnostic tools, increased clinical awareness, an aging population, and increased sun exposed skin [2,3]. In the United States, approximately 50–80% of MCC tumors are Merkel cell polyomavirus-positive (VP-MCC), with clonal integration of viral DNA into the host genome [6–12]. VP-MCC tumors carry a low somatic mutation burden, suggesting that tumorigenesis is driven by viral T antigen oncogenes [11,13–19]. The remaining 20–50% of MCC tumors are polyomavirus-negative (VN-MCC) and possess numerous ultraviolet signature mutations in genes such as *p53* and *RB1* [6,8–16]. Although a number of molecular and cytogenetic alterations have been reported for MCC, no unique signatures have been identified [11,13–16,20,21].

Genomic instability can initiate cancers, contribute to disease progression and impact patient response to treatment [22–24]. Several factors promote genomic instability leading to genomic structural variants in the form of amplifications or deletions, such as telomere damage, epigenetic modifications and DNA damage [22,23]. Here we use Nanostring's nCounter copy number variant (CNV) analysis to identify commonly amplified or deleted cancer-related genes in MCC. Unsupervised clustering identified three tumor groups with distinct genomic structural variant signatures. On average VP-MCC tumors had fewer copy number changes than VN-MCC. Furthermore, the cluster of tumors characterized by very few structural variants were all VP-MCC. In contrast, the tumors with numerous copy number variants clustered independently of virus status, suggesting a shared genomic instability among VN-MCC and a subset of VP-MCC.

2. Results

2.1. Three Genomic Structural Variant Signatures Identified in MCC Tumors

Despite the importance of genomic integrity in cancer, little is known about the genomic structural variants that lead to MCC. Therefore, we sought to identify commonly amplified or deleted cancer genes in MCC. We obtained 31 MCC tumors from Memorial Sloan Kettering (MSK), Marshfield Clinic (MF), and the University of Pennsylvania (UP) (Table 1). The patients ranged from 53 to 100 years of age. Most of the tumors analyzed were obtained from primary tumor lesions on sun exposed skin (Table 1). Genomic DNA from tumor samples and control tissues was analyzed using Nanostring Technologies' copy number variant cancer panel assay. Fresh-frozen tumors from MSK were normalized to fresh-frozen adjacent tissue samples, whereas FFPE tumors were normalized to FFPE normal spleen samples. As depicted in Figure 1, unsupervised clustering identified three distinct structural variant groups. Tumors clustered in group 1 (Del) displayed numerous recurrent deletions in a number of genes, including genes involved in cycle regulation such as *RB1* (Figure 1 and Table S1). Tumors in group 2 (Low) showed very few genomic structural variations. In the third group (Amp), tumors carried very few deletions but contained numerous recurrent amplifications in several genes, including *MDM4*, *AKT3*, *BCL2L1* and *MYCL1* (Figure 1 and Table S1). In this cohort of 31 MCC tumors, most of the tumors (18, 58%) have the structural group 1 Del signature dominated by deletions in cancer related genes. Group 2 Low with few changes accounted for 8 (26%) tumors, whereas only 5 (16%) tumors had the amplification-heavy group 3 Amp signature.

Table 1. Patient Summary.

MCC Sample	Sex	MCPyV Status	Age	Site of MCC	Specimen Code	Cluster	Tissue Source
MF1	Female	Negative	68	right upper arm	primary	Del	FFPE
MF2	Male	Negative	72	left hand	primary	Del	FFPE
MF3	Female	Negative	80	right gluteal	primary	Del	FFPE
MF4	Male	Negative	64	abdominal wall	primary	Del	FFPE
MF5	Male	Negative	89	left ala of nose	primary	Del	FFPE
MF6	Male	Negative	-	frontal scalp	primary	Amp	FFPE
MF7	Female	Negative	94	right scalp	primary	Del	FFPE
MF8	Female	Negative	-	lymph node	metastasis	Del	FFPE
MF9	Male	Positive	58	left thigh	primary	Del	FFPE
MF10	Male	Positive	67	left index finger	primary	Del	FFPE
MF11	Male	Negative	72	left cheek	primary	Del	FFPE
MF12	Male	Positive	-	right neck	primary	Del	FFPE
MF13	Female	Negative	-	right leg	primary	Del	FFPE
MF14	Female	Negative	100	right forehead	primary	Del	FFPE
MF15	Male	Negative	93	left cheek, nose	primary	Amp	FFPE
MF16	Female	Negative	74	left buttock	primary	Del	FFPE
MF18	Female	Negative	-	right forearm	primary	Del	FFPE
MF19	Male	Negative	77	right face	metastasis	Del	FFPE
MF20	Male	Positive	75	top of head	primary	Del	FFPE
MF21	Male	Negative	87	right wrist	primary	Amp	FFPE
MF22	Female	Negative	88	forehead	primary	Amp	FFPE
MF23	Male	Positive	81	left cheek	primary	Del	FFPE
UP1	Female	Positive	75	left brow	-	Amp	FFPE
MSK1	Female	Positive	80	lymph nodes	metastasis	Low	Frozen
MSK2	Male	Positive	73	pancreas	metastasis	Low	Frozen
MSK6	Male	Positive	53	groin	metastasis	Low	Frozen
MSK13	Female	Positive	62	skin	primary	Low	Frozen
MSK19	Male	Positive	59	skin	primary	Low	Frozen
MSK20	Female	Positive	63	skin	primary	Low	Frozen
MSK21	Male	Positive	87	skin	primary	Low	Frozen
MSK24	Male	Positive	82	lymph nodes	metastasis	Low	Frozen

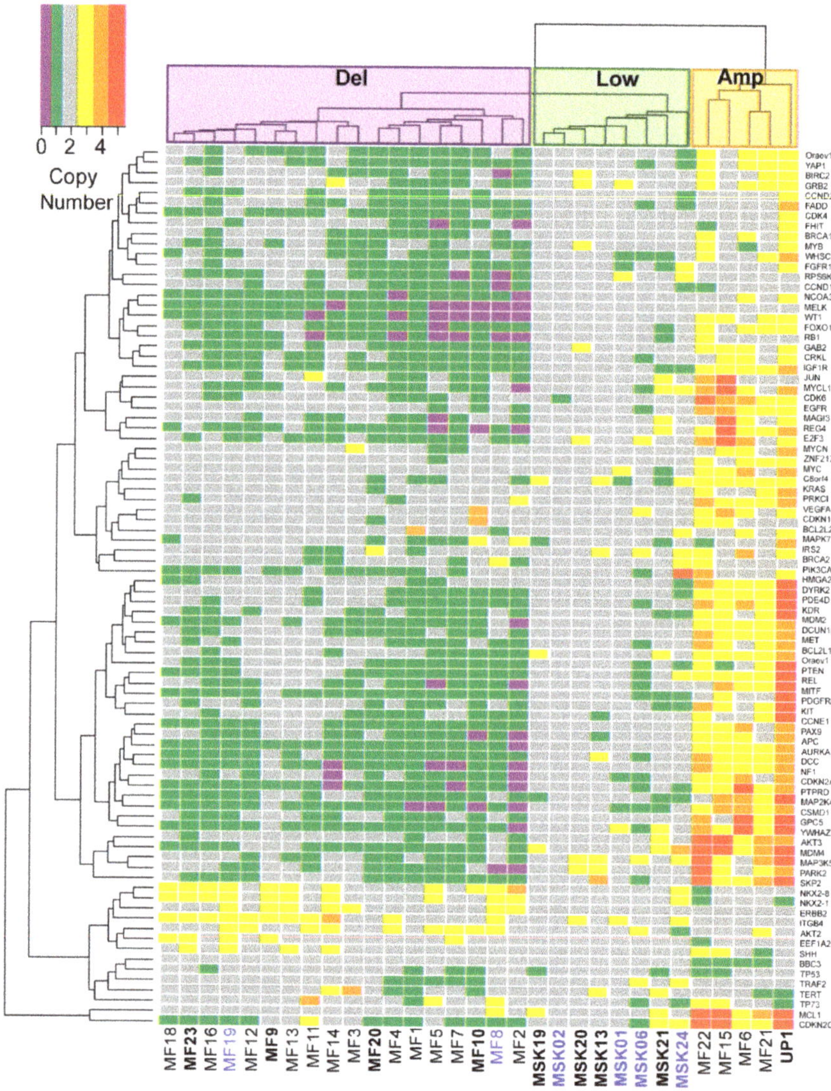

Figure 1. Three genomic structural variant signatures detected in MCC tumors by NanoString nCounter. Tumor DNA from 31 patients with MCC from Memorial Sloan Kettering (MSK), Marshfield Clinic (MF), and the University of Pennsylvania (UP) were subjected to Nanostring nCounter CNV analysis. CNV alterations for 86 gene loci commonly altered in cancer were ascertained and plotted as a heatmap. Three cluster groups denoted as Del (deletion) for group 1, Low for group 2, and Amp (amplification) for group 3. Bold indicates virus positive MCC (VP-MCC) tumors. Blue indicates metastatic tumor.

2.2. MCC Structural Variant Signatures Are Characterized by Deletions, Absence of Copy Changes, or Amplifications

To characterize the differences between MCC structural variant groups we compared the total number of copy number variations per tumor for each cluster. Tumors in both the Del and Amp groups had significantly more CNVs per tumor than tumors in the Low group ($p < 0.0001$, Figure 2A). We then compared the average sum of the allelic variation relative to diploid (-1 for each allelic deletion, +1 for each amplification, total of 86 genes) for the tumors in each cluster. As seen in Figure 2B, the Del group tumors had the lowest average sum of variation (−37 copies), reflective of their numerous deletions. Similarly, the Low group tumors' average sum of variation was 0.125 copies, close to the zero-value seen in the control samples; and the Amp group had an average sum of 82 copies. The significant difference in the average sums of variation ($p < 0.0001$) support there being 3 distinct CNV profiles for MCC rather than random distributions of deletions and amplifications in the tumors.

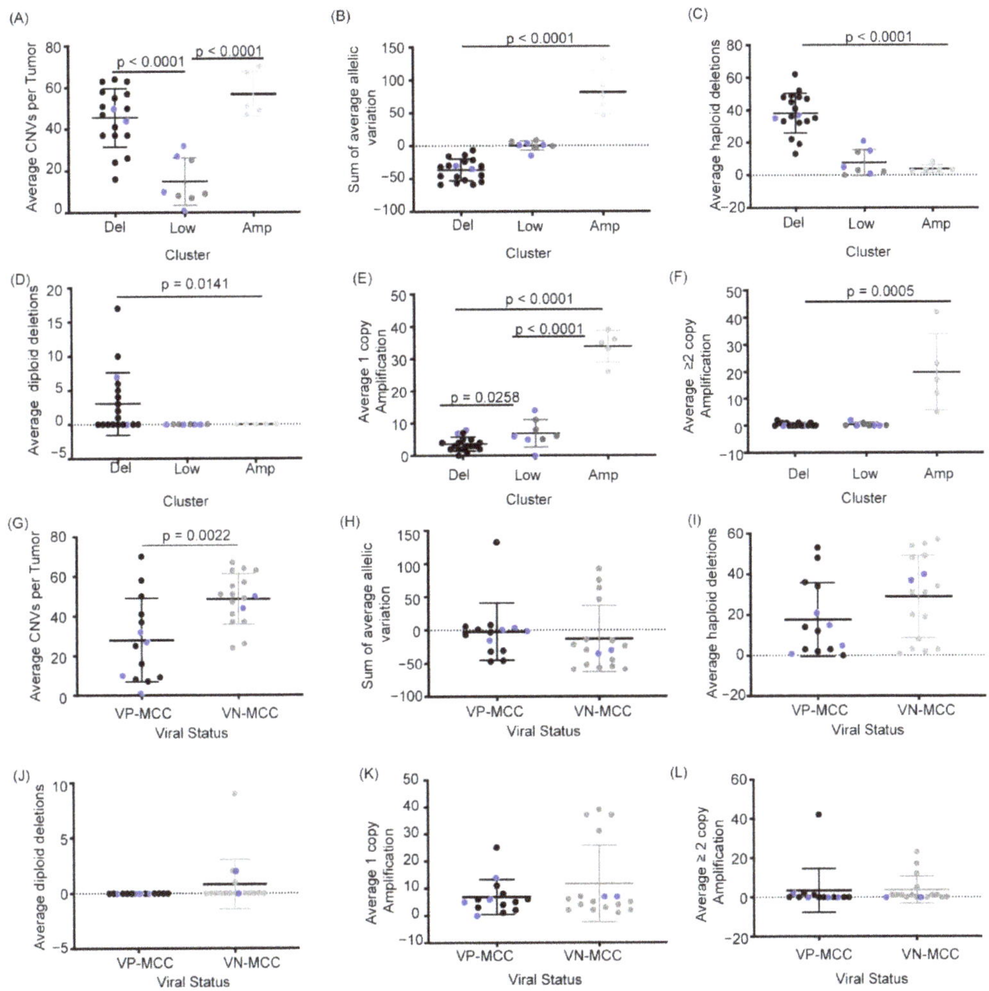

Figure 2. Structural variant clusters show distinct levels and types of CNVs whereas VN-MCC show more structural variation than VP-MCC. Comparison of clusters for the (**A**) average number of CNVs per tumor (one-way ANOVA), (**B**) average sum of allelic variations (−1 for each haploid deletion, +1 for each amplification, Kruskal-Wallis test), (**C**) average haploid deletions (Kruskal-Wallis test), (**D**) average diploid deletions (Kruskal-Wallis test), (**E**) average single-copy amplification (one-way ANOVA), and (**F**) average two or greater copy amplifications (Kruskal-Wallis test). Comparison between VP-MCC and VN-MCC for the (**G**) average number of CNVs per tumor (unpaired *T*-test), (**H**) average s um of allelic variations (Mann-Whitney test), (**I**) average haploid deletions (unpaired *T*-test), (**J**) average diploid deletions (Mann-Whitney test), (**K**) average single-copy amplification (Mann-Whitney test), and (**L**) average two or greater copy amplifications (Mann-Whitney test). Uncolored dots are primary tumor samples, blue dots indicate metastatic tumor samples.

A similar trend between signature groups was seen when quantifying CNV on a per gene level with haploid deletions being more common than diploid deletions, and single copy amplifications being more common than multi copy amplifications (Figure 2C–F). Taken together, quantitative comparisons between the three structural variant signatures suggest that MCC tumors with genetic instability are dominated by either recurrent haploid deletions or recurrent amplifications in cancer associated genes.

2.3. VN-MCC Contains More Structural Variants Than VP-MCC

Comparing the average CNV per tumor for VP-MCC and VN-MCC samples we found that VN-MCC tumor samples contained significantly more structural variation per tumor (48.6) than VP-MCC samples (27.9) (Figure 2G). This is consistent with prior studies that also found higher rates of CNV in VN-MCC [21,25]. Interestingly, VP-MCC and VN-MCC showed no difference in the average sum of the variation, haploid deletions, diploid deletions, single copy amplifications, or multi copy amplifications (Figure 2H–L). Thus, although VN-MCC have more structural variants than VP-MCC on average, each virus status subtype contains similar frequencies of amplifications and deletions. The decreased average CNV count for VP-MCC was largely due to the fact that the eight tumors with the Low variant signature were exclusively VP-MCC. Accordingly, the structural variant signature of MCC tumors correlated with tumor virus status (two-tailed Fisher exact test, $p < 0.005$), with VP-MCC being more likely in the Low variant group and VN-MCC more likely in the deleted or amplified group. However, if a tumor was not in the Low variant group, the likelihood of having a deletion or amplification signature was independent of virus status ($p = 1.0$).

2.4. MCC Structural Variant Signatures Are Not Predictors of Survival

The three distinct CNV signatures observed in MCC tumors suggest differences in their biology that might impact disease progression. We used non-parametric Kaplan–Meier estimate to test for overall survival differences in patients based on the CNV signatures of their MCC tumors. Survival data was available for 29 of the 31 patients in the study. As shown in Figure S1, Kaplan–Meier survival estimates indicate that there is no statistical difference in survival between the three signature groups ($p < 0.8857$). Taken together, although the three signature groups reflect distinct patterns of genomic instability, any difference in survival was not detected in this cohort of patients.

3. Discussion

Merkel cell carcinoma generally arises on sun exposure skin, giving rise to the notion that UV mediated damage induces MCC [26–29]. UV-induced DNA damage is frequently seen in skin cancer and has been shown to cause genomic instability [30–36]. Oncoviruses also leads to genomic instability via virus integration or through the expression of viral oncogenes that alters the fidelity of replication [37]. Interestingly, although VP-MCC tumors do not have a significant enrichment of UV-induced sequence mutations, these tumors primarily occur on sun exposed regions of the skin and these tumors, like VN-MCC, also show genomic instability [21,25,38]. Here we used Nanostring Technologies nCounter system to examine the frequency of structural variation in 31 MCC tumors by quantifying amplifications and deletions in 86 cancer related genes. A number of the alterations found in our data are predicted to disrupt cell cycle regulation, including deletions of *RB1*. Deletions in the *RB1* locus or mutations that functionally inactive RB have been previously identified in MCC [11,13,15,21]. Loss of RB function is a well-established phenotype in a variety of cancers [39–44]. In VP-MCC the MCPyV large T antigen binds and inhibits RB, thereby releasing E2F to promote G1 to S phase transition through the cell cycle [45–47]. Interestingly, 5 (36%) of 14 VP-MCC also showed deletions in *RB1*, suggesting redundant inactivation of RB may play a role in either MCC onset or progression. Future studies to determine whether CNVs in *RB1* correlate with the presence or absence of specific sequence mutations may lead to a better understand of the pathophysiology of this disease.

Genomic amplifications can also dysregulate the cell cycle leading to tumorigenesis [23,48,49]. Our data shows numerous amplifications within the group 3 Amp cluster, some of which are well established proto-oncogenes known to be involved in the onset and progression of many different cancers. Specifically, we observed amplifications in *MYCL1* which was previously shown to me amplified in MCC [20,21,50]. Furthermore, the protein levels of MYC, which was also a gene loci amplified in the group 3 cluster, are stabilized in VP-MCC by the small T antigen binding and inhibiting the function of the F-box protein FBW7 [17]. In small cell lung cancer (SCLC), another neuroendocrine carcinoma, L-myc is thought to induce pre-rRNA synthesis and transcriptional pathways concomitant with ribosomal biogenesis [51]. A similar pathogenesis may be exploited in L-myc amplified MCC. Another interesting finding in our data is that tumors in the Amp signature cluster showed amplifications in *AKT3* whereas tumors with the Del signature had amplifications in *AKT2*, suggesting that both tumor types may utilize the AKT survival pathway for tumorigenesis. Inhibition of the AKT downstream target mTOR has already been implicated as a potential target for the treatment of MCC [52,53]. Moreover, gene mutations and amplification in AKT1 have been found in MCC through next generation sequencing studies [14,21]. Multiple lines of evidence suggestion that both L-myc and AKT could potentially be druggable targets in the treatment of MCC [14,50,51], and assessing CNV signature type may help predict which MCC tumors are more likely to respond to these treatments.

We observed a number of genomic structural variants previously unreported in MCC. Most notably, recurrent deletions of fragile histidine triad (*FHIT*) and recurrent amplifications in integrin β4 (*ITGβ4*). Interestingly, FHIT was shown to inhibit AKT activation leading to one mechanism by which FHIT decreases lung cancer cell survival [54,55]. Additionally, FHIT was shown to transcriptionally repress β-catenin [56], which is a downstream target of not only AKT but also of the WNT signaling pathway [57,58]. The deletions observed in *FHIT* could further implicate AKT in MCC. Intriguingly, ITGβ4 promotes metastasis through the induction of epithelial-mesenchymal transition in pancreatic ductal adenocarcinoma [59]. In addition, expression of ITGβ4, CD24 and Notch were shown to confer non-small cell lung carcinoma (NSCLC) propagation in clonogenic and othotoptic transplantation assays [60]. In MCC, amplification in *ITGβ4* might similarly promote proliferation and metastasis. Taken together, structural alterations in MCC tumors potentially alter a number of different pathways to increase tumor cells survival such as AKT, L-myc, RB, and β-catenin. Additionally, structural variation in *ITGβ4* could play a role in MCC metastases. Further work will be needed to test these potential associations.

The Nanostring technology used in this study allows for direct quantification of fragmented genomic DNA based on hybridization to barcoded probes for genes commonly amplified or deleted in cancers. The technology uses an average of 3 probes per gene, internal control probes to 54 invariant genomic regions, as well as spike-in process controls. Thus, copy number variants relative to similarly processed diploid control tissues can be reproducibly quantified from either FFPE or fresh-frozen tumor samples [61]. The heterogeneity of analyzing both FFPE and fresh-frozen samples from different institutions is a limitation of our study. It is noteworthy that the 8 tumors comprising the Low CNV cluster were all fresh-frozen samples from Memorial Sloan-Kettering, whereas the tumors in the Del and Amp clusters were FFPE samples. Although the fresh-frozen samples and controls met the same quality control endpoints as the FFPE samples and controls, it is possible that there were batch effects related to sample acquisition or fixation. Formalin fixation can cause DNA fragmentation, degradation, crosslinking, and adduct formation that can theoretically impact molecular studies [62]. In addition, unlike the FFPE controls, the fresh-frozen controls were normal adjacent tissues collected at the time of tumor excisions. Normal adjacent tissue has limitations as a control, but it is generally found to contain diploid DNA [63] and thus its use is unlikely to impact a pooled reference for CNV normalization. Despite these concerns, as discussed above, many of the recurrent CNVs observed in our study were previously reported in MCC tumors based on studies using other copy number assays [11,13–15,20,21,50], suggesting some accuracy in our data. Similarly, the observation that the average CNV load in VN-MCC is higher than the average for VP-MCC has also been reported in other studies [11,13,21]. Finally, our finding that individual VP-MCC tumors can have very few structural variants or can contain multiple amplifications or deletions is consistent with the results of Starrett et al., 2020 [21]. Nonetheless, the possibility that some artifact is contributing to the clustering of our data must be considered. Therefore, further studies will be needed to validate our findings and explore the functional implications of MCC tumors with different structural variant signatures.

4. Materials and Methods

4.1. Inclusion Criteria and Patient Samples

Archival cases of MCC were identified by a retrospective search for the diagnosis of Merkel cell carcinoma or neuroendocrine skin tumor in the Pathology Departments of the institutions. For cases where adequate tissue was available for analysis, the diagnosis was confirmed by an expert dermatopathologist (DK, KJB, or EYC) based on histopathology and immunostaining for diagnostic markers. After confirming a diagnosis of MCC and ensuring the sample was >75% tumor, tissue was cut for DNA extraction. For each case, available patient information was retrieved by clinical chart review. De-identified tissue samples and clinical data were sent to the NIH for analysis.

We performed CNV analysis on 23 formalin-Fixed Paraffin-Embedded (FFPE) MCC tumor samples and 8 fresh-frozen MCC tumor samples. FFPE tumors were collected from patients at the University of Pennsylvania and Marshfield clinic between August 1996 and April 2012. Fresh-frozen tumors and normal tissues were collected from patients at Memorial Sloan Kettering between July 1995 and August 2010. Control tissues used for normalization consisted of 2 FFPE normal spleen samples (controls for the FFPE tumors) and 3 fresh-frozen normal tissues adjacent to excised tumors (controls for the fresh-frozen tumors).

4.2. Genomic DNA Isolation

Genomic DNA (gDNA) was extracted from FFPE or fresh-frozen tumor with QIAamp DNA FFPE Tissue kit (Qiagen, Hilden, Germany) or the DNeasy Blood and Tissue kit (Qiagen, Hilden, Germany) respectively, according to the manufacturer protocols. Samples were treated with RNase A (Qiagen, Hilden, Germany) per manufacturer protocol. DNA concentration, 260/280 and 260/230 nm ratios were measured on a DeNovix DS-11 spectrophotometer (DeNovix Inc., Wilmington, DE, USA) prior to DNA fragmentation with Alu1 restriction endonuclease. Following Alu1 restriction digestion, fragmented DNA was analyzed on a 2100 Agilent Bioanalyzer (Agilent Technologies, Santa Clara, CA, USA). Representative electropherograms and gel images of Alu1 digested DNA from FFPE and fresh frozen samples can be found in Figure S2.

4.3. Virus Detection

Nested qPCR was used to detect the presence of the Merkel cell polyomavirus from gDNA. For step one, 20 ng of gDNA underwent 15 cycles of amplification with forward primer GGCAACATCCCTCTGATGAAAGC 3′ and reverse primer 5′ CCACCAGTCAAAACTTTCCCAAGTAGG 3′ using the KAPA2G Fast HotSStart PCR kit according to the manufacture protocol (Kapa Biosystems, Wilmington, MA, USA). Step two, 2 μL of step one product was amplified for 25 cycles in a OneStep Real-Time PCR System with forward primer 5′ CTTAAAGCATCACCCTGATAAAGG 3′ and reverse primer 5′ AAACCAAAGAATAAAGCACTGATAGCA 3′ using Power SYBR green master mix as per the manufacture protocol (ThermoFisher, Carlsbad, CA, USA). Primer set (forward 5′ CCACACTGCCCATCTCGGAGAC 3′ and reverse 5′ GCGGTGAGGTCCCTACGGCCTG 3′) for TPO was used as an endogenous control for quantitative PCR. gDNA from the VP-MCC cell line MKL1 and VN-MCC cell line UISO were used as controls to determine the presence or absences of the polyomavirus.

4.4. Cell Lines

UISO-MCC-1 [64] and MKL-1 [65] were previously described and grown in RPMI-1640 supplemented with 10% fetal bovine serum and 1% streptomycin/penicillin. Cell lines are sent out annually to be tested for authenticity via the Hum 16-Marker STR profile, interspecies contamination test and PCR evaluation for viruses and *Mycoplasma* which was performed by Idexx Bio Research.

4.5. Nanostring Prep and Run

A total of 600 ng of gDNA was process for Nanostring nCounter copy number variants as per the manufacturer protocol (Nanostring Technologies, Seattle, WA, USA).

4.6. Nanostring Data Analysis

Copy number for 86 genes for each tumor sample compared to the appropriate control samples were determined in nSolver according to manufacturer instructions. A heatmap of the normalized copy number data was generated in R using the gplots package and heatmap.2 code.

All statistical analyses were performed in GraphPad Prism (GraphPad Software, La Jolla, CA, USA). For each comparison, Grubbs' Method was used to detect statistical outliers. For populations with normal distributions, *T*-test or one-way ANOVA were performed to assess differences between VP-MCC and VN-MCC or between clusters. For populations with statistical outliers, Mann–Whitney or Kruskal–Wallis test were performed to assess differences between VP-MCC and VN-MCC or between clusters. Significance was based on a *p*-value of less than 0.05.

5. Conclusions

We identified three distinct CNV signatures in MCC tumors. The observation that majority of MCC tumors demonstrate the Del structural signature, independent of virus status, suggests a shared pattern of genomic instability among most VP-MCC and VN-MCC tumors that promotes allelic deletions. In contrast, a subset of MCC tumors appear to be associated with mechanisms that promote genomic amplifications. A further subset of VP-MCC tumors are capable of progression with very few genomic structural alterations. As VP-MCC are known to have a very low somatic mutational burden, observing VP-MCC tumors with few CNVs suggests that viral oncogenes and epigenetic changes may be sufficient for tumorigenesis. Although the different CNV signatures were not associated with survival differences in MCC patients, the signatures were associated with recurrent changes in specific cancer pathways. It is possible that testing genomic structural signatures may help identify MCC patients more likely to respond to targeted therapeutic approaches.

Supplementary Materials: The following are available online at https://www.mdpi.com/2072-6694/13/5/1134/s1, Figure S1: The three genomic structural variant clusters are not predictors of overall survival, Figure S2: Bioanalyzer analysis of Alu1 cut DNA from FFPE and fresh frozen MCC tumor samples, Table S1: Normalized genomic copy number at 86 gene loci for MCC tumor samples.

Author Contributions: Conceptualization, C.G. and I.B.; Formal analysis, N.T.H.; Funding acquisition, I.B.; Investigation, N.T.H.; Methodology, N.T.H., D.K., K.J.B. and E.Y.C.; Project administration, I.B.; Resources, D.K., K.J.B., E.Y.C. and C.G.; Supervision, I.B.; Visualization, N.T.H.; Writing—original draft, N.T.H.; Writing—review and editing, N.T.H., C.G. and I.B. All authors have read and agreed to the published version of the manuscript.

Funding: This work was supported by funding from the Intramural Research Program, Center for Cancer Research, National Cancer Institute (ZIA BC 011394 to I.B.) and the Marshfield Clinic Research Institute.

Institutional Review Board Statement: This retrospective investigation of deidentified data was carried out following the rules of the Declaration of Helsinki of 1975, revised in 2013.

Informed Consent Statement: Analysis of patient samples from the three institutions was conducted under National Cancer Institute (NCI) Protocol 13CN024 without obtaining further consent as the samples were analyzed anonymously.

Data Availability Statement: Data is contained within the article or supplementary material.

Acknowledgments: The opinions expressed in this article are the authors' own and do not reflect the view of the National Institutes of Health, the Department of Health and Human Services, or the United States government. We would like to thank Emily Andrea for her editorial guidance.

Conflicts of Interest: We have no conflict of interest to declare.

References

1. Albores-Saavedra, J.; Batich, K.; Chable-Montero, F.; Sagy, N.; Schwartz, A.M.; Henson, D.E. Merkel cell carcinoma demographics, morphology, and survival based on 3870 cases: A population-based study. *J. Cutan. Pathol.* **2010**, *37*, 20–27. [CrossRef] [PubMed]
2. Harms, P.W.; Harms, K.L.; Moore, P.S.; DeCaprio, J.A.; Nghiem, P.; Wong, M.K.; Brownell, I. International workshop on Merkle cell carcinoma research working. The biology and treatment of Merkel cell carcinoma: Current understanding and research priorities. *Nat. Rev. Clin. Oncol.* **2018**, *15*, 763–776. [CrossRef] [PubMed]
3. Schwartz, R.A.; Lambert, W.C. The Merkel cell carcinoma: A 50-year retrospect. *J. Surg. Oncol.* **2005**, *89*, 5. [CrossRef] [PubMed]
4. Fitzgerald, T.L.; Dennis, S.; Kachare, S.D.; Vohra, N.A.; Wong, J.H.; Zervos, E.E. Dramatic increase in the incidence and mortality from Merkel cell carcinoma in the United States. *Am. Surg.* **2015**, *81*, 802–806. [CrossRef]
5. van Veenendaal, L.M.; van Akkooi, A.C.J.; Verhoef, C.; Grunhagen, D.J.; Klop, W.M.C.; Valk, G.D.; Tesselaar, M.E.T. Merkel cell carcinoma: Clinical outcome and prognostic factors in 351 patients. *J. Surg. Oncol.* **2018**, *117*, 1768–1775. [CrossRef] [PubMed]
6. Feng, H.; Shuda, M.; Chang, Y.; Moore, P.S. Clonal integration of a polyomavirus in human Merkel cell carcinoma. *Science* **2008**, *319*, 1096–1100. [CrossRef] [PubMed]
7. Harms, P.W. Update on Merkel cell carcinoma. *Clin. Lab. Med.* **2017**, *37*, 485–501. [CrossRef]
8. Al-Rohil, R.N.; Milton, D.R.; Nagarajan, P.; Curry, J.L.; Feldmeyer, L.; Torres-Cabala, C.A.; Ivan, D.; Prieto, V.G.; Tetzlaff, M.T.; Aung, P.P. Intratumoral and peritumoral lymphovascular invasion detected by D2-40 immunohistochemistry correlates with metastasis in primary cutaneous Merkel cell carcinoma. *Hum. Pathol.* **2018**, *77*, 98–107. [CrossRef]
9. Aung, P.P.; Parra, E.R.; Barua, S.; Sui, D.; Ning, J.; Mino, B.; Ledesma, D.A.; Curry, J.L.; Nagarajan, P.; Torres-Cabala, C.A.; et al. B7-H3 Expression in Merkel cell carcinoma-associated endothelial cells correlates with locally aggressive primary tumor features and increased vascular density. *Clin. Cancer Res.* **2019**, *25*, 3455–3467. [CrossRef] [PubMed]
10. Feldmeyer, L.; Hudgens, C.W.; Ray-Lyons, G.; Nagarajan, P.; Aung, P.P.; Curry, J.L.; Torres-Cabala, C.A.; Mino, B.; Rodriguez-Canales, J.; Reuben, A.; et al. Density, distribution, and composition of immune infiltrates correlate with survival in Merkel cell carcinoma. *Clin. Cancer Res.* **2016**, *22*, 5553–5563. [CrossRef] [PubMed]
11. Harms, P.W.; Patel, R.M.; Verhaegen, M.E.; Giordano, T.J.; Nash, K.T.; Johnson, C.N.; Daignault, S.; Thomas, D.G.; Gudjonsson, J.E.; Elder, J.T.; et al. Distinct gene expression profiles of viral- and nonviral-associated merkel cell carcinoma revealed by transcriptome analysis. *J. Investig. Dermatol.* **2013**, *133*, 936–945. [CrossRef] [PubMed]
12. Wang, L.; Harms, P.W.; Palanisamy, N.; Carskadon, S.; Cao, X.; Siddiqui, J.; Patel, R.M.; Zelenka-Wang, S.; Durham, A.B.; Fullen, D.R.; et al. Age and gender associations of virus positivity in Merkel cell carcinoma characterized using a novel RNA in situ hybridization assay. *Clin. Cancer Res.* **2017**, *23*, 5622–5630. [CrossRef]
13. Goh, G.; Walradt, T.; Markarov, V.; Blom, A.; Riaz, N.; Doumani, R.; Stafstrom, K.; Moshiri, A.; Yelistratova, L.; Levinsohn, J.; et al. Mutational landscape of MCPyV-positive and MCPyV-negative Merkel cell carcinomas with implications for immunotherapy. *Oncotarget* **2016**, *7*, 3403–3415. [CrossRef]
14. Harms, P.W.; Collie, A.M.; Hovelson, D.H.; Cani, A.K.; Verhaegen, M.E.; Patel, R.M.; Fullen, D.R.; Omata, K.; Dlugosz, A.A.; Tomlins, S.A.; et al. Next generation sequencing of Cytokeratin 20-negative Merkel cell carcinoma reveals ultraviolet-signature mutations and recurrent TP53 and RB1 inactivation. *Mod. Pathol.* **2016**, *29*, 240–248. [CrossRef] [PubMed]
15. Wong, S.Q.; Waldeck, K.; Vergara, I.A.; Schroder, J.; Madore, J.; Wilmott, J.S.; Colebatch, A.J.; De Paoli-Iseppi, R.; Li, J.; Lupat, R.; et al. UV-associated mutations underlie the etiology of MCV-negative merkel cell carcinomas. *Cancer Res.* **2015**, *75*, 5228–5234. [CrossRef] [PubMed]
16. Veija, T.; Sarhadi, V.K.; Koljonen, V.; Bohling, T.; Knuutila, S. Hotspot mutations in polyomavirus positive and negative Merkel cell carcinomas. *Cancer Genet.* **2016**, *209*, 30–35. [CrossRef]
17. Kwun, H.J.; Shuda, M.; Feng, H.; Camacho, C.J.; Moore, P.S.; Chang, Y. Merkel cell polyomavirus small T antigen controls viral replication and oncoprotein expression by targeting the cellular ubiquitin ligase SCFFbw7. *Cell Host Microbe* **2013**, *14*, 125–135. [CrossRef] [PubMed]
18. Shuda, M.; Kwun, H.J.; Feng, H.; Chang, Y.; Moore, P.S. Human Merkel cell polyomavirus small T antigen is an oncoprotein targeting the 4E-BP1 translation regulator. *J. Clin. Investig.* **2011**, *121*, 3623–3634. [CrossRef] [PubMed]
19. Abdul-Sada, H.; Muller, M.; Mehta, R.; Toth, R.; Arthur, J.S.C.; Whitehouse, A.; Macdonald, A. The PP4R1 sub-unit of protein phosphatase PP4 is essential for inhibition of NF-kappaB by merkel polyomavirus small tumour antigen. *Oncotarget* **2017**, *8*, 25418–25432. [CrossRef]

20. Van Gele, M.; Leonard, J.H.; Van Roy, N.; Van Limbergen, H.; Van Belle, S.; Cocquyt, V.; Salwen, H.; De Paepe, A.; Speleman, F. Combined karyotyping, CGH and M-FISH analysis allows detailed characterization of unidentified chromosomal rearrangements in Merkel cell carcinoma. *Int. J. Cancer* **2002**, *101*, 137–145. [CrossRef] [PubMed]
21. Starrett, G.J.; Thakuria, M.; Chen, T.; Marcelus, C.; Cheng, J.; Nomburg, J.; Thorner, A.R.; Slevin, M.K.; Powers, W.; Burns, R.T.; et al. Clinical and molecular characterization of virus-positive and virus-negative Merkel cell carcinoma. *Genome Med.* **2020**, *12*, 30. [CrossRef] [PubMed]
22. Evan, G.I.; Vousden, K.H. Proliferation, cell cycle and apoptosis in cancer. *Nature* **2001**, *411*, 342–348. [CrossRef] [PubMed]
23. Hartwell, L. Defects in a cell cycle checkpoint may be responsible for the genomic instability of cancer cells. *Cell* **1992**, *71*, 543–546. [CrossRef]
24. Limpose, K.L.; Trego, K.S.; Li, Z.; Leung, S.W.; Sarker, A.H.; Shah, J.A.; Ramalingam, S.S.; Werner, E.M.; Dynan, W.S.; Cooper, P.K.; et al. Overexpression of the base excision repair NTHL1 glycosylase causes genomic instability and early cellular hallmarks of cancer. *Nucleic Acids Res.* **2018**, *46*, 4515–4532. [CrossRef] [PubMed]
25. Garneski, K.M.; DeCaprio, J.A.; Nghiem, P. Does a new polyomavirus contribute to Merkel cell carcinoma? *Genome Biol.* **2008**, *9*, 228. [CrossRef] [PubMed]
26. Popp, S.; Waltering, S.; Herbst, C.; Moll, I.; Boukamp, P. UV-B-type mutations and chromosomal imbalances indicate common pathways for the development of Merkel and skin squamous cell carcinomas. *Int. J. Cancer* **2002**, *99*, 352–360. [CrossRef]
27. Hartschuh, W.; Schulz, T. Merkel cell hyperplasia in chronic radiation-damaged skin: Its possible relationship to fibroepithelioma of Pinkus. *J. Cutan. Pathol.* **1997**, *24*, 477–483. [CrossRef]
28. Miller, R.W.; Rabkin, C.S. Merkel cell carcinoma and melanoma: Etiological similarities and differences. *Cancer Epidemiol. Biomark. Prev.* **1999**, *8*, 153–158.
29. Agelli, M.; Clegg, L.X. Epidemiology of primary Merkel cell carcinoma in the United States. *J. Am. Acad. Dermatol.* **2003**, *49*, 832–841. [CrossRef]
30. Brenner, B.; Sulkes, A.; Rakowsky, E.; Feinmesser, M.; Yukelson, A.; Bar-Haim, E.; Katz, A.; Idelevich, E.; Neuman, A.; Barhana, M.; et al. Second neoplasms in patients with Merkel cell carcinoma. *Cancer* **2001**, *91*, 1358–1362. [CrossRef]
31. Ratner, D.; Nelson, B.R.; Brown, M.D.; Johnson, T.M. Merkel cell carcinoma. *J. Am. Acad. Dermatol.* **1993**, *29*, 143–156. [CrossRef]
32. Cerroni, L.; Kerl, H. Primary cutaneous neuroendocrine (Merkel cell) carcinoma in association with squamous- and basal-cell carcinoma. *Am. J. Dermatopathol.* **1997**, *19*, 610–613. [CrossRef] [PubMed]
33. Aydin, A.; Kocer, N.E.; Bekerecioglu, M.; Sari, I. Cutaneous undifferentiated small (Merkel) cell carcinoma, that developed synchronously with multiple actinic keratoses, squamous cell carcinomas and basal cell carcinoma. *J. Dermatol.* **2003**, *30*, 241–244. [CrossRef] [PubMed]
34. Pulitzer, M.P.; Brannon, A.R.; Berger, M.F.; Louis, P.; Scott, S.N.; Jungbluth, A.A.; Coit, D.G.; Brownell, I.; Busam, K.J. Cutaneous squamous and neuroendocrine carcinoma: Genetically and immunohistochemically different from Merkel cell carcinoma. *Mod. Pathol.* **2015**, *28*, 1023–1032. [CrossRef]
35. Iacocca, M.V.; Abernethy, J.L.; Stefanato, C.M.; Allan, A.E.; Bhawan, J. Mixed Merkel cell carcinoma and squamous cell carcinoma of the skin. *J. Am. Acad. Dermatol.* **1998**, *39*, 882–887. [CrossRef]
36. Panich, U.; Sittithumcharee, G.; Rathviboon, N.; Jirawatnotai, S. Ultraviolet radiation-induced skin aging: The role of DNA damage and oxidative stress in epidermal stem cell damage mediated skin aging. *Stem. Cells Int.* **2016**, *2016*, 7370642. [CrossRef]
37. Morales-Sanchez, A.; Fuentes-Panana, E.M. Human viruses and cancer. *Viruses* **2014**, *6*, 4047–4079. [CrossRef]
38. Heath, M.; Jaimes, N.; Lemos, B.; Mostaghimi, A.; Wang, L.C.; Penas, P.F.; Nghiem, P. Clinical characteristics of Merkel cell carcinoma at diagnosis in 195 patients: The AEIOU features. *J. Am. Acad. Dermatol.* **2008**, *58*, 375–381. [CrossRef] [PubMed]
39. Cancer Genome Atlas Network. Comprehensive molecular portraits of human breast tumours. *Nature* **2012**, *490*, 61–70. [CrossRef] [PubMed]
40. Phillips, S.M.; Barton, C.M.; Lee, S.J.; Morton, D.G.; Wallace, D.M.; Lemoine, N.R.; Neoptolemos, J.P. Loss of the retinoblastoma susceptibility gene (RB1) is a frequent and early event in prostatic tumorigenesis. *Br. J. Cancer* **1994**, *70*, 1252–1257. [CrossRef]
41. Phillips, S.M.; Morton, D.G.; Lee, S.J.; Wallace, D.M.; Neoptolemos, J.P. Loss of heterozygosity of the retinoblastoma and adenomatous polyposis susceptibility gene loci and in chromosomes 10p, 10q and 16q in human prostate cancer. *Br. J. Urol.* **1994**, *73*, 390–395. [CrossRef]
42. Bettendorf, O.; Schmidt, H.; Staebler, A.; Grobholz, R.; Heinecke, A.; Boecker, W.; Hertle, L.; Semjonow, A. Chromosomal imbalances, loss of heterozygosity, and immunohistochemical expression of TP53, RB1, and PTEN in intraductal cancer, intraepithelial neoplasia, and invasive adenocarcinoma of the prostate. *Genes Chromosom. Cancer* **2008**, *47*, 565–572. [CrossRef]
43. Dimaras, H.; Khetan, V.; Halliday, W.; Orlic, M.; Prigoda, N.L.; Piovesan, B.; Marrano, P.; Corson, T.W.; Eagle, R.C., Jr.; Squire, J.A.; et al. Loss of RB1 induces non-proliferative retinoma: Increasing genomic instability correlates with progression to retinoblastoma. *Hum. Mol. Genet.* **2008**, *17*, 1363–1372. [CrossRef]
44. Xing, F.; Persaud, Y.; Pratilas, C.A.; Taylor, B.S.; Janakiraman, M.; She, Q.B.; Gallardo, H.; Liu, C.; Merghoub, T.; Hefter, B.; et al. Concurrent loss of the PTEN and RB1 tumor suppressors attenuates RAF dependence in melanomas harboring (V600E)BRAF. *Oncogene* **2012**, *31*, 446–457. [CrossRef] [PubMed]
45. Hesbacher, S.; Pfitzer, L.; Wiedorfer, K.; Angermeyer, S.; Borst, A.; Haferkamp, S.; Scholz, C.J.; Wobser, M.; Schrama, D.; Houben, R. RB1 is the crucial target of the Merkel cell polyomavirus Large T antigen in Merkel cell carcinoma cells. *Oncotarget* **2016**, *7*, 32956–32968. [CrossRef] [PubMed]

46. Nakamura, T.; Sato, Y.; Watanabe, D.; Ito, H.; Shimonohara, N.; Tsuji, T.; Nakajima, N.; Suzuki, Y.; Matsuo, K.; Nakagawa, H.; et al. Nuclear localization of Merkel cell polyomavirus large T antigen in Merkel cell carcinoma. *Virology* **2010**, *398*, 273–279. [CrossRef] [PubMed]
47. Dresang, L.R.; Guastafierro, A.; Arora, R.; Normolle, D.; Chang, Y.; Moore, P.S. Response of Merkel cell polyomavirus-positive merkel cell carcinoma xenografts to a survivin inhibitor. *PLoS ONE* **2013**, *8*, e80543. [CrossRef]
48. Xu, X.; Weaver, Z.; Linke, S.P.; Li, C.; Gotay, J.; Wang, X.W.; Harris, C.C.; Ried, T.; Deng, C.X. Centrosome amplification and a defective G2-M cell cycle checkpoint induce genetic instability in BRCA1 exon 11 isoform-deficient cells. *Mol. Cell* **1999**, *3*, 389–395. [CrossRef]
49. Keyomarsi, K.; Pardee, A.B. Redundant cyclin overexpression and gene amplification in breast cancer cells. *Proc. Natl. Acad. Sci. USA* **1993**, *90*, 1112–1116. [CrossRef]
50. Paulson, K.G.; Lemos, B.D.; Feng, B.; Jaimes, N.; Penas, P.F.; Bi, X.; Maher, E.; Cohen, L.; Leonard, J.H.; Granter, S.R.; et al. Array-CGH reveals recurrent genomic changes in Merkel cell carcinoma including amplification of L-Myc. *J. Investig. Dermatol.* **2009**, *129*, 1547–1555. [CrossRef]
51. Kim, D.W.; Wu, N.; Kim, Y.C.; Cheng, P.F.; Basom, R.; Kim, D.; Dunn, C.T.; Lee, A.Y.; Kim, K.; Lee, C.S.; et al. Genetic requirement for Mycl and efficacy of RNA Pol I inhibition in mouse models of small cell lung cancer. *Genes Dev.* **2016**, *30*, 1289–1299. [CrossRef] [PubMed]
52. Shao, Q.; Byrum, S.D.; Moreland, L.E.; Mackintosh, S.G.; Kannan, A.; Lin, Z.; Morgan, M.; Stack, B.C., Jr.; Cornelius, L.A.; Tackett, A.J.; et al. A proteomic study of human Merkel cell carcinoma. *J. Proteom. Bioinform.* **2013**, *6*, 275–282. [CrossRef] [PubMed]
53. Lin, Z.; McDermott, A.; Shao, L.; Kannan, A.; Morgan, M.; Stack, B.C., Jr.; Moreno, M.; Davis, D.A.; Cornelius, L.A.; Gao, L. Chronic mTOR activation promotes cell survival in Merkel cell carcinoma. *Cancer Lett.* **2014**, *344*, 272–281. [CrossRef]
54. Wu, D.W.; Lee, M.C.; Hsu, N.Y.; Wu, T.C.; Wu, J.Y.; Wang, Y.C.; Cheng, Y.W.; Chen, C.Y.; Lee, H. FHIT loss confers cisplatin resistance in lung cancer via the AKT/NF-kappaB/Slug-mediated PUMA reduction. *Oncogene* **2015**, *34*, 3882–3883. [CrossRef]
55. Semba, S.; Trapasso, F.; Fabbri, M.; McCorkell, K.A.; Volinia, S.; Druck, T.; Iliopoulos, D.; Pekarsky, Y.; Ishii, H.; Garrison, P.N.; et al. Fhit modulation of the Akt-survivin pathway in lung cancer cells: Fhit-tyrosine 114 (Y114) is essential. *Oncogene* **2006**, *25*, 2860–2872. [CrossRef] [PubMed]
56. Weiske, J.; Albring, K.F.; Huber, O. The tumor suppressor Fhit acts as a repressor of beta-catenin transcriptional activity. *Proc. Natl. Acad. Sci. USA* **2007**, *104*, 20344–20349. [CrossRef]
57. Robertson, B.W.; Chellaiah, M.A. Osteopontin induces beta-catenin signaling through activation of Akt in prostate cancer cells. *Exp. Cell. Res.* **2010**, *316*, 1–11. [CrossRef]
58. He, X.C.; Zhang, J.; Tong, W.G.; Tawfik, O.; Ross, J.; Scoville, D.H.; Tian, Q.; Zeng, X.; He, X.; Wiedemann, L.M.; et al. BMP signaling inhibits intestinal stem cell self-renewal through suppression of Wnt-beta-catenin signaling. *Nat. Genet.* **2004**, *36*, 1117–1121. [CrossRef]
59. Masugi, Y.; Yamazaki, K.; Emoto, K.; Effendi, K.; Tsujikawa, H.; Kitago, M.; Itano, O.; Kitagawa, Y.; Sakamoto, M. Upregulation of integrin beta4 promotes epithelial-mesenchymal transition and is a novel prognostic marker in pancreatic ductal adenocarcinoma. *Lab. Investig.* **2015**, *95*, 308–319. [CrossRef]
60. Zheng, Y.; de la Cruz, C.C.; Sayles, L.C.; Alleyne-Chin, C.; Vaka, D.; Knaak, T.D.; Bigos, M.; Xu, Y.; Hoang, C.D.; Shrager, J.B.; et al. A rare population of CD24(+)ITGB4(+)Notch(hi) cells drives tumor propagation in NSCLC and requires Notch3 for self-renewal. *Cancer Cell* **2013**, *24*, 59–74. [CrossRef]
61. nanoString, Analyzing FFPE Specimens with the nCounter®Copy Number Variation (CNV) Assay. 2019. Available online: http://108.166.79.23/products/CNV, nanoString Tech, https://www.nanostring.com/wp-content/uploads/2020/12/TN_MK1113_FFPE_CNV_R3.pdf (accessed on 16 February 2021).
62. Do, H.; Dobrovic, A. Sequence artifacts in DNA from formalin-fixed tissues: Causes and strategies for minimization. *Clin. Chem.* **2015**, *61*, 64–71. [CrossRef] [PubMed]
63. Jakubek, Y.A.; Chang, K.; Sivakumar, S.; Yu, Y.; Giordano, M.R.; Fowler, J.; Huff, C.D.; Kadara, H.; Vilar, E.; Scheet, P. Large-scale analysis of acquired chromosomal alterations in non-tumor samples from patients with cancer. *Nat. Biotechnol.* **2020**, *38*, 90–96. [CrossRef]
64. Ronan, S.G.; Green, A.D.; Shilkaitis, A.; Huang, T.S.; Das Gupta, T.K. Merkel cell carcinoma: In vitro and in vivo characteristics of a new cell line. *J. Am. Acad. Dermatol.* **1993**, *29*, 715–722. [CrossRef]
65. Rosen, S.T.; Gould, V.E.; Salwen, H.R.; Herst, C.V.; Le Beau, M.M.; Lee, I.; Bauer, K.; Marder, R.J.; Andersen, R.; Kies, M.S.; et al. Establishment and characterization of a neuroendocrine skin carcinoma cell line. *Lab. Investig.* **1987**, *56*, 302–312. [PubMed]

cancers

Article

Mutational Landscape of Virus- and UV-Associated Merkel Cell Carcinoma Cell Lines Is Comparable to Tumor Tissue

Kai Horny [1,2], Patricia Gerhardt [3], Angela Hebel-Cherouny [3], Corinna Wülbeck [3], Jochen Utikal [4,5] and Jürgen C. Becker [1,2,3,*]

1 Translational Skin Cancer Research, German Cancer Consortium (DKTK), 45141 Essen, Germany; k.horny@dkfz.de
2 German Cancer Research Center (DKFZ), 69120 Heidelberg, Germany
3 Department of Dermatology, University Medicine Essen, 45141 Essen, Germany; patricia.gerhardt@uni-due.de (P.G.); angela.cherouny@uk-essen.de (A.H.-C.); c.wuelbeck@dkfz.de (C.W.)
4 Skin Cancer Unit, German Cancer Research Center (DKFZ), 69120 Heidelberg, Germany; j.utikal@dkfz.de
5 Department of Dermatology, Venerology and Allergology, University Medical Center Mannheim, 68167 Mannheim, Germany
* Correspondence: j.becker@dkfz.de; Tel.: +49-201-1836727

Simple Summary: Merkel cell carcinoma (MCC) is an aggressive, rare skin cancer which is caused either by a virus or chronic UV exposure. For both forms, distinct genetic alterations have been described; however, these observations were mostly made in tumor tissue. Since cancer cell lines are frequently used as preclinical models to investigate biological function, we considered it necessary to establish the genomic landscape of MCC cell lines by whole-exome sequencing. We confirmed the presence of UV-induced DNA damage, a high number of mutations and several coding mutations in virus-negative cell lines which were absent in virus-positive cell lines; these, however, harbored characteristic copy number variations, suggesting some virally caused genetic instability. Knowing the genomic features of MCC cell lines validates previous, and facilitates upcoming, experimental studies to discover their biological and translational relevance.

Abstract: Merkel cell carcinoma (MCC) is a rare, highly aggressive cutaneous malignancy that is either associated with the integration of the Merkel cell polyomavirus or chronic UV exposure. These two types of carcinogenesis are reflected in characteristic mutational features present in MCC tumor lesions. However, the genomic characteristics of MCC cell lines used as preclinical models are not well established. Thus, we analyzed the exomes of three virus-negative and six virus-positive MCC cell lines, all showing a classical neuroendocrine growth pattern. Virus-negative cell lines are characterized by a high tumor mutational burden (TMB), UV-light-induced DNA damage, functionally relevant coding mutations, e.g., in *RB1* and *TP53*, and large amounts of copy number variations (CNVs). In contrast, virus-positive cell lines have a low TMB with few coding mutations and lack prominent mutational signatures, but harbor characteristic CNVs. One of the virus-negative cell lines has a local *MYC* amplification associated with high *MYC* mRNA expression. In conclusion, virus-positive and -negative MCC cell lines with a neuroendocrine growth pattern resemble mutational features observed in MCC tissue samples, which strengthens their utility for functional studies.

Keywords: merkel cell carcinoma; merkel cell polyoma virus; UV; cell line; MYC; TP53; RB1; whole-exome; significantly mutated genes; copy number variation

Citation: Horny, K.; Gerhardt, P.; Hebel-Cherouny, A.; Wülbeck, C.; Utikal, J.; Becker, J.C. Mutational Landscape of Virus- and UV-Associated Merkel Cell Carcinoma Cell Lines Is Comparable to Tumor Tissue. *Cancers* **2021**, *13*, 649. https://doi.org/10.3390/cancers13040649

Academic Editor: Virve Koljonen
Received: 23 December 2020
Accepted: 2 February 2021
Published: 5 February 2021

1. Introduction

Merkel cell carcinoma (MCC) is a rare, highly aggressive neuroendocrine skin cancer. It is either associated with chronic Ultraviolet (UV)-light exposure or the genomic integration of the Merkel cell polyomavirus (MCPyV) [1,2]. Virus-associated MCCs are highly prevalent in countries with high latitude, while UV-associated MCCs are more frequent in regions close to the equator [1–3].

The different forms of carcinogenesis of MCC are represented in various genomic features, as demonstrated by targeted [4–10], whole-exome [11–14] and whole-genome sequencing [15], as well as comparative genomic hybridization (CGH) [16–21]. Virus-negative MCCs are characterized by a high tumor mutational burden (TMB), the presence of UV-light-induced DNA damage, functional driver mutations, and high numbers of copy number variations (CNVs). Virus-positive MCCs have a very low TMB and lack known cancer-driving mutations and prominent mutational signatures. Characteristic CNV patterns have repeatedly been reported for virus-associated MCC.

The majority of genomic studies analyzed MCC tissue samples; only a few studies addressed MCC cell lines. CNV patterns of six virus-positive MCC cell lines were previously characterized using CGH [17]. Three virus-negative MCC cell lines, previously characterized by targeted sequencing, have variant growth characteristics [6]. Notably, the origin of "variant" MCC cell lines is controversial, since these have different growth and gene expression patterns to other "classical" virus-negative cell lines [22,23], which share neuroendocrine growth features—i.e., growing in suspension as spheroids—with virus-positive cell lines. Thus, since comprehensive mutational characterization of the MCC cell lines is missing, we analyzed the mutational landscape of cell lines that are frequently used in MCC research by whole-exome sequencing (WES).

2. Results

Whole-exome sequencing of the virus-positive cell lines WaGa, MKL-1, UKE-MCC3b, UM-MCC13, UM-MCC29 and PeTa, as well as the virus-negative cell lines UM-MCC9, UM-MCC32 and UM-MCC34, was performed using the SureSelect Exon V6 Kit on a HiSeq 4000 with, on average, 118 million reads per sample. Moreover, we directly compared the cell line PeTa with cryopreserved tissue from which PeTa has been established to assess possible differences between the cell line and tissue.

2.1. Mutational Burden and Signatures of MCC Cell Lines Are in Accordance with MCC Tissue Characteristics

Virus-negative MCC cell lines have a higher mutational burden with, on average, 44.5 mutations per megabasepairs (mut/Mbp), constituting, on average, 2.693 absolute mutations per cell line than virus-positive MCC cell lines, which contain, on average, 10.5 mut/Mbp (an average of 637 mutations) (Figure 1A, Table S1). Similarly, the number of coding mutations is higher in virus-negative MCC cell lines with, on average, 15.1 mut/Mbp (an average of 913 mutations per cell line, i.e., 33.6% of respective mutations) compared to 2.31 mut/Mbp in virus-positive MCC cell lines (on average, 140 mutations, i.e., 21.9% of mutations). The average fraction of missense (29.3%) and silent (15.6%) mutations in virus-negative cell lines is also higher than in virus-positive cell lines (16.4%/6.8%) (Figure 1A).

Since mutations are called between the respective cell line and the human reference genome hg19, the observed somatic TMB strongly depends on the filtering strategy for potential polymorphisms (Figures 1A and S1). Polymorphisms are identified using the variant allele frequency (VAF) reported in databases covering nonmalignant exomes and genomes. In general, exome databases cover 71.6% of all MCC cell line mutations while genome databases cover either 91.4% in the 1000 genomes database or 97.7% in the "genome aggregation database" (gnomAD) genome database (Figure S1B,C). Therefore, we filtered for VAFs greater than 0.001% with the comprehensive gnomAD genome database. This reduced the presented TMB of virus-positive MCC cell lines by 98.8% from, on average, 54,850 to 637 mutations per cell line. The TMB of virus-negative MCC cell lines shows a smaller reduction by 94.8% from, on average, 52,614, to 913 mutations (Figures 1B and S1A).

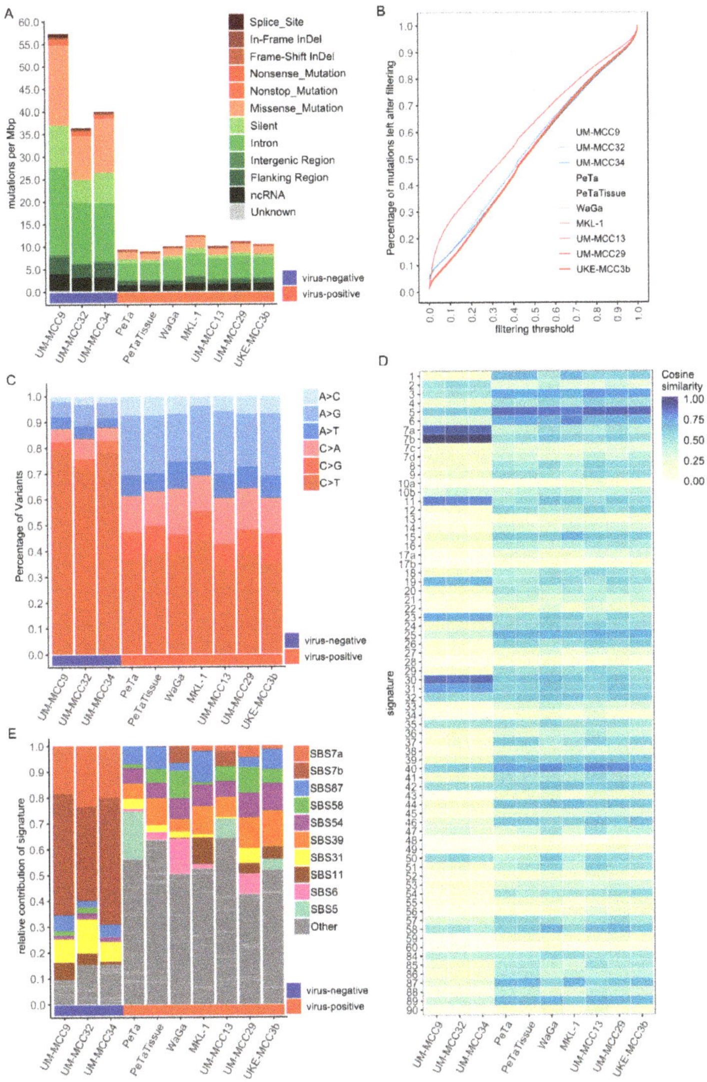

Figure 1. Virus-negative Merkel cell carcinoma (MCC) cell lines show high tumor mutational burden (TMB) and presence of UV-light-induced DNA damage, while virus-positive cell lines have low TMB and lack prominent signatures; (**A**) Mutational burden in mut/Mbp, color-coded by variant classification; (**B**) Filtering of polymorphisms in MCC cell lines showing the relative decrease in TMB (y-axis) with increasing variant allele frequency (VAF) threshold (x-axis) from gnomAD genome database; Figure S1A depicts the same plot with log-transformation of x-axis; (**C**) Contributions of base-pair transitions for single nucleotide variants (SNVs), normalized by total number of SNVs. Complementary transitions are merged in one category (e.g., G > A and C > T as C > T); (**D**) Cosine similarity between trinucleotide context frequencies (TCFs) of MCC cell lines and reference signatures reveals two distinct patterns for virus-positive and -negative cell lines; (**E**) Signature contribution of MCC cell lines after fitting to reference signatures. Signature contributions are normalized to total number of SNVs in the respective cell line. Signatures not reaching at least 10% contribution in at least one sample are summarized as "Other". Abbreviations: mut/Mbp: Mutations per Megabasepair, SBS: Single Base Substitution.

Virus-negative MCC cell lines are characterized by a high fraction of, on average, 77% C > T single-nucleotide variations (SNVs), as compared to 38% in virus-positive MCC cell lines (Figure 1C). This observation already suggests different forms of mutagenesis. Hints regarding the underlying mutagenic process can be retrieved from the first preceding and following basepair of an SNV, i.e., the trinucleotide context frequency (TCF). TCFs for virus-negative MCC cell lines show characteristic C > T transition patterns known to be caused by UV-induced mutagenesis (Figure S2) [24]. In contrast, virus-positive MCC cell lines have a "flat" TCF distribution, i.e., low frequencies for most categories, with only slightly elevated C > T and T > C transitions. For MKL-1 and UKE-MCC3b, there is a higher presence of C > T transitions with guanine as the following basepair; a pattern which often originates from spontaneous deamination of CpGs correlating with progressing age. The systematic comparison of the TCFs of MCC cell lines with reference mutational signatures reflecting defined mutagenic processes reveals distinct patterns for virus-negative and -positive cell lines (Figure 1D). Notably, the aging signature 1 and defective DNA mismatch repair signatures 6 and 15 were very similar to the TCF of MKL-1. Fitting reference signatures to the TCFs demonstrates a high contribution of signatures 7a and 7b for virus-negative MCC cell lines (on average, 67.2%), which are both associated with UV-light-induced DNA damage (Figure 1E). Virus-positive MCC cell lines generally have low individual signature contributions, with no prominent mutational signature present: approximately 50% of the total signature contribution for virus-positive MCC cell lines originates from signatures with less than 10% contribution (Figure 1E). Most of the absolute differences in the mutational burden between virus-negative and -positive MCC cell lines are due to signatures 7a and 7b. However, some mutational signatures have slightly higher signature contributions relative to others, namely signature 31 in virus-negative and signatures 5, 6, 11, 39, 54, 58 and 87 in virus-positive MCC cell lines. The reconstruction efficiency after signature fitting is, on average, higher in virus-negative (99.57%) than in virus-positive (96.95%) cell lines.

To test if the mutational landscape of the MCC cell lines indeed represents that of the original tumor, we compared the MCC cell line PeTa with cryopreserved tissue from which the cell line was derived (Figure 2, Table S2). The respective exomes share almost 80% of mutations, with 21% (120/565) being unique in the cell line and 17% (92/537) unique in the tumor tissue. Somatic variant calling for the cell line using the tissue as reference retrieved 124 variants, of which 38 (31%) were already among the germline-called variants in PeTa. Vice versa, somatic variant calling for the tissue using the cell line as reference resulted in 480 mutations, of which only five (1%) were present in germline-called variants of the tissue (Figure 2).

2.2. Mutations Altering Protein Structure

Next, we investigated mutations predicted to change the amino acid code and likely have an effect on protein function (Figure 3A). Virus-negative MCC cell lines harbor a higher number of nonsense mutations, i.e., mutations introducing a stopcodon, (on average, 53 mutations per cell line, corresponding to 2% of respective mutations) than virus-positive MCC cell lines (on average, six mutations, 0.9% of respective mutations) (Table S3). A total of 21% (≈12 mutations) and 33% (≈2 mutations) of nonsense mutations for virus-negative and -positive cell lines, respectively, are within genes of Hallmark Gene Sets, representing specific biological processes from the molecular signatures database (MSigDB) [25]. Nonsense mutations that are predicted to be pathogenic and cancer-related in ClinVar are in *RB1* in UM-MCC9 (rs794727481) and UM-MCC34 (rs121913304), in *BAP1* in UM-MCC32 (chr3.52437267.G > A), and in the tumor-suppressor gene *CHEK2* in MKL-1 (chr22.29091725.C > T).

Frameshift Insertions and deletions (InDels) are also enriched in absolute numbers in virus-negative MCC cell lines with, on average, 21 mutations (0.8% of respective mutations) compared to 12 mutations (2% of respective mutations) in virus-positive MCC cell lines (Table S3). A total of 18% (≈4 mutations) of virus-negative and 16% (≈2 mutations) of virus-positive frameshift InDels are within Hallmark Gene Sets, among those annotated as pathogenic and cancer-related in ClinVar are *ERCC2* in UM-MCC9 (chr19.45855805.T > -), and *BRCA2* in UM-MCC29 (chr13.32911298.AAAC > -). UM-MCC9 and UM-MCC34 both harbor frameshift deletions in *TP53* (UM-MCC9: chr17.7577070.G > -, UM-MCC34: chr17.7579518.CTTCA > -), and UM-MCC32 a frame-shift deletion in *RB1* (chr13.48919254. CCAGTACCAAAGTTGATAAT > -); the inhibition of both tumor suppressors plays an important role in MCC carcinogenesis [4–9,11–14,26–28]. Besides the frameshift Indels and nonsense mutations, there are missense mutations of *TP53* in UM-MCC29 (rs1057520000) and UM-MCC32 (rs121912651). In UM-MCC9, following the frameshift deletion of *TP53*, are a missense (rs786201059) and a silent (chr17.7577558.G > A) mutation, while UM-MCC34 harbors only a silent mutation (chr17.7579516.G > A) before the frameshift deletion. For *RB1*, UM-MCC9 has a missense mutation (rs137853294) following the nonsense mutation, which, therefore, has no effect on the amino acid sequence. There are several other frameshift InDels that likely contribute to MCC carcinogenesis, for example, a frameshift deletion in *NOTCH1* in UM-MCC9 (chr9.139399867.AG > -). Moreover, only two nonstop mutations are found in this study, the first in the transcription repressor *GMNN* [29] in UM-MCC29 (rs757538616) and the other in chaperonin *TCP1* in WaGa (rs779397332) (Table S3).

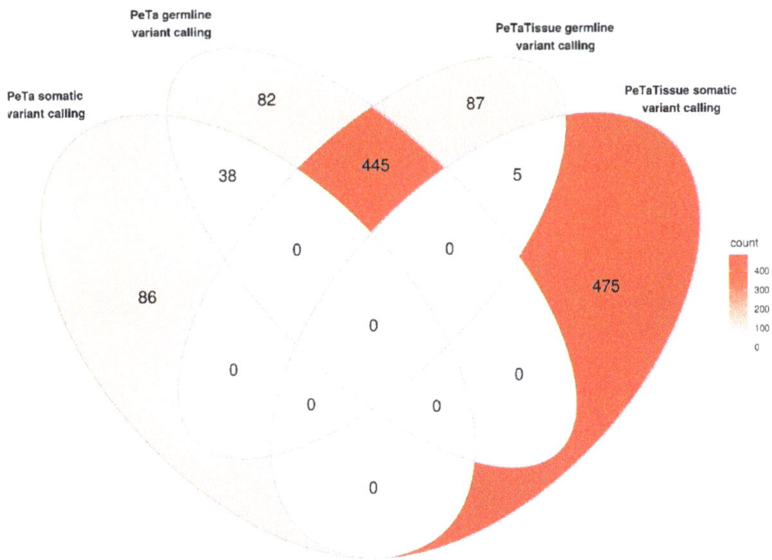

Figure 2. Comparison of unique mutations between PeTa and the respective tumor tissue. Venn-diagram showing how many of the same mutations are shared between the tumor cell line and tissue using either the tissue or the cell line as normal reference for somatic variant calling.

2.3. *Significantly Mutated Genes*

Next, we tested for genes with a significantly higher mutational burden as expected by chance, aka significantly mutated genes (SMGs) (Figure 3B–F, Tables S4–S6) [30]. In this approach, the mutations of several samples are aggregated and compared with a local background model of silent mutations for each respective gene [30]. This analysis was performed separately for all virus-negative (Figure 3B,D) and for all virus-positive (Figure 3C,E) cell lines. Downstream analysis was restricted to Hallmark Gene sets to focus on genes possibly relevant to MCC carcinogenesis. We evaluated the significance of the respective mutational burden by visualizing the distribution of *p*-values (Figure 3B–E). When correcting all *p*-values for multiple testing using Benjamini–Hochberg procedure, the genes *KRT4*, *MDK* and *CACNA1B* remained the only SMGs in the Hallmark Gene Sets present in both MCC types. Virus-negative MCC cell lines harbor more SMGs with a *p*-value lower than 0.01 within Hallmark Gene Sets compared to virus-positive MCC cell lines, i.e., 16 vs. 5 genes, respectively (Figure 3D–F). Among the SMGs for virus-negative cell lines are *TP53* and *RB1*, which are frequently mutated tumor-suppressor genes in virus-negative MCC (Figure 3F) [4–9,11–14]. Of the SMGs found in virus-positive MCC cell lines, UM-MCC29 has a frame-shift deletion in the chromatin modifier *CBX3* (chr7.26248161.A > -), and UKE-MCC3b a falsely annotated nonstop mutation in *NAPA*, the latter composed of an in-frame insertion (chr19.47998837.- > ATTAAA) and deletion (chr19.47998843.GTT > -), resulting in the addition of two and deletion of one amino acid without introducing a stopcodon (Figure 3F).

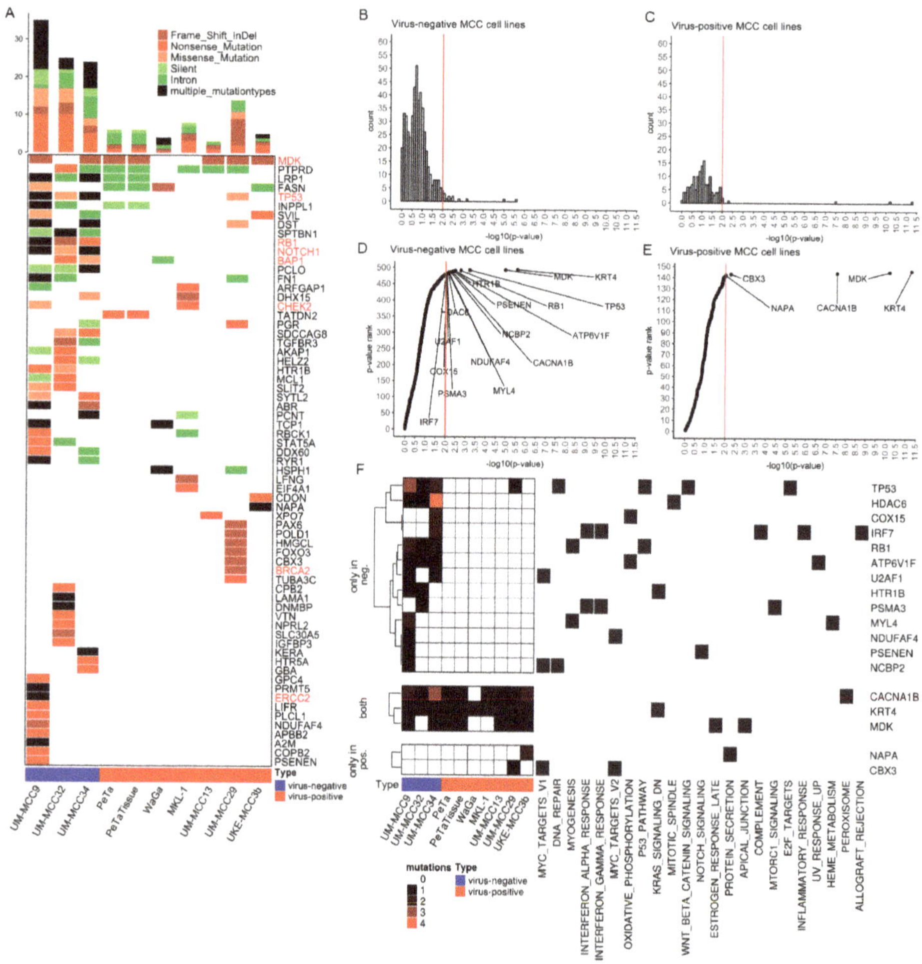

Figure 3. Virus-negative MCC cell lines have high number of coding mutations altering protein structure and significantly mutated genes. (**A**) Oncoplot showing genes selected by the following criteria: (i) containing either a frameshift InDel, nonsense or nonstop mutation and (ii) it is either within a Hallmark Gene Set or its mutation is annotated as pathogenic in ClinVar database. The number of mutations within the selected genes are depicted as bar chart. Both plots are colored by variant classification. Genes emphasized in red are discussed in the results section; (**B**–**E**) Distribution of *p*-values for identification of SMGs as histogram (**B**,**C**) and ranked by *p*-value (**D**,**E**) for virus-negative (**B**,**D**) and -positive (**C**,**E**) MCC cell lines; only genes present in Hallmark Gene Sets were taken into account, red lines indicate a *p*-value of 0.01, genes with a *p*-value of exactly 1 are not shown; (**F**) Mutational burden and involvement in biological processes of SMGs with *p*-value below 0.01 and presence in Hallmark Gene Sets. Abbreviations: MCC: Merkel cell carcinoma, InDel: Insertion and deletion, SMG: Significantly mutated genes.

We identified three SMGs (*KRT4, MDK* and *CACNA1B*) with extraordinary low *p*-values (Figure 3D–F) in all MCC cell lines. *KRT4* contains the exact same large in-frame insertion in all samples (rs11267392), which has a VAF of 87% in the 1000 genomes database. *MDK* comprise the exact same frameshift deletion in a cytosine-rich repeat in six samples, which is actually a mixture of a single-cytosine (chr11.46404342.C > -) and double-cytosine (chr11.46404342.CC > -) deletion. *CACNA1B* has the same large-scale insertion at a splice site in seven samples (chr9.140773612.- > ACGACACGGAGCCC-TATTTCATCGGGATCTTTTGCTTCGAGG CAGGGA, rs370237172).

2.4. Characteristic Copy Number Variation Patterns in Virus-Positive MCC Cell Lines

CNVs were determined from the exome sequencing data (Figure 4A, Table S7). The virus-negative cell lines UM-MCC9 and UM-MCC34, but not UM-MCC32, are characterized by numerous, varying CNVs covering most of the genome. Virus-positive MCC cell lines have less, but more characteristic CNVs, which include whole-chromosome gains of chr1 (UM-MCC29), chr5, chr7, chr8 (UM-MCC29), chr6 (WaGa, PeTa), chr11 (UM-MCC29), chr13 (UM-MCC29, PeTa), chr19, chr20 (UM-MCC29, WaGa) and a complete loss of chr10 (UM-MCC13). Several chromosomes are partially amplified, e.g., chr1q (UM-MCC13, PeTa), chr3q (MKL-1, PeTa) and chr11 (WaGa, PeTa), while others show partial losses, such as chr3p (UM-MCC13), chr8p (UM-MCC13, MKL-1, WaGa) and chr10q (MKL-1, WaGa, PeTa). Only UKE-MCC3b lacks any substantial copy number changes.

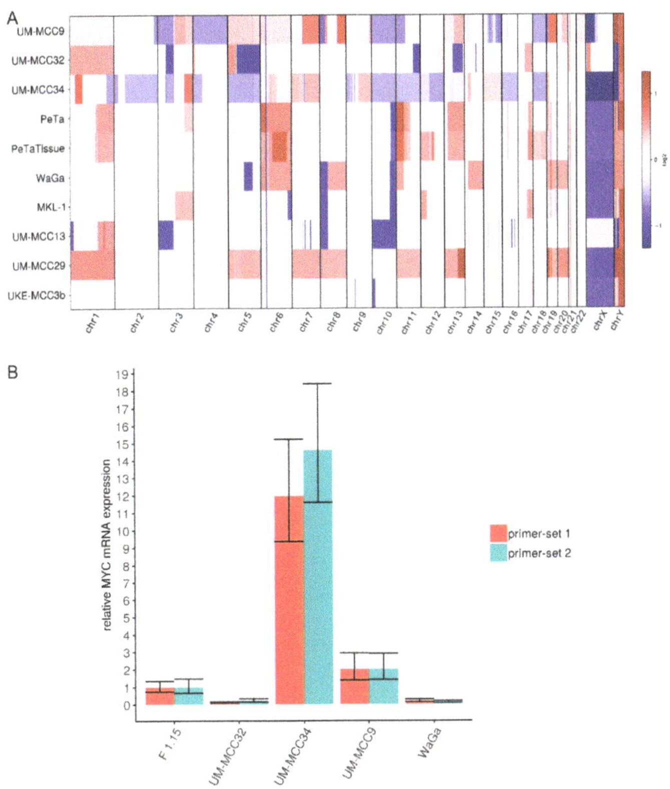

Figure 4. CNVs in MCC cell lines. (**A**) Graphical display of derived CNVs using CNVkit with sex chromosomes relative to haploid reference; (**B**) Expression of *MYC* mRNA in MCC cell lines was determined by qRT-PCR. Cq values were normalized to GAPDH expression and compared to ΔCq value of fibroblasts (F 1.15).

Previous studies reported copy number losses covering *RB1* on chromosome 13 [4,5,11,12,14,16,31]. We also observe large single-copy deletions on chromosome 13 including the loss of *RB1* in virus-negative cell lines (UM-MCC32 and UM-MCC34); in contrast, there are large single-copy gains that include *RB1* in one virus-negative (UM-MCC9) and two virus-positive cell lines (PeTa, UM-MCC29).

Local amplifications of *MYCL* on chromosome 1 have previously been reported for both MCC types [4,5,16,31]. Here, *MYCL* is included in the whole-chromosome gains of UM-MCC29 and UM-MCC32 as well as the partial chromosome gains in UM-MCC34. Interestingly, UM-MCC34 has an extraordinarily high, localized amplification of *MYC* (aka *c-MYC*), with 106 copies covering ~530,000 basepairs on chromosome 8. *MYC* is also included in larger whole- or partial-chromosome gains in UM-MCC29, WaGa and UM-MCC9. These amplifications are associated with a higher *MYC* mRNA expression, which is most pronounced in UM-MCC34 (Figure 4B).

3. Discussion

Due to the lack of suitable genetically engineered mouse models (GEMMs), preclinical functional studies rely on MCC cell lines. However, the detailed genomic characteristics of the applied cell lines are not fully established. Indeed, most studies investigating genomic features of MCC by targeted or WES are based on fresh frozen or formalin-fixed paraffin-embedded (FFPE) tissue samples [4–8,10–14]. Only Wong et al. included three virus-negative cell lines [6] that may be not representative for MCC [22,23]. Here, we present the mutational landscape of three classical virus-negative and six virus-positive MCC cell lines (characteristics are summarized in Table 1). The ratio of virus-positive to -negative cell lines recapitulates the ratio of MCC tumors in countries with high latitude [1]. The genomic features of the MCC cell line cohorts are very similar to those previously reported for the respective MCC tumors. Furthermore, direct comparison of one matched cell line-tissue pair confirmed that genomic alterations accumulated during cell culture only caused minor differences in their mutational landscape. However, expectedly, the cell line did not capture the complete tumor heterogeneity, as many somatic mutations were specific to the tissue.

Table 1. Genomic features of MCC cell lines for both MCC types.

MCC Cell Line Type	Virus-Negative	Virus-Positive
Tumor Mutational Burden	high (on average, 44.5 mut/Mbp)	low (on average, 10.5 mut/Mbp)
Mutagenic processes detected	UV-light-associated DNA damage (SBS7a, SBS7b)	flat mutation profile without prominent signatures.
Coding mutations	many mutations with potential functional effect; many mutated genes	few mutations with potential functional effect; few mutated genes
Copy Number Variations	many widespread CNVs	few, characteristic CNVs

The bold is used to emphasize the row names for the subsequent summary.

Virus-positive MCCs are characterized by very low TMB, a lack of prominent mutational signatures and the absence of functional mutations (Table 1) [4–6,11–13]. Previously reported TMBs for virus-positive MCC, however, show large differences and are inconsistently specified, e.g., regarding normalization. For the WES studies, TMB was reported either as a median of 12.5 SNVs [12], an average of 0.4 mut/Mbp [11] or a median of 1.57 mut/Mbp [13]. We observed, on average, 11 mut/Mbp, which is comparable with studies using targeted sequencing approaches (i.e., an average of 5–10 mut/Mbp [6], a median of 1.2 coding mut/Mbp [5] or up to 16 mut/Mbp [4]). All studies with higher TMB lacked individual normal tissues as a reference for somatic variant calling, hence databases reporting common polymorphisms (e.g., 1000 genomes, exome aggregation consortium (ExAC), gnomAD databases) had to be used for filtering non-somatic variants. Thus, the observed higher TMBs are likely caused by polymorphisms not represented in common databases. This notion is supported by the absence of any prominent mutational signature in virus-positive MCC samples. No single mutational signature has a relevant contribution to the TMB; only "flat" TCF distributions were detected for virus-positive MCC cell lines, which likely represent randomly distributed, unfiltered polymorphisms that may impair the detection of other mutagenic processes. The absence of functional, cancer-related mutations and low signature reconstruction efficiency is in line with this assumption. In contrast, in virus-negative MCC cell lines, TMB is high (on average, 44.5 mut/Mbp), mutational patterns are strongly associated with UV-light-induced DNA damage, and many coding mutations of cancer-related genes exist (Table 1). The primary origin of virus-negative MCC cell lines is associated with UV-exposed areas. UM-MCC9 and UM-MCC32 were derived from primary tumors localized on the scalp, and UM-MCC34 was derived from axillary metastasis presumably originating from a primary tumor on the upper extremity (Table 2) [28]. Some of the virus-positive MCC cell lines were generated from tumors without a clear association with chronic UV-exposure, e.g., PeTa and UKE-MCC3b originated from tumors of the trunk (Table 2) [17]. Interestingly, we did not observe major differences in TMB between cell lines derived from primary tumors (UM-MCC9, UM-MCC32, PeTa) and metastases (UM-MCC34, WaGa, MKL-1, UM-MCC13, UM-MCC29, UKE-MCC3b), which would have been expected from more general observations in cancer (Table 2) [32].

Table 2. Overview of analyzed Merkel cell carcinoma cell lines.

Cell Line	MCPyV Status	Established From	Localization of Primary	Time in Culture	Reference
UM-MCC9	negative	primary, scalp	scalp	>6 years	[28]
UM-MCC32	negative	primary, scalp	scalp	>6 years	[28]
UM-MCC34	negative	axillary lymph node metastasis	presumably arm	>6 years	[28]
PeTa	positive	primary, trunk	trunk	>7 years	[17,26]
WaGa	positive	malignant ascites	head	>10 years	[17,33]
MKL-1	positive	nodal metastasis	unknown	>30 years	[17,34]
UM-MCC13	positive	metastasis, leg	presumably leg	>6 years	[28]
UM-MCC29	positive	inguinal lymph node metastasis	presumably leg	>6 years	[28]
UKE-MCC3b	positive	metastasis, trunk	trunk	>3 years	-

All virus-negative MCC cell lines show *RB1* and *TP53* disruption, either by frameshift deletion, nonsense, missense mutation or, for *RB1*, possibly copy number losses. Alterations in both genes are recurrent mutational features in virus-negative MCC [4–9,11–14,26]. Notably, the exact same nonsense mutations in *RB1* were previously reported for UM-MCC9 (rs794727481 [9,14]) and UM-MCC34 (rs121913304 [4,6]). *RB1* and *TP53* abrogation is also common in other neuroendocrine carcinomas, e.g., in small cell lung, neuroendocrine prostate and pancreatic carcinoma [35]. In this context, it is interesting to note that *MYC* binding motifs are enriched in neuroendocrine genes; thus, it has been proposed that *MYC* overexpression drives the temporal tumor cell evolution [36]. We detected an extraordinarily high *MYC* amplification associated with equally high mRNA expression in UM-MCC34. *MYC* family gene amplification, i.e., 6% for *MYCL* and 4% for *MYC* in virus-negative MCCs [5,16], as well as high *MYC* protein expression, was previously reported [13,37].

The biological importance of SMGs relies on the fact that these may be more prone for mutations due to open chromatin regions, i.e., reflecting the functional state of a cell during mutagenesis, or being positively selected during tumor evolution. The SMGs with extraordinary low *p*-values were *KRT4*, *MDK* and *CACNA1B*, suggesting that these genes may be relevant for MCC carcinogenesis. However, critical examination of these mutations demonstrate that this is very unlikely. The mutations in *KRT4* are present in all cell lines and have been previously identified as a common polymorphism with 87% VAF in the 1000 genomes database. Thus, the *KRT4* mutation is actually the major allele of a single nucleotide polymorphism (SNP) not reflected in the hg19 reference genome. For *MDK*, the detected cytosine deletion is embedded in a sequence of 15 cytosines in close proximity to a stopcodon and is therefore in a region prone to sequencing artifacts. Actually, variations in cytosine counts of this region have already been reported in dbSNP as polymorphisms (rs74916763). Finally, the large-scale insertion in *CACNA1B* is localized at the last basepair of an exon and the inserted sequence is identical to the beginning of the following exon, hence we assume a deletion of an intronic region in between, which has already been reported with 0.3% VAF in the Allele Frequency Aggregator (ALFA) database. Consequently, these three variants likely reflect limitations in the representation and annotation of polymorphisms, which emphasizes the importance of variant filtering and evaluation. In virus-negative MCC cell lines, *TP53* and *RB1* have a relatively low *p*-value compared to other genes (Figure 3D) and, due to their recurrency in MCC, these genes are likely associated with tumorigenesis of virus-negative MCC.

CNVs were previously characterized in MCC using CGH [16–21], genome-wide microarrays [31] or next-generation sequencing [4–7,11,12,15]. We observe higher CNV numbers in virus-negative, and fewer, but characteristic, CNV patterns in virus-positive MCC, indicating a common alteration mechanism for the latter. Notably, the MCPyV-encoded small T antigen was reported to induce centrosome overproduction and to increase the frequency of micronuclei by interaction with E3-ligases, causing chromosome instability [38]. The virus-positive MCC-specific losses and gains may actually affect the tumor suppressor *RB1* and oncogene *MYC*.

APOBEC-mediated mutagenesis is a known feature of viral oncogenesis, e.g., in human-papilloma-virus-associated cancer [39]. In our and previously reported studies, *APOBEC* mutations seem to be absent in MCC [4,5]. However, *APOBEC*-related mutagenesis is restricted to localized, hypermutated regions, aka kataegis, that are difficult to detect by WES and even more so by targeted sequencing. Indeed, in whole-genome analysis, an *APOBEC*-related kataegis was reported in a virus-positive MCC [15]. Thus, to detect *APOBEC*-related mutagenesis with enhanced sensitivity, signature analysis should be restricted to such hypermutated regions [40].

In summary, WES of virus-positive and -negative MCC cell lines with a neuroendocrine growth pattern revealed mutational features resembling those previously observed in MCC tissue samples; hence, our report strengthens the utility of these classical MCC cell lines for functional studies.

4. Materials and Methods

4.1. Cell Lines and Tissues

The MCC cell lines WaGa [33], PeTa [26], MKL-1 [34], UM-MCC13, UM-MCC29, UM-MCC9, UM-MCC32, UM-MCC34 [28] were described before (Table 2). UM-MCC13, UM-MCC29, UM-MCC9, UM-MCC32 and UM-MCC34 were provided by Monique E. Verhaegen, University of Michigan, Ann Arbor, MI, USA. UKE-MCC3b was established at the department of dermatology at the University Medicine Essen, Essen, Germany and the patient gave informed consent (ethics committee approval: 11–4715; 17-7538-BO). WaGa, PeTa, MKL-1, UKE-MCC3b and UM-MCC34 were maintained at 37 °C with 5% CO_2 in RPMI-1640 with 10% fetal bovine serum (FBS), 1% penicillin/streptomycin (PAN Biotech, Aidenbach, Germany), while UM-MCC13, UM-MCC29 and UM-MCC32 were maintained as described previously [28] in self-renewal media [41] including low-glucose DMEM, Neurobasal-A medium, 2-mercaptoethanol, N-2 Supplement (100× (times)), B-27™ Supplement (50×, minus vitamin A), MEM non-essential amino acids solution (100×), Gibco™ Amphotericin (all Thermo Fischer, Dreieich, Germany), retinoic acid (Sigma Aldrich, Darmstadt, Germany), basic fibroblast growth factor, recombinant human IGF-I (Peprotech, Hamburg, Germany), 1% penicillin/streptomycin (PAN Biotech). The self-renewal medium was further supplemented with chicken embryo extract containing HBSS, PBS (PAN Biotech), MEM with Earle's salts and L-glutamine (Thermo Fischer) and Hyaluronidase specs (Sigma Aldrich) [42]. Primary cutaneous fibroblasts (F 1.15) were generated and maintained as previously described [43].

4.2. Library Preparation and Sequencing

DNA was purified using DNeasy Blood & Tissue Kit (Qiagen, Hilden, Germany). Library preparation and sequencing were performed by DKFZ Genomics and Proteomics Core Facility. WES libraries were prepared using SureSelect All Exon V6 Kit (Agilent Technologies, Santa Clara, CA, USA) and subsequently sequenced on HiSeq 4000 (Illumina) paired-end 100bp reads with, on average, 118 million reads per sample.

4.3. Alignment and Variant Calling

Processing of reads in FASTQ format to genomic variations in variant call format (VCF) was performed according to genome analysis toolkit (GATK) best practices of germline short variant discovery for all MCC cell lines. Additionally, for PeTa and PetaTissue, GATK best practices of somatic short variant discovery were used. Paired-end reads in FASTQ Format were aligned to the human reference genome hg19 (GRCh37) using Burrows–Wheeler aligner (BWA) mem v0.7.17 [44]; duplicates were marked using Picard MarkDuplicates and aligned reads sorted using samtools v1.7. GATK Toolkits of version 4.0.12.0 were used. For germline short variant discovery, GATK BaseRecalibrator and ApplyBQSR were applied and, subsequently, variants were called using GATK HaploTypeCaller without normal tissue reference data. For somatic short-variant discovery, the panel of normal (PoN) for PeTa and PeTaTissue were created and variants were called with GATK Mutect2, once with Peta cell line and once with PeTaTissue as normal reference. Variants were annotated using ANNOVAR (Version from 8 June 2020) and databases of Ensembl Gencode v31 (29 September 2019), dbSNP with allelic splitting and left-normalization v150 (29 September 2017) ClinVar (05 March 2015), ExAC (29 November 2015), gnomAD exome and genome collection (v2.1.1, 18 March 2019), 1000 genomes dataset (24 August 2015) and Kaviar database (03 December 2015) were used.

4.4. Variant Filtering

The Maftools R package v2.0.05 was used for VCF to Mutation Annotation Format (MAF) conversion using ensemble genes as gene column, and used for manipulation of MAF files in R [45]. Variants that are not within the probe region of the SureSelect All Exon V6 Kit were removed from analysis. Variants from germline variant calling were filtered and removed from analysis if one of the following criteria was met: SNVs with QD < 2.0, MQ < 50.0, FS > 60.0, SOR > 5.0, MQRankSum < -12.5 or ReadPosRankSum < -8.0 and InDels with QD < 2.0, FS > 200.0, SOR > 10.0, InbreedingCoeff < -0.8 or ReadPosRankSum < -20.0. For evaluation of subsequent filtering of possible polymorphisms, we compared different databases reporting VAFs (Figure S1B,C). Based on this analysis, we filtered a variant as germline polymorphism if it reported a VAF of more than 0.001% in gnomAD v2.1.1 genome. Variants from somatic variant calling of Peta/PeTaTissue were filtered using GATK FilterMutectCalls and not filtered for germline polymorphisms.

4.5. Quantitative Real-Time PCR (qRT-PCR)

RNA of cell lines was extracted using peqlabGold Micro RNA Kit (PEQLAB Biotechnologie GmbH, Erlangen, Germany) and transcribed to cDNA using SuperScript IV Reverse Transcriptase (1000u, Life Technologies GmbH, Darmstadt, Germany). qRT-PCR was performed on the CFX Real-Time PCR system (Bio-Rad Laboratories, Hercules, CA, USA) using LuminoCT SYBR Green qPCR ready Mix (Sigma Aldrich). For *MYC* following primers were used: primer-set 1 forward: GGCTCCTGGCAAAAGGTCA, reverse: CTGCG-

TAGTTGTGCTGATGT; primer-set 2 forward: GTCAAGAGGCGAACACACAAC, reverse: TTGGACGGACAGGATGTATGC. *GAPDH* primer set: forward ACCACAGTCCATGC-CATCAC, reverse TCCACCACCCTGTTGCTGTA. Annealing was performed at 60 °C for 15 s. Relative quantification was performed using the $2^{-\Delta\Delta Cq}$ method implemented in the R package "pcr" [46].

4.6. Bioinformatic Processing

The R Markdown script for analysis of MAF format files is attached in File S1. Normalization to mutations per Megabasepair was done through dividing the number of mutations by the sum of the length of all regions covered by probes (60,456,963 basepairs). Signature analysis was performed using only SNVs and MutationalPatterns R package v1.8.0 [47]. Reference mutational signatures of version 3.1 [48] were downloaded from the Cataloque of Somatic Mutations in Cancer (COSMIC, https://cancer.sanger.ac.uk/cosmic/signatures, (accessed date 1 October 2020)). SMGs were determined using MutSigCV v1.41 [30] and Hallmark Gene Sets v7.1 downloaded from MSigDB was used [25]. Heatmaps were created using ComplexHeatmap R package v2.1.0 [49], other plots with ggplot v3.1.0 [50] and ggVennDiagram using R programming language v3.5.2. CNVs were derived using CNVkit v0.9.6 with default settings.

5. Conclusions

Virus-negative MCC cell lines show high TMB, UV-light DNA damage and several functional coding mutations, while virus-positive MCC cell lines harbor few mutations. Thus, the mutational landscape of MCC cell lines that are frequently used in preclinical research reflect the observations from tumor tissue and confirm their suitability for functional studies.

Supplementary Materials: The following are available online at https://www.mdpi.com/2072-6694/13/4/649/s1, Figure S1: Filtering of polymorphisms in MCC cell lines, Figure S2: TCF for each MCC cell line, Table S1: Mutations found in MCC cell lines, Table S2: Mutations from somatic variant calling of PeTa and tissue of PeTa, Table S3: All nonsense, frameshift and nonstop mutations found in MCC cell lines with respective Hallmark Gene Set, Table S4: Mutations found in SMGs with $p < 0.01$ and within a Hallmark Gene Set, Table S5: SMGs from MutSigCV of virus-negative MCC cell lines, Table S6: SMGs from MutSigCV of virus-positive MCC cell lines, Table S7: CNVs found in MCC cell lines, File S1: R Markdown script used for analysis of MAF files.

Author Contributions: Data curation: K.H.; formal analysis: K.H.; funding acquisition: J.C.B.; investigation: P.G., A.H.-C., C.W.; methodology: K.H.; project administration: J.C.B.; resources: P.G., A.H.-C., C.W., J.U.; supervision: J.C.B.; visualization: K.H.; conceptualization: J.C.B.; writing—original draft preparation: K.H.; writing—review and editing: J.C.B. All authors have read and agreed to the published version of the manuscript.

Funding: This work was funded by the DKTK site budget OE 0460 ED03.

Institutional Review Board Statement: The study was conducted according to the guidelines of the Declaration of Helsinki, and approved by the Institutional Review Board of the University Duisburg-Essen (11-4715, 14 September 2011).

Informed Consent Statement: Informed consent was obtained from all subjects involved in the study.

Data Availability Statement: The data presented in this study is available in Tables S1 and S2. FASTQ Files are available on request from the corresponding author. The FASTQ Files are not publicly available due to privacy reasons.

Acknowledgments: For technical support and sequencing services we thank the German Cancer Research Center (DKFZ) Genomics & Proteomics Core Facility, Heidelberg, Germany. We thank Monique E. Verhaegen, University of Michigan, USA, for providing MCC cell lines. Further, we thank Jan Gravemeyer, Translational Skin Cancer Research, German Cancer Consortium (DKTK), Essen, Germany, for his excellent advice and help in bioinformatic analysis.

Conflicts of Interest: J.C.B. is receiving speaker's bureau honoraria from Amgen, Pfizer, Merck-Serono, Recordati and Sanofi, is a paid consultant/advisory board member/DSMB member for Boehringer Ingelheim, eTheRNA, InProTher, MerckSerono, Pfizer, 4SC, and Sanofi/Regeneron. His group receives research grants from Bristol-Myers Squibb, Merck Serono, HTG, IQVIA, and Alcedis. None of these activities are related to the present manuscript. J.U. is on the advisory board or has received honoraria and travel support from Amgen, Bristol Myers Squibb, GSK, LeoPharma, Merck Sharp and Dohme, Novartis, Pierre Fabre, Roche, Sanofi outside the submitted work. All of the other authors declare no conflict of interest.

References

1. Becker, J.C.; Stang, A.; DeCaprio, J.A.; Cerroni, L.; Lebbé, C.; Veness, M.; Nghiem, P. Merkel cell carcinoma. *Nat. Rev. Dis. Primers* **2017**, *3*, 17077. [CrossRef]
2. Feng, H.; Shuda, M.; Chang, Y.; Moore, P.S. Clonal Integration of a Polyomavirus in Human Merkel Cell Carcinoma. *Science* **2008**, *319*, 1096–1100. [CrossRef] [PubMed]
3. Garneski, K.M.; Warcola, A.H.; Feng, Q.; Kiviat, N.B.; Helen Leonard, J.; Nghiem, P. Merkel Cell Polyomavirus Is More Frequently Present in North American than Australian Merkel Cell Carcinoma Tumors. *J. Investig. Dermatol.* **2009**, *129*, 246–248. [CrossRef] [PubMed]
4. Starrett, G.J.; Thakuria, M.; Chen, T.; Marcelus, C.; Cheng, J.; Nomburg, J.; Thorner, A.R.; Slevin, M.K.; Powers, W.; Burns, R.T.; et al. Clinical and molecular characterization of virus-positive and virus-negative Merkel cell carcinoma. *Genome Med.* **2020**, *12*, 30. [CrossRef] [PubMed]
5. Knepper, T.C.; Montesion, M.; Russell, J.S.; Sokol, E.S.; Frampton, G.M.; Miller, V.A.; Albacker, L.A.; McLeod, H.L.; Eroglu, Z.; Khushalani, N.I.; et al. The Genomic Landscape of Merkel Cell Carcinoma and Clinicogenomic Biomarkers of Response to Immune Checkpoint Inhibitor Therapy. *Clin. Cancer Res.* **2019**, *25*, 5961. [CrossRef]
6. Wong, S.Q.; Waldeck, K.; Vergara, I.A.; Schröder, J.; Madore, J.; Wilmott, J.S.; Colebatch, A.J.; De Paoli-Iseppi, R.; Li, J.; Lupat, R.; et al. UV-Associated Mutations Underlie the Etiology of MCV-Negative Merkel Cell Carcinomas. *Cancer Res.* **2015**, *75*, 5228. [CrossRef] [PubMed]
7. Harms, P.W.; Collie, A.M.B.; Hovelson, D.H.; Cani, A.K.; Verhaegen, M.E.; Patel, R.M.; Fullen, D.R.; Omata, K.; Dlugosz, A.A.; Tomlins, S.A.; et al. Next generation sequencing of Cytokeratin 20-negative Merkel cell carcinoma reveals ultraviolet-signature mutations and recurrent TP53 and RB1 inactivation. *Mod. Pathol.* **2016**, *29*, 240–248. [CrossRef]
8. Veija, T.; Sarhadi, V.K.; Koljonen, V.; Bohling, T.; Knuutila, S. Hotspot mutations in polyomavirus positive and negative Merkel cell carcinomas. *Cancer Genet.* **2016**, *209*, 30–35. [CrossRef] [PubMed]
9. Cohen, P.R.; Tomson, B.N.; Elkin, S.K.; Marchlik, E.; Carter, J.L.; Kurzrock, R. Genomic portfolio of Merkel cell carcinoma as determined by comprehensive genomic profiling: Implications for targeted therapeutics. *Oncotarget* **2016**, *7*, 23454. [CrossRef] [PubMed]
10. Graves, C.A.; Jones, A.; Reynolds, J.; Stuart, J.; Pirisi, L.; Botrous, P.; Wells, J. Neuroendocrine Merkel Cell Carcinoma Is Associated with Mutations in Key DNA Repair, Epigenetic and Apoptosis Pathways: A Case-Based Study Using Targeted Massively Parallel Sequencing. *Neuroendocrinology* **2015**, *101*, 112–119. [CrossRef]
11. Harms, P.W.; Vats, P.; Verhaegen, M.E.; Robinson, D.R.; Wu, Y.-M.; Dhanasekaran, S.M.; Palanisamy, N.; Siddiqui, J.; Cao, X.; Su, F.; et al. The Distinctive Mutational Spectra of Polyomavirus-Negative Merkel Cell Carcinoma. *Cancer Res.* **2015**, *75*, 3720–3727. [CrossRef]
12. Goh, G.; Walradt, T.; Markarov, V.; Blom, A.; Riaz, N.; Doumani, R.; Stafstrom, K.; Moshiri, A.; Yelistratova, L.; Levinsohn, J.; et al. Mutational landscape of MCPyV-positive and MCPyV-negative Merkel cell carcinomas with implications for immunotherapy. *Oncotarget* **2015**, *7*, 3403. [CrossRef] [PubMed]
13. González-Vela, M.d.C.; Curiel-Olmo, S.; Derdak, S.; Beltran, S.; Santibañez, M.; Martínez, N.; Castillo-Trujillo, A.; Gut, M.; Sánchez-Pacheco, R.; Almaraz, C.; et al. Shared Oncogenic Pathways Implicated in Both Virus-Positive and UV-Induced Merkel Cell Carcinomas. *J. Investig. Dermatol.* **2017**, *137*, 197–206. [CrossRef]
14. Cimino, P.J.; Robirds, D.H.; Tripp, S.R.; Pfeifer, J.D.; Abel, H.J.; Duncavage, E.J. Retinoblastoma gene mutations detected by whole exome sequencing of Merkel cell carcinoma. *Mod. Pathol.* **2014**, *27*, 1073–1087. [CrossRef]
15. Starrett, G.J.; Marcelus, C.; Cantalupo, P.G.; Katz, J.P.; Cheng, J.; Akagi, K.; Thakuria, M.; Rabinowits, G.; Wang, L.C.; Symer, D.E.; et al. Merkel Cell Polyomavirus Exhibits Dominant Control of the Tumor Genome and Transcriptome in Virus-Associated Merkel Cell Carcinoma. *mBio* **2017**, *8*, e02079-02016. [CrossRef] [PubMed]
16. Paulson, K.G.; Lemos, B.D.; Feng, B.; Jaimes, N.; Peñas, P.F.; Bi, X.; Maher, E.; Cohen, L.; Helen Leonard, J.; Granter, S.R.; et al. Array-CGH Reveals Recurrent Genomic Changes in Merkel Cell Carcinoma Including Amplification of L-Myc. *J. Investig. Dermatol.* **2009**, *129*, 1547–1555. [CrossRef] [PubMed]
17. Schrama, D.; Sarosi, E.-M.; Adam, C.; Ritter, C.; Kaemmerer, U.; Klopocki, E.; König, E.-M.; Utikal, J.; Becker, J.C.; Houben, R. Characterization of six Merkel cell polyomavirus-positive Merkel cell carcinoma cell lines: Integration pattern suggest that large T antigen truncating events occur before or during integration. *Int. J. Cancer* **2019**, *145*, 1020–1032. [CrossRef]

18. Van Gele, M.; Leonard, J.H.; Van Roy, N.; Van Limbergen, H.; Van Belle, S.; Cocquyt, V.; Salwen, H.; De Paepe, A.; Speleman, F. Combined karyotyping, CGH and M-FISH analysis allows detailed characterization of unidentified chromosomal rearrangements in Merkel cell carcinoma. *Int. J. Cancer* **2002**, *101*, 137–145. [CrossRef]
19. Van Gele, M.; Speleman, F.; Vandesompele, J.; Van Roy, N.; Leonard, J.H. Characteristic pattern of chromosomal gains and losses in Merkel cell carcinoma detected by comparative genomic hybridization. *Cancer Res.* **1998**, *58*, 1503–1508.
20. Larramendy, M.L.; Koljonen, V.; Böhling, T.; Tukiainen, E.; Knuutila, S. Recurrent DNA copy number changes revealed by comparative genomic hybridization in primary Merkel cell carcinomas. *Mod. Pathol.* **2004**, *17*, 561–567. [CrossRef]
21. Härle, M.; Arens, N.; Moll, I.; Back, W.; Schulz, T.; Scherthan, H. Comparative genomic hybridization (CGH) discloses chromosomal and subchromosomal copy number changes in Merkel cell carcinomas. *J. Cutan. Pathol.* **1996**, *23*, 391–397. [CrossRef] [PubMed]
22. Gele, M.V.; Boyle, G.M.; Cook, A.L.; Vandesompele, J.; Boonefaes, T.; Rottiers, P.; Roy, N.V.; De Paepe, A.; Parsons, P.G.; Leonard, J.H.; et al. Gene-expression profiling reveals distinct expression patterns for Classic versus Variant Merkel cell phenotypes and new classifier genes to distinguish Merkel cell from small-cell lung carcinoma. *Oncogene* **2004**, *23*, 2732–2742. [CrossRef] [PubMed]
23. Daily, K.; Coxon, A.; Williams, J.S.; Lee, C.-C.R.; Coit, D.G.; Busam, K.J.; Brownell, I. Assessment of Cancer Cell Line Representativeness Using Microarrays for Merkel Cell Carcinoma. *J. Investig. Dermatol.* **2015**, *135*, 1138–1146. [CrossRef] [PubMed]
24. Alexandrov, L.B.; Nik-Zainal, S.; Wedge, D.C.; Aparicio, S.A.J.R.; Behjati, S.; Biankin, A.V.; Bignell, G.R.; Bolli, N.; Borg, A.; Børresen-Dale, A.-L.; et al. Signatures of mutational processes in human cancer. *Nature* **2013**, *500*, 415–421. [CrossRef]
25. Liberzon, A.; Birger, C.; Thorvaldsdóttir, H.; Ghandi, M.; Mesirov, J.P.; Tamayo, P. The Molecular Signatures Database (MSigDB) hallmark gene set collection. *Cell Syst.* **2015**, *1*, 417–425. [CrossRef] [PubMed]
26. Houben, R.; Dreher, C.; Angermeyer, S.; Borst, A.; Utikal, J.; Haferkamp, S.; Peitsch, W.K.; Schrama, D.; Hesbacher, S. Mechanisms of p53 restriction in Merkel cell carcinoma cells are independent of the Merkel cell polyoma virus T antigens. *J. Investig. Dermatol.* **2013**, *133*, 2453–2460. [CrossRef]
27. Hesbacher, S.; Pfitzer, L.; Wiedorfer, K.; Angermeyer, S.; Borst, A.; Haferkamp, S.; Scholz, C.-J.; Wobser, M.; Schrama, D.; Houben, R. RB1 is the crucial target of the Merkel cell polyomavirus Large T antigen in Merkel cell carcinoma cells. *Oncotarget* **2016**, *7*, 32956. [CrossRef]
28. Verhaegen, M.E.; Mangelberger, D.; Weick, J.W.; Vozheiko, T.D.; Harms, P.W.; Nash, K.T.; Quintana, E.; Baciu, P.; Johnson, T.M.; Bichakjian, C.K.; et al. Merkel cell carcinoma dependence on bcl-2 family members for survival. *J. Investig. Dermatol.* **2014**, *134*, 2241–2250. [CrossRef] [PubMed]
29. Dhar, S.K.; Yoshida, K.; Machida, Y.; Khaira, P.; Chaudhuri, B.; Wohlschlegel, J.A.; Leffak, M.; Yates, J.; Dutta, A. Replication from oriP of Epstein-Barr Virus Requires Human ORC and Is Inhibited by Geminin. *Cell* **2001**, *106*, 287–296. [CrossRef]
30. Lawrence, M.S.; Stojanov, P.; Polak, P.; Kryukov, G.V.; Cibulskis, K.; Sivachenko, A.; Carter, S.L.; Stewart, C.; Mermel, C.H.; Roberts, S.A.; et al. Mutational heterogeneity in cancer and the search for new cancer-associated genes. *Nature* **2013**, *499*, 214–218. [CrossRef]
31. Carter, M.D.; Gaston, D.; Huang, W.-Y.; Greer, W.L.; Pasternak, S.; Ly, T.Y.; Walsh, N.M. Genetic profiles of different subsets of Merkel cell carcinoma show links between combined and pure MCPyV-negative tumors. *Hum. Pathol.* **2018**, *71*, 117–125. [CrossRef]
32. Patel, S.A.; Rodrigues, P.; Wesolowski, L.; Vanharanta, S. Genomic control of metastasis. *Br. J. Cancer* **2021**, *124*, 3–12. [CrossRef]
33. Houben, R.; Shuda, M.; Weinkam, R.; Schrama, D.; Feng, H.; Chang, Y.; Moore, P.S.; Becker, J.C. Merkel cell polyomavirus-infected Merkel cell carcinoma cells require expression of viral T antigens. *J. Virol.* **2010**, *84*, 7064–7072. [CrossRef] [PubMed]
34. Rosen, S.T.; Gould, V.E.; Salwen, H.R.; Herst, C.V.; Le Beau, M.M.; Lee, I.; Bauer, K.; Marder, R.J.; Andersen, R.; Kies, M.S.; et al. Establishment and characterization of a neuroendocrine skin carcinoma cell line. *Lab. Investig.* **1987**, *56*, 302–312.
35. Rickman, D.S.; Beltran, H.; Demichelis, F.; Rubin, M.A. Biology and evolution of poorly differentiated neuroendocrine tumors. *Nat. Med.* **2017**, *23*, 664–673. [CrossRef] [PubMed]
36. Ireland, A.S.; Micinski, A.M.; Kastner, D.W.; Guo, B.; Wait, S.J.; Spainhower, K.B.; Conley, C.C.; Chen, O.S.; Guthrie, M.R.; Soltero, D.; et al. MYC Drives Temporal Evolution of Small Cell Lung Cancer Subtypes by Reprogramming Neuroendocrine Fate. *Cancer Cell* **2020**, *38*, 60–78.e12. [CrossRef] [PubMed]
37. Shao, Q.; Kannan, A.; Lin, Z.; Stack, B.C.; Suen, J.Y.; Gao, L. BET Protein Inhibitor JQ1 Attenuates Myc-Amplified MCC Tumor Growth In Vivo. *Cancer Res.* **2014**, *74*, 7090. [CrossRef]
38. Kwun, H.J.; Wendzicki, J.A.; Shuda, Y.; Moore, P.S.; Chang, Y. Merkel cell polyomavirus small T antigen induces genome instability by E3 ubiquitin ligase targeting. *Oncogene* **2017**, *36*, 6784–6792. [CrossRef] [PubMed]
39. Burns, M.B.; Temiz, N.A.; Harris, R.S. Evidence for APOBEC3B mutagenesis in multiple human cancers. *Nat. Genet.* **2013**, *45*, 977–983. [CrossRef]
40. Maura, F.; Degasperi, A.; Nadeu, F.; Leongamornlert, D.; Davies, H.; Moore, L.; Royo, R.; Ziccheddu, B.; Puente, X.S.; Avet-Loiseau, H.; et al. A practical guide for mutational signature analysis in hematological malignancies. *Nat. Commun.* **2019**, *10*, 2969. [CrossRef]
41. Molofsky, A.V.; He, S.; Bydon, M.; Morrison, S.J.; Pardal, R. Bmi-1 promotes neural stem cell self-renewal and neural development but not mouse growth and survival by repressing the p16Ink4a and p19Arf senescence pathways. *Genes Dev.* **2005**, *19*, 1432–1437. [CrossRef]

42. Stemple, D.L.; Anderson, D.J. Isolation of a stem cell for neurons and glia from the mammalian neural crest. *Cell* **1992**, *71*, 973–985. [CrossRef]
43. Fan, K.; Spassova, I.; Gravemeyer, J.; Ritter, C.; Horny, K.; Lange, A.; Gambichler, T.; Ødum, N.; Schrama, D.; Schadendorf, D.; et al. Merkel cell carcinoma-derived exosome-shuttle miR-375 induces fibroblast polarization by inhibition of RBPJ and p53. *Oncogene* **2020**. [CrossRef]
44. Li, H.; Durbin, R. Fast and accurate short read alignment with Burrows-Wheeler transform. *Bioinformatics* **2009**, *25*, 1754–1760. [CrossRef] [PubMed]
45. Mayakonda, A.; Lin, D.C.; Assenov, Y.; Plass, C.; Koeffler, H.P. Maftools: Efficient and comprehensive analysis of somatic variants in cancer. *Genome Res.* **2018**, *28*, 1747–1756. [CrossRef]
46. Ahmed, M.; Kim, D.R. pcr: An R package for quality assessment, analysis and testing of qPCR data. *PeerJ* **2018**, *6*, e4473. [CrossRef] [PubMed]
47. Blokzijl, F.; Janssen, R.; van Boxtel, R.; Cuppen, E. MutationalPatterns: Comprehensive genome-wide analysis of mutational processes. *Genome Med.* **2018**, *10*, 33. [CrossRef] [PubMed]
48. Alexandrov, L.B.; Kim, J.; Haradhvala, N.J.; Huang, M.N.; Tian Ng, A.W.; Wu, Y.; Boot, A.; Covington, K.R.; Gordenin, D.A.; Bergstrom, E.N.; et al. The repertoire of mutational signatures in human cancer. *Nature* **2020**, *578*, 94–101. [CrossRef]
49. Gu, Z.; Eils, R.; Schlesner, M. Complex heatmaps reveal patterns and correlations in multidimensional genomic data. *Bioinformatics* **2016**, *32*, 2847–2849. [CrossRef] [PubMed]
50. Wickham, H. *Ggplot2: Elegant Graphics for Data Analysis*; Springer: New York, NY, USA, 2016.

cancers

Article

Sex Differences in Overall Survival and the Effect of Radiotherapy in Merkel Cell Carcinoma—A Retrospective Analysis of A Swedish Cohort

Hannah Björn Andtback [1], Viveca Björnhagen-Säfwenberg [2], Hao Shi [1], Weng-Onn Lui [1], Giuseppe V. Masucci [1] and Lisa Villabona [1,*]

[1] Department Oncology-Pathology, Karolinska Institute and BioClinicum, Karolinska University Hospital, 17176 Stockholm, Sweden; Hannah.bjorn-andtback@sll.se (H.B.A.); hao.shi@ki.se (H.S.); weng-onn.lui@ki.se (W.-O.L.); Giuseppe.masucci@ki.se (G.V.M.)

[2] Department of Reconstructive Plastic Surgery, Karolinska University Hospital, 17176 Stockholm, Sweden; Viveca.bjornhagen-safwenberg@sll.se

* Correspondence: lisa.villbona@ki.se; Tel.: +46-736-63-98-98

Simple Summary: Merkel cell carcinoma (MCC) is a rare and aggressive skin cancer which is believed to be partially caused by a virus or ultraviolet exposure. Most previous studies have shown that MCC is more common in men compared to women, virus associated MCC has a better prognosis and surgery followed by radiotherapy gives a better outcome. In this article, we explore these traits in a Swedish cohort of 113 patients and find that MCC is more common in women and female patients have a longer survival compared to male patients. In addition, we found that virus negative MCC has a worse outcome in male patients and radiotherapy after surgery gives a better outcome for patients who are treated with a curative dosage, irrespective of sex.

Abstract: Merkel cell carcinoma (MCC) is a rare and aggressive skin cancer where Merkel cell Polyomavirus (MCPyV) contributes to the pathogenesis. In an adjuvant setting, radiotherapy (RT) is believed to give a survival benefit. The prognostic impact of sex related to MCPyV-status and adjuvant RT were analyzed in patients referred to Karolinska University Hospital. Data were collected from 113 patients' hospital records and MCPyV analyses were made in 54 patients (48%). We found a significantly better overall survival (OS) for women compared to men and a significant difference in OS in patients receiving adjuvant RT. Furthermore, we found that men with virus negative MCC have an increased risk for earlier death (HR 3.6). This indicates that MCPyV positive and negative MCC act as two different diseases, and it might be due to different mechanism in the immune response between male and female patients. This could have significance in tailoring treatment and follow-up in MCC patients in the future.

Keywords: merkel cell carcinoma; merkel cell polyoma virus; sex; radiotherapy

Citation: Björn Andtback, H.; Björnhagen-Säfwenberg, V.; Shi, H.; Lui, W.-O.; Masucci, G.V.; Villabona, L. Sex Differences in Overall Survival and the Effect of Radiotherapy in Merkel Cell Carcinoma—A Retrospective Analysis of A Swedish Cohort. *Cancers* **2021**, *13*, 265. https://doi.org/10.3390/cancers13020265

Received: 30 November 2020
Accepted: 31 December 2020
Published: 12 January 2021

1. Introduction

Merkel cell carcinoma (MCC) is a rare and highly malignant neuroendocrine skin cancer that mainly affects older people. The yearly incidence is 2500 in the United States and Europe and 60 cases in Sweden [1–3]. The disease mortality in MCC is as high as 46% within five years [4]. The rarity of the disease and its tendency to affect the elderly has contributed to MCC being little studied and the needs for novel prognostic and predictive biomarkers and new treatment regimens are substantial. Although rare, MCC has in several reports shown a rise in incidence over the last decades [2,5–7].

Merkel cell polyomavirus (MCPyV) was discovered in 2008, which was shown to be clonally integrated in the DNA of up to 80% of MCC tumors [8,9]. The presence of the virus has since been reported to be a favorable prognostic trait in MCC [10–12]. A trend towards women having a better outcome in MCC has been seen previously [13,14] and a recent finding from a large cohort in the U.S. establishes that women do have a better disease specific survival than men [15].

Several clinical risk factors for developing MCC have been identified, but much is still to be learned about the pathogenesis. Besides MCPyV, other risk factors are advanced age, chronic immunosuppression and prolonged ultraviolet (UV) exposure, therefore the primary tumor most often is found on sun exposed skin [16,17]. Curative treatment for MCC with localized disease consists of surgical resection of the primary tumor and the addition of postoperative (adjuvant) radiotherapy (RT) which in some settings has been shown to give a reduced risk of occurrence and survival advantage in a subgroup of patients [18]. However, a study of clinical outcomes and variables for a Swedish MCC-cohort has to our knowledge not yet been published.

In recent years, with the birth of immune checkpoint inhibitors, a new treatment option was born for patients with MCC [19,20]. Considering the pathogenesis of viral association and/or UV radiation, which is prone to cause a high tumor mutational burden [21], the immunogenicity of MCC should not be surprising. Furthermore, patients with a T cell dysfunction were shown to have an increased risk of developing MCC [22–24].

Our aim with this study was to analyze outcomes with regard to sex, adjuvant RT and MCPyV-status in a Swedish cohort, in order to improve the knowledge of MCC and identify prognostic traits for a better understanding of the possibility to tailor treatment and follow up strategies in the future.

2. Results

2.1. Cohort Characteristics and Overall Survival

In total, 113 patients, mostly living in the Stockholm Region and referred to Karolinska University Hospital between 1 January 1987 and 31 December 2019, diagnosed with MCC and treated with a curative intent, were included in the study. Detailed data on the patients are presented in Table 1. Of the patients, 64 were female (57%) and 49 were male (43%). Median age at surgery was 76 years (range 19–100) for the entire cohort, for women 79 years (range 19–100) and for men 75 years (range 59–94). There was a clear difference in overall survival (OS) between patients aged 19–69 years (younger: 22%) and >70 years (older: 78%) (Figure 1).

Table 1. Patients clinical characteristics and treatments.

Cohort Characteristics		Cohort		Female		Male	
		n	%	n	%	n	%
Cohort		113	100	64	57	49	43
Age	Median, years	76		79		75	
	19–69	25	22	18	28	7	14
	>70	88	78	46	46	42	86
Tumor Location	Head and neck	53	47	30	47	23	47
	Upper extremity	24	21	14	22	10	20
	Lower extremity	20	18	13	20	7	14
	Trunk	12	11	5	8	7	14
	Genital area	4	4	2	3	2	4
Stage	I	64	57	36	56	28	57
	II	35	31	22	34	13	27
	III	14	12	6	9	8	16
MCPyV-Status in Tumor		54		29		25	
	Positive	40	74	21	72	19	76
	Negative	14	26	8	28	6	24
Treatment	Surgery	66	58	36	56	30	61
	Surgery and radiotherapy	47	42	28	44	19	39

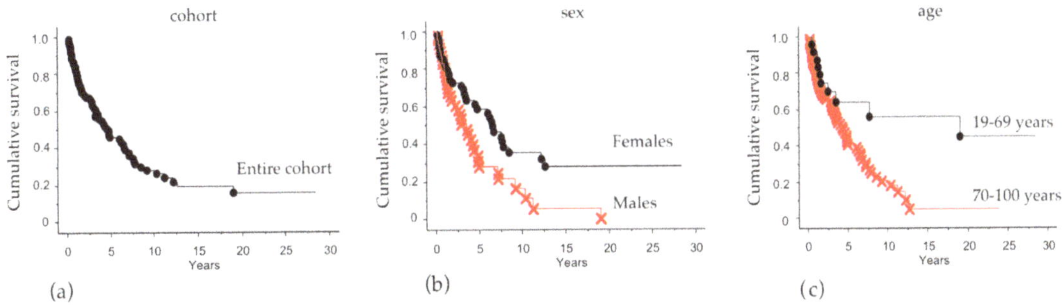

Figure 1. Overall survival analysis in relation to gender and age. Kaplan–Meier plots illustrating overall survival (OS) in: (**a**) the entire cohort; (**b**) Female (black) vs. Male (red) $p = 0.04$; and (**c**) age groups 19–69 years (black) vs. >70 years (red) $p = 0.005$.

The localization of the primary tumor was distributed with a large proportion in the head and neck region (47%) and the others divided among upper extremity (21%), lower extremity (18%), trunk (11%) and genital area (3%) (Table 1).

The patients presented with clinical stages I–III and the majority was stage I (57%) followed by stage II (31%) and stage III (12%).

At the end of the observation period, the probability of survival in the entire cohort was 16% (Figure 1a). There was significantly higher OS for women (30%) compared to men ($p = 0.04$; Figure 1b). Patients under 70 years old had a better outcome (46%) than older patients ($p = 0.005$; Figure 1c). There was no statistically significant difference in outcome between clinical stages.

Patients who had the primary MCC localized in the extremities had a better outcome compared to other anatomical sites (Table 2). This was seen both in entire cohort (HR 0.48) and in the female patients (HR 0.35) for extremities vs. trunk. The comparison of extremities vs. head and neck region was statistically significant when comparing the whole cohort, but only a tendency when separated by sex. There was no significant difference between localization in the head and neck region compared to the trunk (Table 2).

Table 2. Overall survival comparison between primary tumor site.

Parameters	Extremities vs. Trunk			Extremities vs. H&N			H&N vs. Trunk		
	Cohort	Females	Males	Cohort	Females	Males	Cohort	Females	Males
Hazard	0.48	0.35	0.88	0.53	0.52	0.48	0.9	0.65	1.6
C.I. 95%	0.23–0.97	0.12–1.02	0.31–2.4	0.32–0.87	0.24–1.11	0.21–1.08	0.47–1.7	0.24–1.7	0.65–4.3
P	0.03	0.05	ns	0.034	ns	ns	ns	ns	ns

HR, Hazard ratio; CI, confidence interval; ns, not significant.

2.2. MCPyV-Status and Overall Survival

Tumor samples from 54 patients (47%) were available for the detection of MCPyV in tumor tissue (Table 1). In these samples, 74% were positive and 26% negative. The distribution by sex was similar: 72% positive and 28% negative in the female patients and 76% positive and 24% negative in the male patients. A comparison between male and female patients for the risk to die due to MCPyV status is shown in Table 3. Among the 54 MCC patients with MCPyV status, there was no difference in the risk for negative or positive patients. However, male patients with virus-negative MCC had an increased risk for death compared to male patients with virus-positive tumors (HR 3.6; 95% CI, 1.2–10; $p = 0.018$). Using Kaplan–Meier survival analysis, a better survival was also observed in the MCPyV positive male patients (Figure 2). Female patients' viral status had no impact on OS in this analysis (Table 3).

Table 3. Hazard ratios by Cox–Mantel regression analysis comparing MCPyV negative vs. positive filtered by sex.

Sample	MCPyV Negative vs. Positive		
	HR	95% C.I.	*p*-Value
MCPyV cohort (*n* = 54)	1.3	0.65–2.6	ns
Females (*n* = 29)	0.84	0.32–2.2	ns
Males (*n* = 25)	3.6	1.2–10	0.018

HR, hazard ratio; CI, confidence interval; ns, not significant.

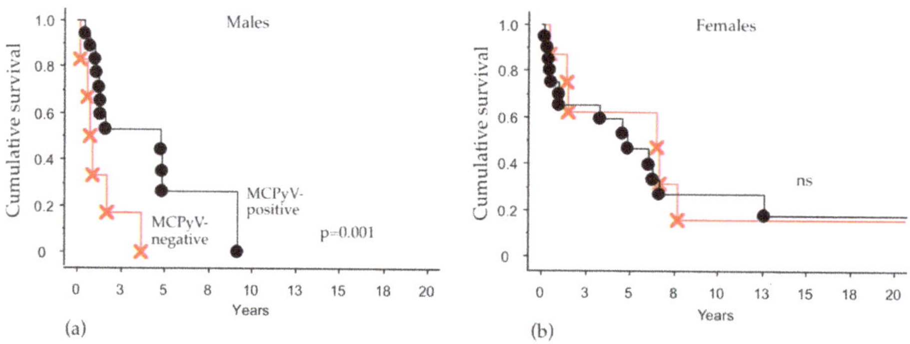

Figure 2. Overall survival analysis in male patients in relation to MCPyV status. Kaplan–Meier plots illustrating overall survival (OS) in (**a**) male and in (**b**) female patients with MCPyV-positive (black) or MCPyV-negative (red) tumor tissue, $p = 0.001$.

2.3. Treatment and Overall Survival

In this cohort, 66 (58%) patients were treated with surgery alone and 47 (42%) patients received radiotherapy in a variety of regimens (Table 1).

Patients who received adjuvant RT after surgery had a significant benefit for survival ($p = 0.0001$) (Figure 3a). No difference was detected between male and female patients (Figure 3b,c).

In addition, we analyzed the efficacy of radiotherapy in cases where relapse was detected. For this reason, the patients were divided into three subgroups considering the total amount of radiation (palliative and adjuvant) received during their disease process (never exposed to RT, RT < 50 Gy and RT ≥ 50 Gy) (Figure 3d–f).

Patients who received ≥50 Gy had a better outcome compared to patients who received a lower dose. The latter group did not differ from patients who never received radiation and this tendency was most explicit in the female group of patients.

Univariate and multivariate analysis of the risk (Cox–Mantel) was performed on the clinical variable collected and summarized in a forest plot (Figure 4). The figure summarizes the findings of the prognostic variables investigated, where younger age, tumor location on extremity and radiotherapy treatment were associated with a better outcome, while male sex was a factor for a worse outcome and increased risk for death.

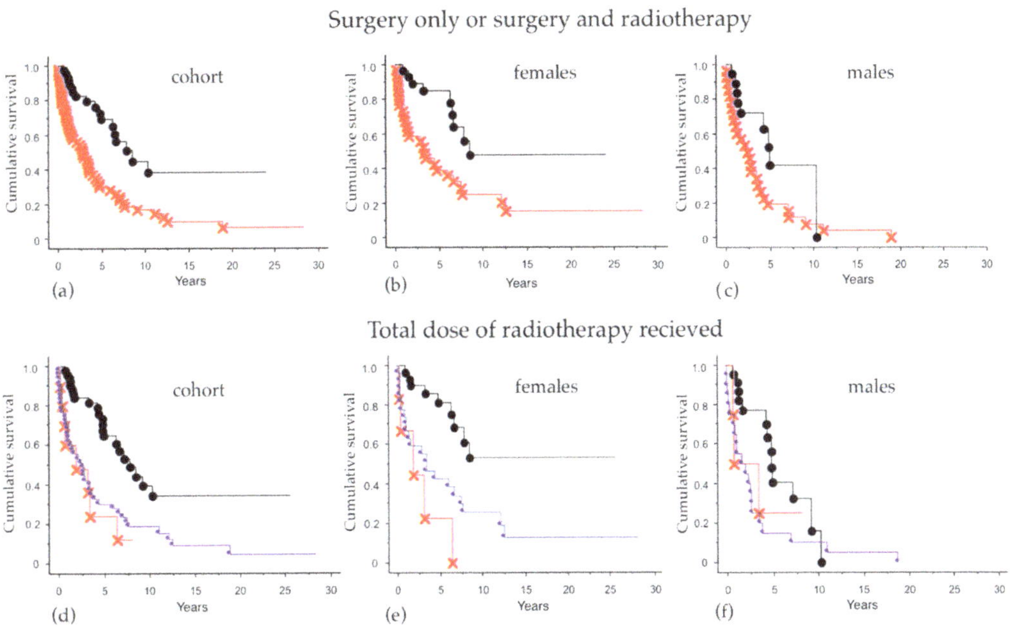

Figure 3. Overall survival in patients receiving surgery alone, a combination of surgery and adjuvant radiotherapy (RT), as well as radiation doses (**a**–**c**). Comparisons of overall survival between patients treated with surgery alone (red) and surgery plus adjuvant RT (black) in: (**a**) the entire cohort, $p = 0.0001$; (**b**) female patients only, $p = 0.002$; and (**c**) male patients only, $p = 0.03$. (**d**–**f**) Comparisons of overall survival among patients treated with radiation ≥50 Gy (black), <50 Gy (red) and patients never exposed to radiotherapy (blue) in: (**d**) the entire cohort, $p = 0.0001$; (**e**) only female, $p = 0.0005$; and (**f**) only male patients, $p = 0.07$.

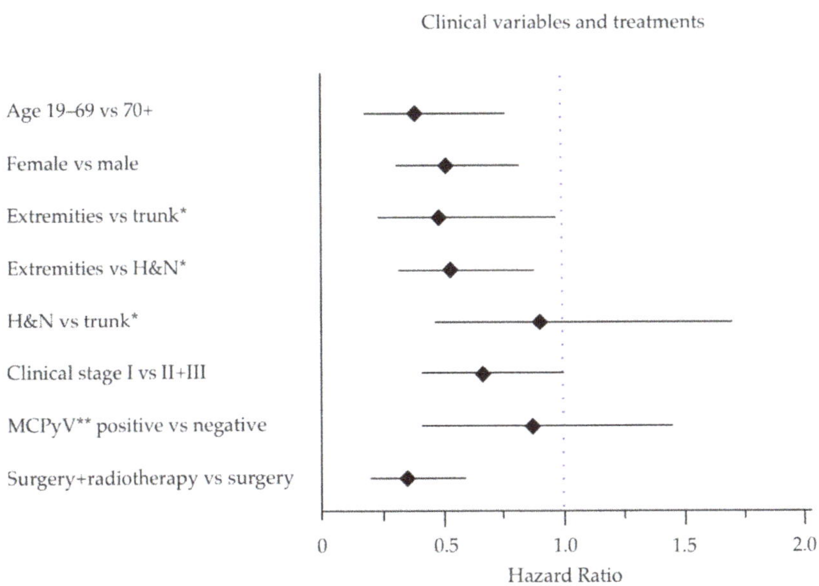

Figure 4. Forest plot for the hazard ratio of the clinical variables and MCPyV-status. * For both sexes, differences are shown in Table 2. ** For both sexes, differences are shown in Table 3.

3. Discussion

In this study, we reviewed clinical data and outcomes for 113 MCC patients from the Stockholm region who were referred to the Karolinska University Hospital in Stockholm, Sweden. The rarity of MCC makes the relatively small number a large cohort by Nordic standards and to our knowledge the largest historical cohort with clinical outcomes described in Sweden. We utilized data from patient hospital records and cause of death registry, as well as MCPyV status available in tumor tissue. The data were correlated to overall survival and sex in addition to treatment received.

In our cohort, we show a better overall survival in patients receiving adjuvant RT after surgery compared to patients who were treated with surgery alone.

Adjuvant RT for MCC has been used in selected cases since the 1970s at Karolinska University Hospital; however, it was only since the late 1980s that a definite treatment schedule has been applied for adjuvant purposes. Very little is presented in the literature in this respect. Despite all the limitations, most retrospective analyses show with relatively clear consensus that adjuvant RT reduces recurrence [25–27], and only two other studies have shown a positive impact on overall survival [28,29]. Both studies are large retrospective MCC cohorts investigating the benefit of adjuvant RT. Chen et al. analyzed 4815 patients with MCC in the head and neck region and showed a survival benefit from adjuvant RT in patients with narrow surgery margins, large tumors and male sex [28]. Bhatia et al. analyzed 6908 patients and reported a benefit both for local recurrence and overall survival in patients with stage I and II disease, but not stage III [29]. Our results show that patients receiving radiotherapy had a clear survival benefit compared to patients who received surgery alone. In our much smaller cohort, we clearly see a survival benefit in both male and female patients receiving adjuvant RT > 50 Gy. Patients who received <50 Gy were most likely offered radiotherapy with palliative intent, which may be the reason for their much worse prognosis. Even though the number of patients was insufficient to analyze any benefit for patients in different clinical stages, our results strengthen the international consensus that MCC patients should be offered adjuvant RT.

Our findings also show that female patients, regardless of MCPyV status, had a significantly improved OS compared to male patients. This finding is also supported by a recent report from an analysis of a large cohort of MCC cases in the US [15].

Previous analysis has shown an inconsistency of the prognostic traits of MCPyV; some studies have shown that patients with MCPyV-positive tumors have a more favorable outcome, whereas others have either found it to be unclear or even prognostically unfavorable [10–12,30–33]. MCPyV-positivity and better outcome was a trend in our material, but the results were non-significant. Interestingly, when we made a multivariate analysis with sex and MCPyV-status, we found that the male patients with MCPyV-negative tumors had the worst outcome and a significantly higher risk for death compared to male patients with MCPyV positive tumors (HR 3.6). The MCPyV status of female patients did not affect outcome in our cohort. This novel finding may serve as a prognostic marker, where male patients and especially virus negative ones, could benefit from closer clinical monitoring and evaluation after primary treatment.

The differences in MCPyV-positive and negative MCC have been extensively researched [34], some even going as far as suggesting that MCPyV-negative MCC does not exist [35]. Our findings in gender differences in outcome may add another dimension to previous findings.

Considering the immunogenicity of MCC, however, one may raise the question of whether these differences in outcome of the patients regarding sex could be due to different immune responses between men and women. Several publications [36–39] have explored both the difference in outcome of immunotherapy treatment between men and women, but also the differences in immune response between the sexes [38]. Given the immunogenicity of MCC, additional studies of immunological markers, such as CD8+ lymphocyte infiltration, MHC class I expression and HLA-genotype would be of interest to further shed light on the sex differences in the immune response. The novel treatment options of immunotherapy for MCC and the reports of the differences in immune response between male and female indicates that sex may play a role in the future treatment options for these patients.

Male sex has been described as an independent risk factor for developing MCC [1,2]. However, in our cohort, we found a shift towards female patients (64%). Similar results have recently been reported from a Finnish study where female patients constituted 65% [40]. The increased incidence in female patients in a Swedish cohort was also previously discussed by Zaar et al. [6] who calculated the age adjusted incidence as higher in male patients. This finding may suggest that there are differences in the sex distribution in the older populations between the Nordic countries compared to the cohorts previously described. It does not, however, explain the differences in outcome between male and female patients discussed above.

Another clinical parameter that had an impact on OS was age, which unsurprisingly showed a better OS in younger patients (19–69) compared to older patients (>70). The distribution of men and women in these groups was somewhat uneven, however the impact on our findings should be limited considering the median age was higher for the female group compared to the male group (79 and 76 years, respectively). Clinical stage could also have an impact on OS; however, these groups were evenly distributed between the sexes (Table 1). Clinical stage in itself did not show a statistically significant difference in OS (not shown), however this may be due to the limited number of stage III patients (n = 14, Table 1) who all received adjuvant RT which may have a positive impact on their outcome.

The most common anatomical location for the primary tumor was in the head and neck region (47%) and the next most common anatomical location was the upper extremities (21%), which are consistent with other publications and no difference between men and women [14,34,41].

We acknowledge several limitations of this historic cohort analysis. The main limitation is the sample size due to the rare nature of MCC, however these findings still add insight to several important prognostic traits in curative patients with MCC.

4. Materials and Methods

4.1. Patient Selection

Patients diagnosed with MCC and referred to the plastic surgery unit at Karolinska University Hospital from 1987 until the end of 2019 were included in the study. Patients underwent primary surgery alone with additional scar excision and with wide margins or were assessed for adjuvant RT at the Onco-Radiation Therapy department of the hospital. Start date was set to the day of surgery. Censor date was set to death date or end-date of the study, 31 December 2019.

The pathology evaluation and diagnosis were mainly performed or reviewed by pathologists at Karolinska University Hospital at the time of diagnosis.

Survival data and given treatment were retrieved from patient hospital records, pathology reports, population registry and the Swedish cause of death registry. Largest diameter of the primary tumor was identified in patient records prior to surgery or from pathology reports. Tumor stage was assessed according to the 8th edition consensus staging system by the American Joint committee on Cancer (AJCC) published in 2017.

The study was conducted in accordance with ethical approval Dnr 2019-05951 approved by the Ethics Review Board (Etikprövningsnämnden) in Sweden.

4.2. Surgery

Patients with MCC stage I and II underwent radical tumor excision, preferably of 1–2 cm in margin down to muscle fascia, pericondrium or periosteum. The aim of surgery is to achieve free margins.

4.3. Radiotherapy Treatment

Radiotherapy as a treatment option for MCC has been a tradition at the Oncology and Radiotherapy department at Karolinska since the 1970s and post-operative radiotherapy has been widely used. Established practice is to offer 2 Gy per fraction, 5 fractions per week up to a total dose of 50 Gy or more. Bolus is used in selected cases to achieve adequate doses in the skin. Common margins have been 1–3 cm. For patients with microscopically or macroscopically positive margins a total dose of 56–66 Gy have been given. When radiotherapy treatment is given after relapse, doses vary depending on indication.

4.4. McPyV Analysis

MCPyV analyses were made by MCPyV LT immunohistochemistry using CM2B4 (Santa Cruz Biotechnology, Dallas, TX) or Ab3 (gift from Dr. J.A. DeCaprio) antibody and PCR detection of MCPyV DNA in tumor samples, as previously described [42]. The virus status of 40 patients was characterized in previous studies [42–44] and 14 patients were characterized in this study.

4.5. Statistical Analysis

Descriptive statistics for nominal or numeric variables was applied. When required distribution differences and correlations between categorical data were compared with the χ^2 test and ordinal data with the Spearman Rank's test. This was used to examine relationships between patient's demographics, clinical variables and biomarkers. Student *t*-test was used to compare mean values. Survival analysis was performed using the Kaplan–Meier method and differences in survival were tested with the log-rank test. Cox–Mantel regression was used in the univariate and multivariate analyses. The results were considered significant if $p \leq 0.05$. Calculations were performed with the program StatView™ for Windows, SAS Institute Inc. Version 5.0.1. The Forest Plot presentation was performed using MedCalc™ program version 19.1.

5. Conclusions

Our data confirm the positive impact of RT on survival in a Swedish MCC cohort. Our findings also show not only that women have a better prognosis, but also that men with virus negative MCC have the worst outcome. Our findings thus indicate that MCPyV positive and negative MCC act as two different diseases and raise questions of whether there is a difference in the disease itself or the immune response towards MCC in male and female patients.

Author Contributions: Conceptualization, L.V., G.V.M. and V.B.-S.; Methodology, H.B.A., G.V.M. and W.-O.L.; formal analysis, G.V.M. and H.B.A.; investigation, H.B.A. and H.S.; resources, W.-O.L. and G.V.M.; writing—original draft preparation, L.V.; writing—review and editing, H.B.A., G.V.M. and W.-O.L.; visualization, H.B.A., G.V.M. and W.-O.L.; supervision, L.V.; project administration, L.V.; and funding acquisition, G.V.M. All authors have read and agreed to the published version of the manuscript.

Funding: This research was funded by the Swedish Cancer Society, Cancer Research Funds of Radiumhemmet, Stockholm County Council and Karolinska Institutet.

Institutional Review Board Statement: The study was conducted according to the guidelines of the Declaration of Helsinki and approved by the National Ethics Committee (Etiksprövningsnämnden) in Sweden (Dnr 2019-05951 approved 2020-05-13).

Informed Consent Statement: Informed consent was obtained from all subjects involved in the study.

Data Availability Statement: The data presented in this study are available on request from the corresponding author. The data are not publicly available due to them containing information that could compromise research participant privacy.

Conflicts of Interest: The authors declare no conflict of interest.

References

1. Paulson, K.G.; Park, S.Y.; Vandeven, N.A.; Lachance, K.; Thomas, H.; Chapuis, A.G.; Harms, K.L.; Thompson, J.A.; Bhatia, S.; Stang, A.; et al. Merkel cell carcinoma: Current US incidence and projected increases based on changing demographics. *J. Am. Acad. Dermatol.* **2018**, *78*, 457–463.e2. [CrossRef]
2. Stang, A.; Becker, J.C.; Nghiem, P.; Ferlay, J. The association between geographic location and incidence of Merkel cell carcinoma in comparison to melanoma: An international assessment. *Eur. J. Cancer* **2018**, *94*, 47–60. [CrossRef] [PubMed]
3. Bjornhagen, V. Nationellt Vårdprogram för Merkelcellscancer. 2020. [Updated 16 November 2020]. Available online: https://www.cancercentrum.se/globalassets/vara-uppdrag/kunskapsstyrning/vardprogram/kommande-vardprogram/2020/200915/nvp-merkelcellscancer-2020-09-15.pdf (accessed on 6 January 2011).
4. Lemos, B.; Nghiem, P. Merkel cell carcinoma: More deaths but still no pathway to blame. *J. Investig. Dermatol.* **2007**, *127*, 2100–2103. [CrossRef] [PubMed]
5. Agelli, M.; Clegg, L.X. Epidemiology of primary Merkel cell carcinoma in the United States. *J. Am. Acad. Dermatol.* **2003**, *49*, 832–841. [CrossRef]
6. Zaar, O.; Gillstedt, M.; Lindelof, B.; Wennberg-Larko, A.M.; Paoli, J. Merkel cell carcinoma incidence is increasing in Sweden. *J. Eur. Acad. Dermatol. Venereol.* **2016**, *30*, 1708–1713. [CrossRef] [PubMed]
7. Hodgson, N.C. Merkel cell carcinoma: Changing incidence trends. *J. Surg. Oncol.* **2005**, *89*, 1–4. [CrossRef]
8. Feng, H.; Shuda, M.; Chang, Y.; Moore, P.S. Clonal integration of a polyomavirus in human Merkel cell carcinoma. *Science* **2008**, *319*, 1096–1100. [CrossRef]
9. Pietropaolo, V.; Prezioso, C.; Moens, U. Merkel Cell Polyomavirus and Merkel Cell Carcinoma. *Cancers* **2020**, *12*, 1774. [CrossRef]
10. Higaki-Mori, H.; Kuwamoto, S.; Iwasaki, T.; Kato, M.; Murakami, I.; Nagata, K.; Sano, H.; Horie, Y.; Yoshida, Y.; Yamamoto, O.; et al. Association of Merkel cell polyomavirus infection with clinicopathological differences in Merkel cell carcinoma. *Hum. Pathol.* **2012**, *43*, 2282–2291. [CrossRef]
11. Sihto, H.; Kukko, H.; Koljonen, V.; Sankila, R.; Bohling, T.; Joensuu, H. Merkel cell polyomavirus infection, large T antigen, retinoblastoma protein and outcome in Merkel cell carcinoma. *Clin. Cancer Res.* **2011**, *17*, 4806–4813. [CrossRef]
12. Moshiri, A.S.; Doumani, R.; Yelistratova, L.; Blom, A.; Lachance, K.; Shinohara, M.M.; Delaney, M.; Chang, O.; McArdle, S.; Thomas, H.; et al. Polyomavirus-Negative Merkel Cell Carcinoma: A More Aggressive Subtype Based on Analysis of 282 Cases Using Multimodal Tumor Virus Detection. *J. Investig. Dermatol.* **2017**, *137*, 819–827. [CrossRef] [PubMed]
13. Reichgelt, B.A.; Visser, O. Epidemiology and survival of Merkel cell carcinoma in the Netherlands. A population-based study of 808 cases in 1993–2007. *Eur. J. Cancer* **2011**, *47*, 579–585. [CrossRef] [PubMed]
14. Kukko, H.; Bohling, T.; Koljonen, V.; Tukiainen, E.; Haglund, C.; Pokhrel, A.; Sankila, R.; Pukkala, E. Merkel cell carcinoma—A population-based epidemiological study in Finland with a clinical series of 181 cases. *Eur. J. Cancer* **2012**, *48*, 737–742. [CrossRef] [PubMed]
15. Tam, M.; Luu, M.; Barker, C.A.; Gharavi, N.M.; Hamid, O.; Shiao, S.L.; Nguyen, A.T.; Lu, D.J.; Ho, A.S.; Zumsteg, Z.S. Improved survival in women versus men with merkel cell carcinoma. *J. Am. Acad. Dermatol.* **2020**. [CrossRef]
16. Heath, M.; Jaimes, N.; Lemos, B.; Mostaghimi, A.; Wang, L.C.; Penas, P.F.; Nghiem, P. Clinical characteristics of Merkel cell carcinoma at diagnosis in 195 patients: The AEIOU features. *J. Am. Acad. Dermatol.* **2008**, *58*, 375–381. [CrossRef]
17. Albores-Saavedra, J.; Batich, K.; Chable-Montero, F.; Sagy, N.; Schwartz, A.M.; Henson, D.E. Merkel cell carcinoma demographics, morphology, and survival based on 3870 cases: A population based study. *J. Cutan. Pathol.* **2010**, *37*, 20–27. [CrossRef]
18. Petrelli, F.; Ghidini, A.; Torchio, M.; Prinzi, N.; Trevisan, F.; Dallera, P.; De Stefani, A.; Russo, A.; Vitali, E.; Bruschieri, L.; et al. Adjuvant radiotherapy for Merkel cell carcinoma: A systematic review and meta-analysis. *Radiother. Oncol.* **2019**, *134*, 211–219. [CrossRef]
19. Kaufman, H.L.; Russell, J.S.; Hamid, O.; Bhatia, S.; Terheyden, P.; D'Angelo, S.P.; Shih, K.C.; Lebbe, C.; Milella, M.; Brownell, I.; et al. Updated efficacy of avelumab in patients with previously treated metastatic Merkel cell carcinoma after ≥1 year of follow-up: JAVELIN Merkel 200, a phase 2 clinical trial. *J. Immunother. Cancer* **2018**, *6*, 7. [CrossRef]
20. Nghiem, P.T.; Bhatia, S.; Lipson, E.J.; Kudchadkar, R.R.; Miller, N.J.; Annamalai, L.; Berry, S.; Chartash, E.K.; Daud, A.; Fling, S.P.; et al. PD-1 Blockade with Pembrolizumab in Advanced Merkel-Cell Carcinoma. *N. Engl. J. Med.* **2016**, *374*, 2542–2552. [CrossRef]
21. Knepper, T.C.; Montesion, M.; Russell, J.S.; Sokol, E.S.; Frampton, G.M.; Miller, V.A.; Albacker, L.A.; McLeod, H.L.; Eroglu, Z.; Khushalani, N.I.; et al. The Genomic Landscape of Merkel Cell Carcinoma and Clinicogenomic Biomarkers of Response to Immune Checkpoint Inhibitor Therapy. *Clin. Cancer Res.* **2019**, *25*, 5961–5971. [CrossRef]
22. Engels, E.A.; Frisch, M.; Goedert, J.J.; Biggar, R.J.; Miller, R.W. Merkel cell carcinoma and HIV infection. *Lancet* **2002**, *359*, 497–498. [CrossRef]
23. Lanoy, E.; Engels, E.A. Skin cancers associated with autoimmune conditions among elderly adults. *Br. J. Cancer* **2010**, *103*, 112–114. [CrossRef] [PubMed]
24. Clarke, C.A.; Robbins, H.A.; Tatalovich, Z.; Lynch, C.F.; Pawlish, K.S.; Finch, J.L.; Hernandez, B.Y.; Fraumeni, J.F., Jr.; Madeleine, M.M.; Engels, E.A. Risk of merkel cell carcinoma after solid organ transplantation. *J. Natl. Cancer Inst.* **2015**, *107*, dju382. [CrossRef]
25. Harrington, C.; Kwan, W. Radiotherapy and Conservative Surgery in the Locoregional Management of Merkel Cell Carcinoma: The British Columbia Cancer Agency Experience. *Ann. Surg. Oncol.* **2016**, *23*, 573–578. [CrossRef] [PubMed]

26. Finnigan, R.; Hruby, G.; Wratten, C.; Keller, J.; Tripcony, L.; Dickie, G.; Rischin, D.; Poulsen, M. The impact of preradiation residual disease volume on time to locoregional failure in cutaneous Merkel cell carcinoma–a TROG substudy. *Int. J. Radiat. Oncol. Biol. Phys.* **2013**, *86*, 91–95. [CrossRef] [PubMed]
27. Strom, T.; Carr, M.; Zager, J.S.; Naghavi, A.; Smith, F.O.; Cruse, C.W.; Messina, J.L.; Russell, J.; Rao, N.G.; Fulp, W.; et al. Radiation Therapy is Associated with Improved Outcomes in Merkel Cell Carcinoma. *Ann. Surg. Oncol.* **2016**, *23*, 3572–3578. [CrossRef] [PubMed]
28. Chen, M.M.; Roman, S.A.; Sosa, J.A.; Judson, B.L. The role of adjuvant therapy in the management of head and neck merkel cell carcinoma: An analysis of 4815 patients. *JAMA Otolaryngol. Head Neck Surg.* **2015**, *141*, 137–141. [CrossRef] [PubMed]
29. Bhatia, S.; Storer, B.E.; Iyer, J.G.; Moshiri, A.; Parvathaneni, U.; Byrd, D.; Sober, A.J.; Sondak, V.K.; Gershenwald, J.E.; Nghiem, P. Adjuvant Radiation Therapy and Chemotherapy in Merkel Cell Carcinoma: Survival Analyses of 6908 Cases From the National Cancer Data Base. *J. Natl. Cancer Inst.* **2016**, *108*, djw042. [CrossRef] [PubMed]
30. Bhatia, K.; Goedert, J.J.; Modali, R.; Preiss, L.; Ayers, L.W. Merkel cell carcinoma subgroups by Merkel cell polyomavirus DNA relative abundance and oncogene expression. *Int. J. Cancer* **2010**, *126*, 2240–2246. [CrossRef] [PubMed]
31. Handschel, J.; Muller, D.; Depprich, R.A.; Ommerborn, M.A.; Kubler, N.R.; Naujoks, C.; Reifenberger, J.; Schafer, K.L.; Braunstein, S. The new polyomavirus (MCPyV) does not affect the clinical course in MCCs. *Int. J. Oral Maxillofac. Surg.* **2010**, *39*, 1086–1090. [CrossRef]
32. Nardi, V.; Song, Y.; Santamaria-Barria, J.A.; Cosper, A.K.; Lam, Q.; Faber, A.C.; Boland, G.M.; Yeap, B.Y.; Bergethon, K.; Scialabba, V.L.; et al. Activation of PI3K signaling in Merkel cell carcinoma. *Clin. Cancer Res.* **2012**, *18*, 1227–1236. [CrossRef] [PubMed]
33. Laude, H.C.; Jonchere, B.; Maubec, E.; Carlotti, A.; Marinho, E.; Couturaud, B.; Peter, M.; Sastre-Garau, X.; Avril, M.F.; Dupin, N.; et al. Distinct merkel cell polyomavirus molecular features in tumour and non tumour specimens from patients with merkel cell carcinoma. *PLoS Pathog.* **2010**, *6*, e1001076. [CrossRef]
34. Harms, K.L.; Healy, M.A.; Nghiem, P.; Sober, A.J.; Johnson, T.M.; Bichakjian, C.K.; Wong, S.L. Analysis of Prognostic Factors from 9387 Merkel Cell Carcinoma Cases Forms the Basis for the New 8th Edition AJCC Staging System. *Ann. Surg. Oncol.* **2016**, *23*, 3564–3571. [CrossRef] [PubMed]
35. Rodig, S.J.; Cheng, J.; Wardzala, J.; DoRosario, A.; Scanlon, J.J.; Laga, A.C.; Martinez-Fernandez, A.; Barletta, J.A.; Bellizzi, A.M.; Sadasivam, S.; et al. Improved detection suggests all Merkel cell carcinomas harbor Merkel polyomavirus. *J. Clin. Investig.* **2012**, *122*, 4645–4653. [CrossRef]
36. Botticelli, A.; Onesti, C.E.; Zizzari, I.; Cerbelli, B.; Sciattella, P.; Occhipinti, M.; Roberto, M.; Di Pietro, F.; Bonifacino, A.; Ghidini, M.; et al. The sexist behaviour of immune checkpoint inhibitors in cancer therapy? *Oncotarget* **2017**, *8*, 99336–99346. [CrossRef]
37. Wu, Y.; Ju, Q.; Jia, K.; Yu, J.; Shi, H.; Wu, H.; Jiang, M. Correlation between sex and efficacy of immune checkpoint inhibitors (PD-1 and CTLA-4 inhibitors). *Int. J. Cancer* **2018**, *143*, 45–51. [CrossRef]
38. Conforti, F.; Pala, L.; Goldhirsch, A. Different effectiveness of anticancer immunotherapy in men and women relies on sex-dimorphism of the immune system. *Oncotarget* **2018**, *9*, 31167–31168. [CrossRef] [PubMed]
39. Grassadonia, A.; Sperduti, I.; Vici, P.; Iezzi, L.; Brocco, D.; Gamucci, T.; Pizzuti, L.; Maugeri-Sacca, M.; Marchetti, P.; Cognetti, G.; et al. Effect of Gender on the Outcome of Patients Receiving Immune Checkpoint Inhibitors for Advanced Cancer: A Systematic Review and Meta-Analysis of Phase III Randomized Clinical Trials. *J. Clin. Med.* **2018**, *7*, 542. [CrossRef] [PubMed]
40. Sahi, H.; Their, J.; Gissler, M.; Koljonen, V. Merkel Cell Carcinoma Treatment in Finland in 1986–2016—A Real-World Data Study. *Cancers* **2020**, *12*, 1224. [CrossRef] [PubMed]
41. Asgari, M.M.; Sokil, M.M.; Warton, E.M.; Iyer, J.; Paulson, K.G.; Nghiem, P. Effect of host, tumor, diagnostic, and treatment variables on outcomes in a large cohort with Merkel cell carcinoma. *JAMA Dermatol.* **2014**, *150*, 716–723. [CrossRef]
42. Xie, H.; Lee, L.; Caramuta, S.; Hoog, A.; Browaldh, N.; Bjornhagen, V.; Larsson, C.; Lui, W.O. MicroRNA expression patterns related to merkel cell polyomavirus infection in human merkel cell carcinoma. *J. Investig. Dermatol.* **2014**, *134*, 507–517. [CrossRef] [PubMed]
43. Xie, H.; Liu, T.; Wang, N.; Bjornhagen, V.; Hoog, A.; Larsson, C.; Lui, W.O.; Xu, D. TERT promoter mutations and gene amplification: Promoting TERT expression in Merkel cell carcinoma. *Oncotarget* **2014**, *5*, 10048–10057. [CrossRef] [PubMed]
44. Kumar, S.; Xie, H.; Shi, H.; Gao, J.; Juhlin, C.C.; Bjornhagen, V.; Hoog, A.; Lee, L.; Larsson, C.; Lui, W.O. Merkel cell polyomavirus oncoproteins induce microRNAs that suppress multiple autophagy genes. *Int. J. Cancer* **2020**, *146*, 1652–1666. [CrossRef] [PubMed]

cancers

MDPI

Article

The Merkel Cell Polyomavirus T Antigens Function as Tumor Promoters in Murine Skin

Megan E. Spurgeon [1,*], Amy Liem [1], Darya Buehler [2], Jingwei Cheng [3], James A. DeCaprio [4,5] and Paul F. Lambert [1,*]

1 McArdle Laboratory for Cancer Research, Department of Oncology, University of Wisconsin School of Medicine and Public Health, Madison, WI 53705, USA; aliem@wisc.edu
2 Department of Pathology and Laboratory Medicine, University of Wisconsin School of Medicine and Public Health, Madison, WI 53705, USA; buehler2@wisc.edu
3 Department of Molecular, Cellular and Biomedical Sciences, University of New Hampshire, Durham, NH 03824, USA; Jingwei.Cheng@unh.edu
4 Department of Medical Oncology, Dana-Farber Cancer Institute, Boston, MA 02215, USA; james_decaprio@dfci.harvard.edu
5 Department of Medicine, Brigham and Women's Hospital, Harvard Medical School, Boston, MA 02215, USA
* Correspondence: megan.spurgeon@wisc.edu (M.E.S.); plambert@wisc.edu (P.F.L.)

Simple Summary: Merkel cell polyomavirus, a recently discovered human virus, is linked to the development of a rare form of skin cancer called Merkel cell carcinoma. The virus does not replicate in cancer cells, yet there is continued expression of viral proteins known as T antigens. The T antigens are believed to contribute to Merkel cell carcinoma development, yet how they do so remains an active area of research. In this study, we used transgenic mice expressing the viral T antigens in their skin to determine at which stage of skin cancer development these viral proteins function. We discovered that the Merkel cell polyomavirus T antigens function as tumor promoters, rather than tumor initiators, in the skin. These findings suggest that other tumor-initiating events may cooperate with the tumor-promoting activities of the viral T antigens, thus providing important insight into how Merkel cell polyomavirus can cause cancer in human skin.

Abstract: Merkel cell polyomavirus (MCPyV) causes the majority of human Merkel cell carcinomas (MCC), a rare but highly aggressive form of skin cancer. We recently reported that constitutive expression of MCC tumor-derived MCPyV tumor (T) antigens in the skin of transgenic mice leads to hyperplasia, increased proliferation, and spontaneous epithelial tumor development. We sought to evaluate how the MCPyV T antigens contribute to tumor formation in vivo using a classical, multi-stage model for squamous cell carcinoma development. In this model, two chemical carcinogens, DMBA and TPA, contribute to two distinct phases of carcinogenesis—initiation and promotion, respectively—that are required for tumors to develop. By treating the MCPyV transgenic mice with each chemical carcinogen, we determined how the viral oncogenes contributed to carcinogenesis. We observed that the MCPyV T antigens synergized with the tumor initiator DMBA, but not with the tumor promoter TPA, cause tumors. Therefore, the MCPyV tumor antigens function primarily as tumor promoters, similar to that seen with human papillomavirus (HPV) oncoproteins. These studies provide insight into the role of MCPyV T antigen expression in tumor formation in vivo and contribute to our understanding of how MCPyV may function as a human DNA tumor virus.

Keywords: Merkel cell polyomavirus; Merkel cell carcinoma; skin carcinogenesis; T antigens; human polyomaviruses; DNA tumor viruses; viral oncoproteins

Citation: Spurgeon, M.E.; Liem, A.; Buehler, D.; Cheng, J.; DeCaprio, J.A.; Lambert, P.F. The Merkel Cell Polyomavirus T Antigens Function as Tumor Promoters in Murine Skin. *Cancers* **2021**, *13*, 222. https://doi.org/10.3390/cancers13020222

Received: 30 November 2020
Accepted: 8 January 2021
Published: 9 January 2021

1. Introduction

Viruses are the etiological agents of at least 15% of human cancers worldwide [1]. Several viruses with a double-stranded DNA genome, including adenoviruses, papillomaviruses, herpesviruses, and polyomaviruses, possess oncogenic activities in a variety of in vitro and in vivo settings [2]. These DNA tumor viruses have significantly contributed to our understanding of viral oncogenesis and have facilitated many advances in molecular biology and cancer research (reviewed in [2,3]). Given the strong causal relationship of high-risk human papillomaviruses (HPVs) and gamma herpesviruses with human cancers [4,5], these viruses have largely dominated much of the DNA tumor virus research landscape in recent decades. However, the recent discoveries of multiple human polyomaviruses, including one that causes a human skin cancer, have reinvigorated interest in the study of this family of DNA tumor viruses [6].

Merkel cell carcinoma (MCC) was first described in 1972 [7]. Although relatively rare, MCC is one of the most aggressive skin cancers, with a high mortality rate [8]. MCC incidence is predicted to rise dramatically in the coming years due to an increase in the aging population [9]. In addition to advanced age, there are several known risk factors for MCC development, including light-colored skin, ultraviolet (UV) light exposure, and immunosuppression [10,11]. The association with immunosuppression prompted researchers to question whether MCC has a viral etiology. Merkel cell polyomavirus (MCPyV or MCV) was discovered in 2008 through digital transcriptome subtraction, which identified unique, non-human sequences present in human MCC tissues that were subsequently determined to be those of a previously unidentified polyomavirus [12]. MCPyV infection is ubiquitous, largely asymptomatic, and occurs during early childhood [13–20]. While the exact cell targeted for infection remains undetermined, MCPyV most likely exhibits cutaneous tropism for a cell type residing in the skin. MCPyV can be detected in skin swabs of healthy individuals [21] and there is current in vitro evidence that MCPyV can infect and/or replicate in keratinocytes and dermal fibroblasts [22–26].

In at least 80% of MCCs, the MCPyV genome is clonally integrated into the genomic DNA of tumor cells with an integration pattern that frequently preserves the early region of the viral genome [12]. This integration pattern facilitates continued expression of the MCPyV early viral proteins known as tumor (T) antigens, and more specifically the small tumor (ST) and large tumor (LT) antigens [12], both of which have been implicated in transformation/tumorigenesis using in vitro and in vivo studies [27–34]. Within MCC cells, mutations are consistently found within integrated MCPyV DNA genomes that result in expression of truncated forms of LT protein [30,35]. While these C-terminal LT truncations prevent the virus from replicating in the tumor cells, truncated LT proteins retain the LXCXE motif that mediates binding to and inactivation of the cellular tumor suppressor pRb critical in cell cycle regulation [30,36]. ST is also expressed in MCC [37,38] and has oncogenic activity in a variety of assays. MCPyV ST, either alone or in combination with LT, can transform rodent and human fibroblasts in vitro [27–29,38] and is tumorigenic in murine skin [31–34]. Therefore, MCPyV viral genome integration in MCC cells preserves expression of viral proteins with oncogenic activity. To date, MCPyV is the only known polyomavirus to cause cancer in humans.

Several lines of evidence point to MCPyV causing MCC. Integration of the MCPyV genome in MCC cells appears to be an early event in neoplastic progression, as evidenced by clonal integration patterns within individual MCCs and shared integration patterns with distant metastases [12,37,39]. Within MCC tumor cells, viral genome copy numbers generally average at least 1 viral genome copy per cell [39–43], a finding that further supports causality. The importance of the MCPyV ST and truncated LT antigens to MCC oncogenesis is underscored by their retained expression in MCC tumors [37,38,44,45]. Continued expression of the MCPyV T antigens is required for MCC survival and optimal cell growth and proliferation [27,41,46], and the truncated LT and its ability to bind pRb appears to be particularly important in this regard [27,36,47,48]. Recent reports suggest that MCPyV-truncated LT antigen helps drive transdifferentiation of presumably MCPyV-infected MCC precursor cells, at least in part through an ability to increase expression of the Merkel cell specification factor atonal homolog 1 (ATOH1) [49,50]. Expression of survivin (BIRC5), an anti-apoptotic gene, is increased by LT in vitro [51] and is also elevated in the skin of transgenic mice expressing the MCPyV T antigens [32]. In transgenic mouse models, combined expression of MCPyV ST and ATOH1 in squamous epithelial cells induces intraepidermal MCC-like lesions [33]. In our transgenic mouse model, epithelial expression of the MCPyV ST and truncated LT antigens induces hyperplasia, proliferation, and spontaneous tumor development [32]. Furthermore, MCPyV T antigen expression in epithelial cells induces Merkel cell-related gene expression and Merkel cell phenotypes [52]. There are also several other potentially oncogenic functions of the MCPyV T antigens [53]. Overall, the collective evidence that MCPyV contributes to MCC carcinogenesis is ample and continues to grow.

In this study, we sought insight into how MCPyV T antigens cause tumors using a well-validated and widely used multi-stage model of skin carcinogenesis. This model involves the topical application of chemical carcinogens that function as tumor initiators or tumor promoters and thus allow one to define the role of genes or factors in discrete stages of cutaneous tumor development [54,55]. The skin of experimental laboratory animals is treated with a subcarcinogenic dose of a tumor initiator, 7,12-dimethylbenz[a]-anthracene (DMBA), or a tumor promoter, 12-O-tetradecanoylphorbol-13-acetate (TPA), or a combination of both. Skin tumorigenesis requires both tumor initiation and promotion, and malignancy requires progression. If transgenic mice develop tumors after only being treated with TPA (a promoter), this indicates that the transgene product functions as a tumor initiator. Likewise, if tumors develop after only being treated with DMBA (an initiator), this indicates that the transgene product functions as a tumor promoter. Treatment with both DMBA and TPA allows investigation into the role of the transgene in malignant progression. Our laboratory has used this approach, sometimes referred to as a 'skin painting' or DMBA/TPA model, to determine the tumorigenic functions of high-risk human papillomavirus oncogenes E5, E6, and E7 [56,57]. In this study, we used MCPyV transgenic mice that express ST and truncated LT antigens in the stratified epithelia and discovered that the MCPyV T antigens can synergize with DMBA, a tumor initiator, to promote tumorigenesis, but do not synergize with TPA, a tumor promoter. Therefore, our results indicate that the MCPyV T antigens function primarily as tumor promoters, and not tumor initiators, in murine skin. These results provide insight into potential mechanisms by which the MCPyV T antigens contribute to MCC neoplastic progression and carcinogenesis.

2. Results

2.1. Model Validation and Experimental Overview of Studies to Determine the Role of the MCPyV T Antigens in Skin Carcinogenesis

We previously reported that keratin 14 (K14) promoter-driven expression of MCC tumor-derived MCPyV small T and truncated LT antigens in murine skin promotes severe epithelial phenotypes [32]. These *K14Cre-MCPyV168* transgenic mice also spontaneously develop benign epithelial tumors on their skin. Prior to beginning our studies, we sought to verify that our current colony of *K14Cre-MCPyV168* transgenic mice replicate the epithelial tumorigenesis phenotype that we previously observed. Therefore, *K14Cre-MCPyV168* transgenic mice (n = 42) were monitored over the course of 28 weeks and scored for tumor development. Consistent with previous observations, *K14Cre-MCPyV168* mice developed overt phenotypes and approximately 38% (n = 16/42) of mice developed benign epithelial skin tumors at some point during the 28 week period (Figure 1A, left). Consistent with our previous observations, MCPyV T antigen-induced skin tumors contained hyperplastic epithelia and histopathology indicative of non-invasive exophytic lesions with varying degrees of dysplasia (benign papillomas) (Figure 1A, right). While the observed incidence of spontaneous tumor development was slightly lower than our previously reported observations (46% incidence; n = 16/35) [32], it was not significantly different ($p = 0.64$; Fisher's Exact Test). These findings indicated that the MCPyV T antigens expressed in *K14Cre-MCPyV168* transgenic mice were functioning as expected and allowed us to move forward with our study.

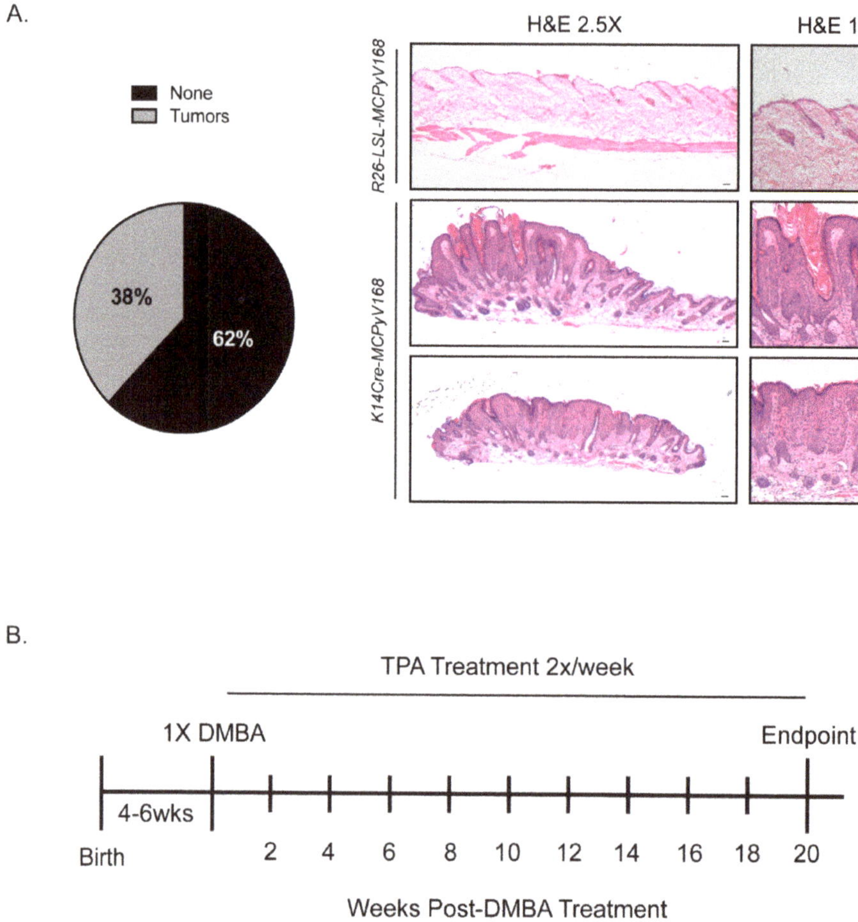

Figure 1. Model validation and experimental overview to determine the role of the MCPyV T antigens in skin carcinogenesis. **(A)** Spontaneous tumor development in untreated *K14Cre-MCPyV168* transgenic mice. The pie chart on the left reflects the percentage of *K14Cre-MCPyV168* mice that did (gray) or did not (black) develop spontaneous tumors. Representative H&E-stained images are shown on the right. Normal skin from *Rosa26-LSL-MCPyV168* mice is shown on top, and sections from two representative spontaneous squamous papillomas that developed on *K14Cre-MCPyV168* mice are shown in the middle and bottom panels. All scale bars = 100 μM. **(B)** Experimental overview of DMBA and TPA treatment regimens in skin carcinogenesis studies. At 4–6 weeks of age, areas of dorsal skin were shaved and prepared in three groups of mice: *Rosa26-LSL-MCPyV168*, *K14E6/E7*, and *K14Cre-MCPyV168*. For mice treated with DMBA only, a one-time topical treatment was applied to the shaved dorsal skin. For mice treated with TPA only, topical treatment was performed twice a week for 20 weeks. For DMBA+TPA treatment, these topical treatments were combined. Mice were monitored for tumor development every 2 weeks during the 20 week treatment period. The average number of tumors per mouse per group was quantified at each time point.

To determine the role of the MCPyV T antigens in skin cancer development, we utilized a multi-stage model of skin carcinogenesis that allows one to determine the role of genes in three stages: initiation, promotion, and progression [54,55]. In this model, murine skin is treated with a subcarcinogenic dose of a tumor initiator, 7,12-dimethylbenz[a]-anthracene (DMBA), or a tumor promoter, 12-O-tetradecanoylphorbol-13-acetate (TPA), or a combination of both. Skin tumorigenesis requires both tumor initiation and promotion, and malignancy requires progression. The ability of a gene or transgene to act as a tumor promoter or initiator in the skin can be elucidated by treating mice with DMBA only or TPA only, respectively. For instance, if transgenic mice develop tumors after only being treated with TPA (a promoter), this indicates that the transgene product functions as a tumor initiator. Likewise, if transgenic mice develop tumors after only being treated with DMBA (an initiator), this indicates that the transgene product functions as a tumor promoter. Transgenic mice treated with both DMBA and TPA allows investigation into the role of the transgene in malignant progression.

In these studies, we included three groups of mice: (1) *Rosa26-LSL-MCPyV168* that do not express Cre recombinase and therefore do not express the MCPyV T antigens as a negative control, (2) *K14E6/E7* HPV16 transgenic mice that have been previously tested using the DMBA/TPA model and therefore serve as a positive control [57], and (3) *K14Cre-MCPyV168* transgenic mice expressing the MCPyV T antigens in K14-positive cells of the stratified epithelia. These mice were separated into three different treatment groups: (1) TPA only, (2) DMBA only, and (3) DMBA+TPA (Table 1). We also included a group of untreated *K14Cre-MCPyV168* mice to monitor the background level of spontaneous tumor development. At 4–6 weeks old, a region of the dorsal skin was shaved to prepare an area for topical treatment. Following the experimental design, we previously used for skin painting studies of HPV16 transgenic mice [56,57], the shaved area of skin of mice in the DMBA-only treatment group was treated topically one time with 0.3 μmol DMBA, and mice in the DMBA+TPA treatment group were treated once with 0.01 μmol DMBA. One week later, mice in the TPA-only and DMBA+TPA treatment groups were treated topically with 15 nmol TPA twice a week for 20 weeks (Figure 1B). All mice were evaluated every 2 weeks for the development of squamous papillomas within the treatment area.

Table 1. Overview of treatment groups and number of mice per group at study onset and endpoint. Starting number of mice per each treatment group and genotype are shown. The number of mice remaining at 20 weeks post-DMBA treatment is shown in parentheses. Any change in the number of mice per group is accounted for in statistical tests at each time point.

Experimental Group	Treatment			
	TPA Only	**DMBA Only**	**DMBA+TPA**	**No Treatment**
ROSA26-LSL-MCPyV168	17	22 (18)	23 (19)	0
K14Cre-MCPyV168	12	14	17 (16)	28 (22)
K14E6/E7	13	10	22 (16)	0

2.2. The MCPyV T Antigens Function as Tumor Promoters, Not Initiators, in Murine Skin

We hypothesized that, if the MCPyV T antigens are able to function as tumor initiators in murine skin, then TPA-treated *K14Cre-MCPyV168* mice will develop significantly more tumors than their untreated counterparts and TPA-treated *Rosa26-LSL-MCPyV168* mice. To determine whether the MCPyV T antigens function as tumor initiators, we treated groups of *Rosa26-LSL-MCPyV168* (n = 17), *K14E6/E7* (n = 13), and *K14Cre-MCPyV168* (n = 12) with TPA only (15 nmol) twice a week for 20 weeks (Figure 1B). We also included a group of untreated *K14Cre-MCPyV168* mice (n = 28) to monitor the incidence of spontaneous tumor development over time. As expected, the negative control *Rosa26-LSL-MCPyV168* mice did not develop any tumors throughout the course of treatment (Figure 2A; blue line). In TPA only treated *K14E6/E7* mice, we observed three total tumors over the course of treatment (1 each at 4, 18, and 20 weeks post-treatment; Figure 2A; red line), a tumor incidence level that was not significantly higher from that observed in the *Rosa26-LSL-MCPyV168* mice (Wilcoxon rank-sum test; all p-values >0.2). These findings indicate that the HPV16 E6 and E7 oncogenes do not function as tumor initiators, consistent with our previous studies [57]. In TPA only treated *K14Cre-MCPyV168* mice, we observed a significantly higher tumor incidence compared to *Rosa26-LSL-MCPyV* control mice only at 20 weeks post-treatment (Figure 2A; p-value = 0.05). However, this significant increase in tumor incidence likely reflects spontaneous tumor development in MCPyV transgenic mice, as there was no significant difference in tumor incidence between untreated and TPA-treated *K14Cre-MCPyV168* mice at this time point ($p = 0.43$). At no point during the treatment period did the tumor incidence in TPA only treated *K14Cre-MCPyV168* differ significantly from the spontaneous tumor development observed in untreated *K14Cre-MCPyV168* mice (Figure 2A; black line; all p-values > 0.13). Together, these data indicate that MCC tumor-derived MCPyV T antigens do not function as tumor initiators in murine skin.

We next tested whether the MCPyV T antigens could function as tumor promoters in murine skin. We hypothesized that, if the MCPyV T antigens are able to function as tumor promoters in murine skin, then DMBA-treated *K14Cre-MCPyV168* mice will develop significantly more tumors than their untreated counterparts and DMBA-treated *Rosa26-LSL-MCPyV168* mice. We previously found that the HPV16 E6 and E7 oncoproteins function as tumor promoters in murine skin [57]. Therefore, tumor incidence in *K14E6/E7* mice should be significantly higher than in DMBA-treated *Rosa26-LSL-MCPyV168* mice and similar to the number of tumors in the DMBA-treated *K14Cre-MCPyV168* transgenic mice, should the MCPyV T antigens function as tumor promoters. Groups of *Rosa26-LSL-MCPyV168* (n = 22), *K14E6/E7* (n = 10), and *K14Cre-MCPyV168* (n = 14) mice were treated with a single dose (0.3 μmol) of the tumor initiator DMBA and monitored for tumor development every 2 weeks for 20 weeks (Figure 1B, Figure 2B). Tumor incidence data from the group of untreated *K14Cre-MCPyV168* mice (n = 28) was included in order to compare the incidence of spontaneous tumor development over time with DMBA-induced tumors. There were no significant differences among any of the groups between 0 and 12 weeks post-DMBA treatment (Figure 2B). At 14 weeks post-treatment, the number of tumors in *K14E6/E7* mice rose to a level significantly higher than *Rosa26-LSL-MCPyV168* mice ($p = 0.02$). Starting at 16 weeks and continuing until the endpoint of 20 weeks post-treatment, the average number of tumors per mouse in DMBA-treated *K14E6/E7* mice (red line) and DMBA-treated *K14Cre-MCPyV168* mice (green line) increased significantly over DMBA-treated *Rosa26-LSL-MCPyV168* mice (blue line) (Figure 2B; *K14E6/E7* and *K14Cre-MCPyV168* versus *Rosa26-LSL-MCPyV168* mice, all p-values < 0.0005). The tumor numbers in the DMBA-treated *K14Cre-MCPyV168* mice at 16, 18, and 20 weeks post-infection were all significantly higher than the number of spontaneously arising tumors in untreated *K14Cre-MCPyV168* mice (black line; 16 weeks, $p = 0.0004$; 18 weeks, $p = 1 \times 10^{-5}$; 20 weeks, $p = 1.5 \times 10^{-5}$), indicating that the elevated average number of tumors in DMBA-treated MCPyV transgenic mice was a consequence of carcinogen treatment. At no point during the treatment period did the number of tumors differ significantly between DMBA-treated *K14E6/E7* and DMBA-treated *K14Cre-MCPyV168* mice (all p-values >0.08). Taken together, these results indicate that the MCPyV T antigens, like the high-risk HPV16 oncoproteins, function as tumor promoters in murine skin.

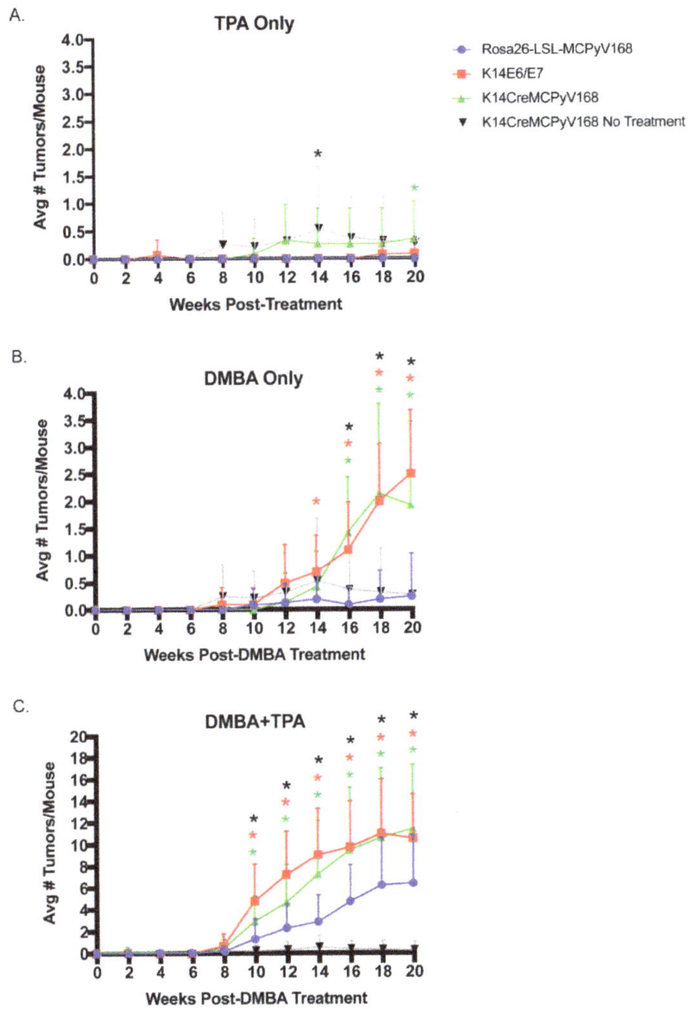

Figure 2. The MCPyV T antigens function as tumor promoters, not initiators, in murine skin and synergize with chemical carcinogens to exacerbate skin tumorigenesis. Groups of *Rosa26-LSL-MCPyV168* (blue data points), *K14E6/E7* (red data points), and *K14Cre-MCPyV168* (green data points) mice were treated topically with (**A**) 15 nmol of TPA twice a week for 20 weeks, (**B**) one time with 0.3 μmol DMBA, or (**C**) treated one time with 0.01 μmol DMBA and then twice a week for 20 weeks with 15 nmol TPA. Tumor incidence over time in an untreated group of *K14Cre-MCPyV168* mice is also included (black data points). At each time point, the average number of tumors/mouse in each group was calculated by dividing the total number of tumors by the total number of mice in each group. Group sizes were adjusted when necessary. A two-sided Wilcoxon rank-sum test was performed on data from each time point to compare the average number of tumors per mouse. Statistical significance is indicated with an asterisk for the comparisons indicated in the figure. Statistical significance indicated in (**A**) black asterisk: *K14Cre-MCPyV168* No Treatment vs. *Rosa26-LSL-MCPyV168* $p = 0.05$; green asterisk: *K14Cre-MCPyV168* vs. *Rosa26-LSL-MCPyV16* $p = 0.05$; (**B**) black asterisk: *K14Cre-MCPyV168* No Treatment vs. *K14Cre-MCPyV168* Treated $p < 0.0004$; red asterisk: *K14E6/E7* vs. *Rosa26-LSL-MCPyV168* $p < 0.03$; green asterisk: *K14Cre-MCPyV168* vs. *Rosa26-LSL-MCPyV168* $p < 2 \times 10^{-5}$; (**C**) black asterisk: *K14Cre-MCPyV168* No Treatment vs. *K14Cre-MCPyV168* Treated $p < 1.6\text{x}10^{-9}$; red asterisk: *K14E6/E7* vs. *Rosa26-LSL-MCPyV168* $p < 0.008$; green asterisk: *K14Cre-MCPyV168* vs. *Rosa26-LSL-MCPyV168* $p < 0.04$. The number of mice per group is indicated in Table 1. Error bars indicate standard deviation.

2.3. The MCPyV T Antigens Synergize with Chemical Carcinogens to Exacerbate Skin Tumorigenesis

To determine the relative contribution of viral genes and carcinogens to malignant progression, animals were treated with both tumor-initiating and -promoting chemicals (DMBA and TPA, respectively). This dual treatment will induce tumor formation even in non-transgenic murine skin. Tumor-bearing mice were then held for an additional period of time to monitor for progression of benign tumors to cancer. However, prior to this holding period, we analyzed whether the MCPyV T antigens can synergize with DMBA and TPA to exacerbate tumorigenesis. Groups of *Rosa26-LSL-MCPyV168* (n = 23), *K14E6/E7* (n = 17), and *K14Cre-MCPyV168* (n = 12) were treated once with 0.01 μmol DMBA and then twice a week with 15 nmol TPA for 20 weeks (dual carcinogen-treated, Figure 1B). The number of tumors was recorded every two weeks and the average number of tumors per mouse calculated at each time point (Figure 2C). In all groups of dual carcinogen-treated mice, the number of tumors began to rise at 10 weeks post-treatment. The average number of tumors in dual carcinogen-treated *K14Cre-MCPyV168* mice was significantly higher than in the dual carcinogen-treated *Rosa26-LSL-MCPyV168* mice at all time points between 10 and 20 weeks post-treatment (all p-values < 0.04). This significant increase in the average number of tumors compared to dual carcinogen-treated *Rosa26-LSL-MCPyV168* mice was similar in the dual carcinogen-treated *K14E6/E7* mice (all p-values < 0.008). However, despite these similarities, the difference between dual carcinogen-treated *Rosa26-LSL-MCPyV168* and dual carcinogen-treated *K14E6/E7* mice was generally more highly significant, especially between 10 and 14 weeks post-treatment, than the difference between dual carcinogen-treated *Rosa26-LSL-MCPyV168* and dual carcinogen-treated *K14Cre-MCPyV168* mice. At 12 weeks post-DMBA treatment, the average number of tumors in dual carcinogen-treated *K14E6/E7* mice was significantly higher than the number in dual carcinogen-treated *K14Cre-MCPyV168* mice (p = 0.03), and the difference between these two groups trended towards significance at both 10 weeks (p = 0.07) and 12 weeks (0.09) post-DMBA treatment. However, towards the end of the treatment period at 16, 18, and 20 weeks, the average number of tumors present in dual carcinogen-treated *K14E6/E7* and dual carcinogen-treated *K14Cre-MCPyV168* mice was statistically indistinguishable (16 weeks, p = 0.51; 18 weeks, p = 0.38; 20 weeks, p = 0.79). At all time points between 10 and 20 weeks, the number of tumors that developed in dual carcinogen-treated *K14Cre-MCPyV168* mice was significantly higher than in untreated controls (all p-values $<1.6 \times 10^{-9}$). These results indicate that the MCPyV T antigens synergize with chemical carcinogens to promote tumorigenesis, and do so at a level similar to that of the high-risk HPV16 E6 and E7 oncoproteins.

2.4. Assessment of Malignant Progression

To evaluate malignant progression, mice in the dual carcinogen (DMBA + TPA)-treated group normally are held for an additional 20 weeks following completion of the TPA treatment to allow time for malignant progression to occur [57]. At the end of this 20 week holding period, tissues were harvested and evaluated for histopathological disease and progression to squamous cell carcinoma (SCC). In this study, several mice developed excessive tumor burden that necessitated humane euthanasia of many of the mice, particularly in the dual carcinogen-treated *K14E6/E7* and dual carcinogen-treated *K14Cre-MCPyV168* groups, well ahead of the 20 week hold period. This compromised our ability to monitor for malignant progression. Nevertheless, we did analyze the histopathology and scored for worst disease in tumors harvested 5 weeks into the holding period from dual carcinogen-treated *Rosa26-LSL-MCPyV168* mice (n = 24 foci from n = 8 mice), dual carcinogen-treated *K14E6/E7* mice (n = 20 foci from n = 6 mice), and dual carcinogen-treated *K14Cre-MCPyV168* mice (n = 26 foci from n = 7 mice). Each tumor/foci was scored as having either No Disease, Squamous Dysplasia Grade 1 (mild), Squamous Dysplasia Grade 2 (moderate), Squamous Dysplasia Grade 3 (severe), Squamous Cell Carcinoma (SCC) Grade 1 (well differentiated), SCC Grade 2 (moderately differentiated), or SCC Grade 3 (poorly differentiated) (Table 2, Figure 3).

Table 2. Histopathological scoring of disease in tumors arising on murine skin after treatment with chemical carcinogens. After completing 20 weeks of TPA treatment following one-time treatment with DMBA, mice were held for an additional 5 weeks to monitor malignant progression. The number of mice and total tumors/foci evaluated is indicated in parentheses. The number of foci scored as having each disease grade is indicated for each group of mice.

Disease Grade	**Experimental Groups**		
	R26-LSL-MCPyV168 **(n = 8 mice, n = 24 foci)**	***K14E6/E7*** **(n = 6 mice, n = 20 foci)**	***K14Cre-MCPyV168*** **(n = 7 mice, n = 26 foci)**
Dysplasia Grade 1	12	6	14
Dysplasia Grade 2	4	6	4
Dysplasia Grade 3	0	2	2
SCC Grade 1	8	4	6
SCC Grade 2	0	1	0
SCC Grade 3	0	1	0

In dual carcinogen-treated *Rosa26-LSL-MCPyV168* mice, 67% of tumors (n = 16/24) progressed to precancerous dysplasia (n = 12 Dysplasia Grade 1, n = 4 Dysplasia Grade 2) and the remaining 33% of tumors (n = 8/24) had progressed to SCC Grade 1 within the 5 week holding period (Figure 3A). There was little difference between the disease that developed in dual carcinogen-treated *Rosa26-LSL-MCPyV168* mice and dual carcinogen-treated *K14Cre-MCPyV168* mice. Nearly 77% of tumors evaluated from dual carcinogen-treated *K14Cre-MCPyV168* mice had progressed to dysplasia (n = 14 Dysplasia Grade 1, n = 4 Dysplasia Grade 2, n = 2 Dysplasia Grade 3) and the remaining 23% (n = 6/26) of tumors had progressed to SCC Grade 1 (Figure 3A). There was no statistically significant difference between the overall disease severity in dual carcinogen-treated *Rosa26-LSL-MCPyV168* and dual carcinogen-treated *K14Cre-MCPyV168* mice ($p = 0.68$). In dual carcinogen-treated *K14E6/E7* mice, 70% of tumors progressed to dysplasia (n = 6 Dysplasia Grade 1, n = 6 Dysplasia Grade 2, n = 2 Dysplasia Grade 3). While around the same percentage of tumors from dual carcinogen-treated *K14E6/E7* mice progressed to SCC (30%; n = 6/20) as in the other groups of mice, the SCC grade of severity was slightly elevated such that we identified some tumors that progressed to SCC Grade 1 (n = 4/20), SCC Grade 2 (n = 1/20) and SCC Grade 3 (n = 1/20) (Figure 3A). However, the overall disease severity in dual carcinogen-treated *K14E6/E7* mice was not significantly higher than that in dual carcinogen-treated *Rosa26-LSL-MCPyV168* mice ($p = 0.33$) or dual carcinogen-treated *K14Cre-MCPyV168* mice ($p = 0.17$). Representative histology of worst disease that developed after 5 weeks hold in each group of mice is shown in Figure 3B. We conclude that the MCPyV T antigens do not acutely contribute to significant malignant progression, at least within 5 weeks following DMBA+TPA treatment. We were able to hold one dual carcinogen-treated *K14Cre-MCPyV168* mouse for a total of 8 weeks post-treatment (Figure 3C). Of the 5 total foci we evaluated from this single mouse, 60% (n = 3/5) were dysplastic and 40% (n = 2) developed into invasive SCCs. One cancer was scored as SCC Grade 1 and the other was scored as SCC Grade 3. Therefore, we have reason to believe that, had we been able to hold the mice longer, we might have observed a contribution of MCPyV T antigens to malignant progression.

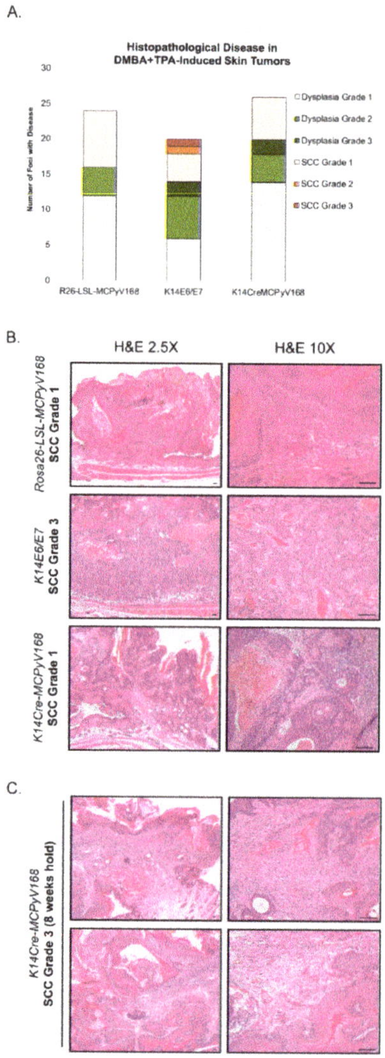

Figure 3. Assessment of malignant progression. (**A**) After the 20 week DMBA+TPA treatment period, tumor-bearing *Rosa26-LSL-MCPyV168, K14E6/E*, and *K14Cre-MCPyV168* mice were held without further treatment for an additional 5 weeks. Skin was harvested, sectioned into 5 µM sections and placed on glass slides, and stained with hematoxylin and eosin (H&E). Tissues were evaluated for histopathological disease. Each foci/tumor evaluated was given a score for worst disease among the following grades: Dysplasia Grade 1, Dysplasia Grade 2, Dysplasia Grade 3, Squamous Cell Carcinoma (SCC) Grade 1, SCC Grade 2, or SCC Grade 3. The number of foci with each disease score in each group of mice is indicated in the bar graph. The overall disease severity between groups was not statistically significant (two-sided Wilcoxon rank-sum test; all *p*-values > 0.17 for all comparisons). (**B**) Representative H&E-stained images of tissue sections from tumors harvested from DMBA+TPA-treated *Rosa26-LSL-MCPyV168, K14E6/E7*, and *K14Cre-MCPyV168* mice after 5 weeks hold. Images show representative examples of worst disease state present in each treatment group. All scale bars = 100 µM. (**C**) Representative H&E-stained images of tissue sections from tumors harvested from a DMBA+TPA-treated *K14Cre-MCPyV168* mouse after 8 weeks hold. Images show representative examples of worst disease state present. All scale bars = 100 µM.

3. Discussion

In this study, we evaluated how the MCPyV T antigens function within discrete stages of tumorigenesis in murine skin using a classical multi-stage model of squamous cell carcinoma development. In this model, both tumor-initiating and -promoting functions are needed for skin tumorigenesis in murine skin and resulting tumors can undergo malignant progression to cancer [55]. We found that MCPyV transgenic mice treated with the tumor promoter TPA failed to give rise to an increased number of tumors relative to the number that spontaneously develop on the skin of *K14Cre-MCPyV168* mice (Figure 1, Figure 2A). Conversely, there was a significant increase in tumor development in MCPyV transgenic mice treated with the tumor initiator DMBA and tumor development occurred at a level similar to that observed in HPV16 transgenic mice (Figure 2B). We also found that the MCPyV T antigens synergized with dual carcinogen treatment to significantly exacerbate skin tumor development, again to a level similar to HPV16 oncoproteins (Figure 2C). Finally, although we were unable to complete our studies on malignant progression as planned, we found some evidence that MCPyV T antigen expression in *K14Cre-MCPyV168* mice could contribute to malignant progression but did not significantly do so within the abbreviated time period of our study (Figure 3). In total, our results demonstrate that MCC tumor-derived MCPyV T antigens function as tumor promoters, and not tumor initiators, in murine skin.

Our finding that MCPyV T antigens function in tumor promotion, but not initiation, raises the question: what factor(s) drive tumor initiation to give rise to MCPyV-positive MCC? We believe there are several possibilities for initiating events during MCPyV-induced MCC pathogenesis. One possibility is MCPyV genome integration, a hypothesis presented in the initial report on MCPyV's discovery [12]. In a recent study, Starrett and colleagues found that MCPyV genome integration is associated with host genome amplifications and copy number variations (CNVs), similar to what is observed with other integrated oncogenic viruses like HPV [58,59], which the authors speculated could cause genomic instability. Therefore, MCPyV viral genome integration may initiate tumorigenesis by potentiating these potentially mutagenic events. In light of this proposed model of integration-induced tumor initiation by DNA tumor viruses, it is interesting that we also saw no evidence of tumor-initiating roles for the high-risk HPV16 oncogenes E5 [56], E6, or E7 [57] in our transgenic models. Notably, none of our HPV transgenic models nor our MCPyV transgenic model are infection models and do not involve the process of viral genome integration. Our murine models, therefore, effectively obscure any role of viral genome integration in genomic instability and tumor initiation. Nevertheless, viral genome integration is one possible mechanism of tumor initiation during MCC pathogenesis.

Another possibility is that the MCPyV T antigens themselves may contribute to tumor initiation, either through inherent functions or within certain contexts. For instance, the C terminus of the MCPyV full-length LT antigen induces DNA damage and genomic instability during in vitro MCPyV infection [60]. It is unclear when this DNA damage-inducing domain is lost during MCC pathogenesis, although there is evidence that LT truncation occurs before or during integration [61], thus allowing a window of time for LT-induced host genomic instability. ST can induce several chromosomal abnormalities, such as aneuploidy, chromosomal breaks, and micronuclei, in transduced fibroblasts and in ST transgenic mice [62]. Theoretically, ST could also function as a tumor initiator through its association with MYCL and the chromatin remodeling complex EP400, which leads to increased transcription of several pro-oncogenic genes [63]. One such gene is MDM4, which acts with MDM2 to cause p53 ubiquitination and subsequent degradation [64]. There is also evidence that ST expression in the epidermis initiates MCC-like lesions in murine skin, but only when expressed in keratinocytes that are also expressing the Merkel cell specification factor ATOH1 [33]. It is also possible that there is a heretofore unappreciated co-carcinogenic factor that contributes to the initiation events in MCPyV+ MCC. Therefore, the MCPyV T antigens may require cooperation with additional co-factors and/or precise conditions to function as tumor initiators in the complex environment of the skin. This hypothesis seems plausible given the ubiquitous nature of MCPyV infection, yet rare incidence of MCC, in the human population.

It is possible that host factors contribute to tumor initiation during MCC pathogenesis. It is now well established that MCPyV-negative and MCPyV-positive MCC tumor cells contain significantly different mutational landscapes in their host genomes [58,59,65–67]. MCPyV-negative MCCs are characterized by an abundance of mutations that bear signatures consistent with UV-mediated mutagenesis, whereas MCPyV-positive MCCs have a low overall mutational burden. While these findings seem to imply that the contribution of somatic mutations to tumor initiation in MCPyV-positive MCC is minor, there may be random or rare mutational events in driver genes whose likelihood to arise over time is increased by the hyperproliferative effects of the MCPyV T antigens. Potential driver mutations have been identified in MCPyV-positive MCCs, with one study finding such mutations in approximately 30% of tumors [67]. For instance, activating gene mutations in the phosphatidyl-3-kinase/AKT/mammalian target of rapamycin (PI3K/AKT/mTOR) pathway are found in MCCs. Activating mutations in *AKT1*, *PIK3CA*, and *HRAS* and loss of function mutations in genes that act as negative regulators of the pathway, such as *PTEN* and *TSC1*, have been detected by multiple groups in MCPyV-positive MCCs [65,67–71]. Along these lines, it is interesting to note that DMBA, the tumor-initiating chemical carcinogen used in this study, is thought to act as a tumor initiator by inducing activating mutations in *HRAS* [72]. We therefore speculate that such DMBA-induced mutations may imitate at least one of the tumor-initiating somatic driver mutations found in MCPyV-induced MCCs, which could then cooperate with the tumor-promoting functions of the MCPyV T antigens to drive tumorigenesis. Interestingly, we have evidence that activating PI3K mutations are sufficient to drive tumorigenesis in HPV16 transgenic mice [73]. In a model of HPV-positive anal cancer, DMBA treatment and the HPV oncogenes serve the tumor initiator and tumor promoter roles, respectively [74]. However, an activating mutation in *PIK3CA* was sufficient to drive anal carcinogenesis in HPV16 transgenic mice without DMBA administration [73]. Therefore, it is reasonable to hypothesize that such driver mutations can also cooperate with the tumor-promoting activity of the MCPyV T antigens in a similar way during MCC pathogenesis.

The multi-stage carcinogenesis model used in this study is an eminent model used to investigate the development of tumors of epithelial origin, most notably squamous cell carcinoma [54,75]. The cellular origin of MCC remains unclear, as does the cell type infected by MCPyV, and both are active areas of research (reviewed in [76]). Despite similarities in gene expression signatures, biomarkers, and histology, Merkel cells are post-mitotic [77] and quite rare within the skin and therefore themselves seem an unlikely precursor to MCC. The current paradigm suggests that MCPyV-positive MCCs derive from an infected precursor cell type that has yet to be identified. Some possibilities include dermal fibroblasts, which can support MCPyV infection and replication in vitro [24], lymphoid pre/pro-B cells [78], and neuronal cells [50]. However, growing evidence supports an epithelial cell of origin. Merkel cells arise from epidermal progenitor cells [77,79] and ectopic expression of the Merkel cell specification factor atonal homolog 1 (ATOH1) in epithelial cells induces Merkel cell development [80]. Interestingly, MCPyV LT expression in keratinocytes has been observed to not only increase ATOH1 expression [50] but also prevent its degradation [52]. Furthermore, combined expression of the MCPyV T antigens and cellular genes GLI1 or ATOH1 induces Merkel cell-like phenotypes in vitro [52] and MCC-like lesions in vivo [33], respectively. A recent study reported sequencing evidence that a MCPyV-positive MCC was derived from an epithelial lineage [81]. While it is possible that the T antigen functions differ in other potential MCC precursor cell types, our evaluation of MCPyV T antigen function in epithelial tumorigenesis determined that these viral proteins function as tumor promoters when their expression is targeted to one likely MCC precursor cell population, K14-positive epithelial cells.

Throughout the course of our studies with *K14Cre-MCPyV168* transgenic mice, we continue to observe considerable similarities between the activities of keratin 14-driven expression of the MCPyV T antigens and the high-risk HPV16 oncoproteins E6 and E7. During our initial characterization of MCPyV transgenic mice, we found that the MCPyV T antigens induced epithelial hyperplasia, cellular proliferation, and E2F-dependent gene expression to the same extent as HPV16 E6 and E7 [32]. Here, we have observed that the MCPyV T antigens and HPV16 oncoproteins act in very similar ways in a model of skin carcinogenesis. The HPV oncoproteins, particularly E5 and E7, function as tumor promoters in murine skin [56,57], while E6 primarily functions in malignant progression [57]. We included *K14E6/E7* transgenic mice in our study, to compare findings to those obtained with the MCPyV transgenic mice, and discovered that the MCPyV T antigens also function as tumor promoters. The tumor incidence in *K14Cre-MCPyV168* mice largely mirrored that in *K14E6/E7* mice following DMBA-only treatment (Figure 2B), and the same was true when both transgenic lines were treated with DMBA+TPA (Figure 2C). These similarities seem to reflect the often overlapping, parallel functions of oncogenic DNA tumor virus proteins [2]. A few observations may suggest that the HPV16 oncoproteins are slightly more potent tumor promoters than the MCPyV T antigens. For instance, tumor incidence in DMBA only treated animals rose to a level significantly higher than the negative control group slightly faster in *K14E6/E7* mice than in *K14Cre-MCPyV168* mice (14 versus 16 weeks, respectively; Figure 2B). While the number of tumors in both groups were significantly higher than in *Rosa26-LSL-MCPyV168* mice, the number of tumors in DMBA+TPA-treated *K14E6/E7* mice were more highly significant than in DMBA+TPA-treated *K14Cre-MCPyV168* mice (Figure 2C). Finally, some of the tumors present in DMBA+TPA-treated *K14E6/E7* mice progressed to a more advanced grade of SCC within the limited 5 week holding period than did tumors in DMBA+TPA-treated *K14Cre-MCPyV168* mice, though the overall severity of disease was not significantly different (Figure 3), which may reflect the actions of E6 in the progression stage [57]. Therefore, while there are functional similarities between epithelial expression of the MCPyV T antigens and HPV16 oncoproteins, there may still be important differences in their underlying mechanisms and/or potency.

There remain several outstanding questions related to the specific role of the MCPyV T antigens in MCC pathogenesis that are raised by our study. For instance, there is still a great deal to learn about the individual and cooperative roles of the truncated LT and ST antigens in MCCs. Both T antigens are expressed in the *K14Cre-MCPyV168* mice used in this study, and it would be interesting to test the role of the individual MCC tumor-derived T antigens using the multi-stage model of skin carcinogenesis. The difficulty we encountered in evaluating malignant progression in our study left us with an unsatisfying level of insight regarding the role of MCPyV T antigens in this stage of carcinogenesis. Because this model is so well studied and utilized, there are several identified areas for optimization related to treatment dose and duration that we can explore in future studies that may increase our ability to study malignant progression [54]. Given its association with MCC [11], we could also adapt this model to study the role of ultraviolet light exposure in MCPyV-associated tumorigenesis and whether it functions more as a mutagen [58,65,67,71] or as an immunosuppressive or immunomodulatory agent [82] in this process. The development of MCPyV transgenic mice provide ample opportunities to further explore the underlying mechanisms of the MCPyV T antigens in MCC pathogenesis and neoplastic progression.

4. Materials and Methods

4.1. Animals

The MCPyV transgenic mice, named *ROSA26-LSL-MCPyV168*, have been described previously [32]. Briefly, the MCPyV early region, isolated from MCC tumor specimen MCCw168 (MCPyV168; GenBank: KC426954.1), was cloned into vectors containing a LoxP-stop-LoxP cassette (LSL) and the pROSA26PA plasmid. These conditional *ROSA26-LSL-MCPyV168* mice were crossed with transgenic mice expressing Cre recombinase driven by the human keratin 14 (*Krt14* or K14) promoter (*K14Cre*) to generate *K14Cre-MCPyV168* mice. The *K14E6/E7* bitransgenic mice included in this study express the HPV16 E6 and E7 oncogenes driven by the K14 promoter and have been described previously [83,84].

All mice were maintained on the *FVB/N* genetic background. All animal experiments were performed in full compliance with standards outlined in the Guide for the Care and Use of Laboratory Animals by the Laboratory Animal Resources (LAR) as specified by the Animal Welfare Act (AWA) and Office of Laboratory Animal Welfare (OLAW) and approved by the Governing Board of the National Research Council (NRC). Mice were housed at the McArdle Laboratory Animal Care Unit in strict accordance with guidelines approved by the Association for Assessment of Laboratory Animal Care (AALAC), at the University of Wisconsin Medical School. All protocols for animal work were approved by the University of Wisconsin Medical School Institutional Animal Care and Use Committee (IACUC; protocol number M005871).

4.2. Genotyping

All transgenic mice used in these studies were verified by PCR genotyping. Genomic DNA was isolated from tail snips and resuspended in water. Separate PCR reactions were used to identify the wild-type or recombined ROSA26 allele, presence of the Cre recombinase gene, and the E6/E7 transgenes. PCR products were evaluated using agarose gel electrophoresis. The following primers were used for genotyping: P1 (5′-AAA GTC GCT CTG AGT TGT TAT-3′), P2 (5′-GCG AAG AGT TTG TCC TCA-3′) and P3 (5′-AGC GGG AGA AAT GGA TAT-3′) specific for the ROSA26 allele; 3069 (5′-TTC CTC AGG AGT GTC TTC GC-3′) and 3070 (5′-GTC CAT GTC CTT CCT GAA GC-3′) for K14Cre; Oligo-2 (5′-GCA TGA CAG CTG GGT TTC TCT ACG-3′) and E6TTL (5′-GCT TAG TTA ACT AAT GCA AAC-3′) for E7, and E7TTL (5′- AGC CTT AGT TAA CTA ACA TTA C-3′) and 709-4 (5′-CCC GGA TCC TAC CTG CAG GAT CAG CCA TG-3′) for E6. All primers were synthesized by Integrated DNA Technologies (Coralville, IA, USA).

4.3. Skin Carcinogenesis Studies

At 4 to 6 weeks of age, the dorsal area of mice was shaved to create an area for topical carcinogen treatment and the animals divided into three groups. One group of mice (TPA only) was treated with 12-O-tetradecanoylphorbol-13-acetate (TPA), a promoting agent. The skin of these mice was topically treated with 15 nmol TPA, dissolved in acetone, twice weekly for 20 weeks. Another group of mice (DMBA only) were topically treated once with 0.3 µmol 7,12-dimethylbenz[a]-anthracene (DMBA), an initiating carcinogen, also dissolved in acetone. The third group of mice (DMBA+TPA) was topically treated with 0.01 µmol DMBA and one week later, the same skin region was topically treated with 15 nmol TPA twice weekly for 20 weeks. We also included an additional group of *K14Cre-MCPyV168* mice that were left untreated to serve as a baseline control for the number of tumors these mice spontaneously develop over time [32]. All mice were examined every 2 weeks for tumors, and the average number of gross tumors per mouse was calculated at each time point.

To evaluate malignant progression, mice that had been treated with DMBA+TPA were held following treatment completion and mice were not treated in any way during this time. While the original protocol used in our laboratory to study malignant progression prescribes a 20 week hold, we were only able to hold mice for 5 weeks due to excessive tumor burden in *K14E6/E7* and *K14Cre-MCPyV168* mice. At the end of 5 weeks, skin was harvested, fixed in 4% paraformaldehyde, and embedded in paraffin. All tissue sections were prepared by an experienced histotechnologist by cutting 5 µM serial sections from the paraffin blocks and placing on glass slides. Every 10th section was stained with hematoxylin and eosin (H&E) to facilitate histopathological analysis. At least 20 tumors, or foci, from each group of DMBA+TPA-treated mice held for 5 weeks were selected for histopathological analysis. Each tumor/foci was assessed for squamous dysplasia and keratinizing invasive squamous cell carcinoma according to the standard histopathologic criteria as having No Disease, Squamous Dysplasia Grade 1 (mild), Squamous Dysplasia Grade 2 (moderate), Squamous Dysplasia Grade 3 (severe), Invasive Squamous Cell Carcinoma (SCC) Grade

1 (well differentiated), SCC Grade 2 (moderately differentiated), or SCC Grade 3 (poorly differentiated).

4.4. Statistical Analysis

The average number of tumors per mouse was calculated by dividing the total number of tumors present within a given group of animals by the total number of animals present per group at the indicated time point. If animals dropped out of the experiment, either by morbidity due to advanced age or required euthanasia for excessive tumor burden, the group size was adjusted accordingly. Data were compiled and graphs generated using the GraphPad Prism program (Version 8.4.3; last accessed 30 October 2020). A two-sided Wilcoxon rank-sum test was used to compare the average number of tumors per mouse between groups at each time point. To compare disease severity, each histopathological grade was assigned a rank (No Disease = 0, Dysplasia Grade 1 = 1, Dysplasia Grade 2 = 2, Dysplasia Grade 3 = 3, SCC Grade 1 = 4, SCC Grade 2 = 5, and SCC Grade 3 = 6) and then analyzed using a two-sided Wilcoxon rank-sum test. Statistical analysis was performed using MSTAT statistical software version 6.6.1 (https://oncology.wisc.edu/mstat/; last accessed 21 October 2020).

5. Conclusions

The causal relationship between MCPyV and MCC represents the first association of a human polyomavirus with human cancer. Understanding the role of this newly discovered DNA tumor virus and its viral proteins in the pathogenesis of MCC is critical to identifying prevention and therapeutic approaches to this aggressive and lethal cutaneous cancer. We have recently developed and characterized a MCPyV transgenic murine model that involves the targeted expression of MCC tumor-derived MCPyV-truncated LT and ST antigens to epithelial cells of murine skin [32]. In the study presented here, we utilized these transgenic mice to determine the role of the MCPyV T antigens in different stages of skin carcinogenesis. This well-validated, multi-stage model of skin cancer development uses topical application of chemical carcinogens to identify the role of genes and factors in tumor initiation, tumor promotion, and malignant progression. We found that the MCPyV T antigens function as tumor promoters, and not tumor initiators, in murine skin. The functions of the MCPyV T antigens in this in vivo assay closely mirrored the actions of viral oncoproteins from another DNA tumor virus, HPV16. It is possible that the MCPyV T antigens function differently in other potential MCC precursor cell types or when expressed under certain conditions within the milieu of human skin. However, in the context of epithelial tumorigenesis, these observations suggest that other factors likely contribute to tumor initiation and cooperate with the tumor-promoting functions of the MCPyV T antigens during MCC development.

Author Contributions: Conceptualization, M.E.S. and P.F.L.; data curation, M.E.S., D.B. and P.F.L.; formal analysis, M.E.S. and D.B.; funding acquisition, M.E.S. and P.F.L.; investigation, M.E.S. and P.F.L.; methodology, M.E.S., A.L. and D.B.; project administration, M.E.S. and P.F.L.; resources, M.E.S., A.L., J.C., J.A.D. and P.F.L.; supervision, M.E.S. and P.F.L.; validation, M.E.S.; visualization, M.E.S.; writing—original draft, M.E.S.; writing—review and editing, M.E.S., D.B., J.C., J.A.D. and P.F.L. All authors have read and agreed to the published version of the manuscript.

Funding: National Institutes of Health: CA211246, DE026787, CA022443, CA210807, CA228543, CA243777, CA232128, CA203655.

Institutional Review Board Statement: Not applicable.

Informed Consent Statement: Not applicable.

Data Availability Statement: The data presented in this study are available within this article.

Acknowledgments: The authors would like to thank all Lambert Laboratory members for constructive feedback and advice. We would also like to acknowledge Harlene Edwards and Ella Ward-Shaw for histotechnological assistance and Ruth Sullivan for early pathology consultations. Megan E.

Spurgeon is supported by a grant from the National Institutes of Health (R50CA211246). Darya Buehler is supported by grants from the National Institutes of Health (P50DE026787). Paul F. Lambert is supported by grants from the National Institutes of Health (P01CA022443, R35CA210807, R01CA228543, and P50DE026787). Jingwei Cheng is supported by a grant from the National Institutes of Health (R50CA243777). James A. DeCaprio is supported by grants from the National Institutes of Health (R35CA232128 and P01CA203655).

Conflicts of Interest: J.A.D. received research funding from Constellation Pharmaceuticals. J.A.D. has served as a consultant to Merck & Co. and EMD Serono. The remaining authors have no conflicts to declare.

References

1. De Martel, C.; Ferlay, J.; Franceschi, S.; Vignat, J.; Bray, F.; Forman, D.; Plummer, M. Global burden of cancers attributable to infections in 2008: A review and synthetic analysis. *Lancet Oncol.* **2012**, *13*, 607–615. [CrossRef]
2. Lambert, P.F. The interwoven story of the small DNA tumor viruses. *Virology* **2009**, *384*, 255. [CrossRef] [PubMed]
3. Pipas, J.M. DNA tumor viruses and their contributions to molecular biology. *J. Virol.* **2019**, *93*, e01524-18. [CrossRef] [PubMed]
4. Jha, H.C.; Banerjee, S.; Robertson, E.S. The role of gammaherpesviruses in cancer pathogenesis. *Pathogens* **2016**, *5*, 18. [CrossRef] [PubMed]
5. Hausen, H.Z. Papillomaviruses in the causation of human cancers-A brief historical account. *Virology* **2009**, *384*, 260–265. [CrossRef] [PubMed]
6. DeCaprio, J.A.; Garcea, R.L. A cornucopia of human polyomaviruses. *Nat. Rev. Genet.* **2013**, *11*, 264–276. [CrossRef]
7. Toker, C. Trabecular carcinoma of the skin. *Arch. Dermatol* **1972**, *105*, 107–110. [CrossRef]
8. Hodgson, N.C. Merkel cell carcinoma: Changing incidence trends. *J. Surg. Oncol.* **2004**, *89*, 1–4. [CrossRef]
9. Paulson, K.G.; Park, S.Y.; Vandeven, N.A.; Lachance, K.; Thomas, H.; Chapuis, A.G.; Harms, K.L.; Thompson, J.A.; Bhatia, S.; Stang, A.; et al. Merkel cell carcinoma: Current US incidence and projected increases based on changing demographics. *J. Am. Acad. Dermatol.* **2018**, *78*, 457–463.e2. [CrossRef]
10. Harms, P.W.; on behalf of the International Workshop on Merkel Cell Carcinoma Research (IWMCC) Working Group; Harms, K.L.; Moore, P.S.; DeCaprio, J.A.; Nghiem, P.; Wong, M.K.K.; Brownell, I. The biology and treatment of Merkel cell carcinoma: Current understanding and research priorities. *Nat. Rev. Clin. Oncol.* **2018**, *15*, 763–776. [CrossRef]
11. Heath, M.; Jaimes, N.; Lemos, B.; Mostaghimi, A.; Wang, L.C.; Peñas, P.F.; Nghiem, P. Clinical characteristics of Merkel cell carcinoma at diagnosis in 195 patients: The AEIOU features. *J. Am. Acad. Dermatol.* **2008**, *58*, 375–381. [CrossRef] [PubMed]
12. Feng, H.; Shuda, M.; Chang, Y.; Moore, P.S. Clonal integration of a polyomavirus in human Merkel cell carcinoma. *Science* **2008**, *319*, 1096–1100. [CrossRef] [PubMed]
13. Carter, J.J.; Paulson, K.G.; Wipf, G.C.; Miranda, D.; Madeleine, M.M.; Johnson, L.G.; Lemos, B.D.; Lee, S.; Warcola, A.H.; Iyer, J.G.; et al. Association of Merkel cell polyomavirus-specific antibodies with Merkel cell carcinoma. *J. Natl. Cancer Inst.* **2009**, *101*, 1510–1522. [CrossRef] [PubMed]
14. Chen, T.; Hedman, L.; Mattila, P.S.; Jartti, T.; Ruuskanen, O.; Söderlund-Venermo, M.; Hedman, K. Serological evidence of Merkel cell polyomavirus primary infections in childhood. *J. Clin. Virol.* **2011**, *50*, 125–129. [CrossRef] [PubMed]
15. Kean, J.M.; Rao, S.; Wang, M.; Garcea, R.L. Seroepidemiology of human polyomaviruses. *PLoS Pathog.* **2009**, *5*, e1000363. [CrossRef]
16. Pastrana, D.V.; Tolstov, Y.L.; Becker, J.C.; Moore, P.S.; Chang, Y.; Buck, C.B. Quantitation of Human Seroresponsiveness to Merkel Cell Polyomavirus. *PLoS Pathog.* **2009**, *5*, e1000578. [CrossRef]
17. Tolstov, Y.L.; Knauer, A.; Chen, J.-G.; Kensler, T.W.; Kingsley, L.A.; Moore, P.S.; Chang, Y. Asymptomatic primary Merkel cell polyomavirus infection among adults. *Emerg. Infect. Dis.* **2011**, *17*, 1371–1380. [CrossRef]
18. Tolstov, Y.L.; Pastrana, D.V.; Feng, H.; Becker, J.C.; Jenkins, F.J.; Moschos, S.; Chang, Y.; Buck, C.B.; Moore, P.S. Human Merkel cell polyomavirus infection II. MCV is a common human infection that can be detected by conformational capsid epitope immunoassays. *Int. J. Cancer* **2009**, *125*, 1250–1256. [CrossRef]
19. Touzé, A.; Gaitan, J.; Arnold, F.; Cazal, R.; Fleury, M.J.; Combelas, N.; Sizaret, P.-Y.; Guyetant, S.; Maruani, A.; Baay, M.; et al. Generation of Merkel cell polyomavirus (MCV)-like particles and their application to detection of MCV antibodies. *J. Clin. Microbiol.* **2010**, *48*, 1767–1770. [CrossRef]
20. Viscidi, R.P.; Rollison, D.E.; Sondak, V.K.; Silver, B.; Messina, J.L.; Giuliano, A.R.; Fulp, W.; Ajidahun, A.; Rivanera, D. Age-specific seroprevalence of Merkel cell polyomavirus, BK virus, and JC virus. *Clin. Vaccine Immunol.* **2011**, *18*, 1737–1743. [CrossRef]
21. Schowalter, R.M.; Pastrana, D.V.; Pumphrey, K.A.; Moyer, A.L.; Buck, C.B. Merkel cell polyomavirus and two previously unknown polyomaviruses are chronically shed from human skin. *Cell Host Microbe* **2010**, *7*, 509–515. [CrossRef] [PubMed]
22. Becker, M.; Dominguez, M.; Greune, L.; Soria-Martinez, L.; Pfleiderer, M.M.; Schowalter, R.; Buck, C.B.; Blaum, B.S.; Schmidt, M.A.; Schelhaas, M. Infectious entry of Merkel cell polyomavirus. *J. Virol.* **2019**, *93*, 93. [CrossRef] [PubMed]
23. Kwun, H.J.; Chang, Y.; Moore, P.S. Protein-mediated viral latency is a novel mechanism for Merkel cell polyomavirus persistence. *Proc. Natl. Acad. Sci. USA* **2017**, *114*, E4040–E4047. [CrossRef] [PubMed]
24. Liu, W.; Krump, N.A.; Buck, C.B.; You, J. Merkel cell polyomavirus infection and detection. *J. Vis. Exp.* **2019**, e58950. [CrossRef]

25. Neumann, F.; Borchert, S.; Schmidt, C.; Reimer, R.; Hohenberg, H.; Fischer, N.; Grundhoff, A. Replication, gene expression and particle production by a consensus Merkel cell polyomavirus (MCPyV) genome. *PLoS ONE* **2011**, *6*, e29112. [CrossRef]
26. Schowalter, R.M.; Reinhold, W.C.; Buck, C.B. Entry tropism of BK and Merkel cell polyomaviruses in cell culture. *PLoS ONE* **2012**, *7*, e42181. [CrossRef]
27. Angermeyer, S.; Hesbacher, S.; Becker, J.C.; Schrama, D.; Houben, R. Merkel Cell polyomavirus-Positive Merkel cell carcinoma cells do not require expression of the viral small T antigen. *J. Investig. Dermatol.* **2013**, *133*, 2059–2064. [CrossRef]
28. Cheng, J.; Rozenblatt-Rosen, O.; Paulson, K.G.; Nghiem, P.; DeCaprio, J.A. Merkel cell polyomavirus large T antigen has growth-promoting and inhibitory activities. *J. Virol.* **2013**, *87*, 6118–6126. [CrossRef]
29. Nwogu, N.; Ortiz, L.E.; Kwun, H.J. Surface charge of Merkel cell polyomavirus small T antigen determines cell transformation through allosteric FBW7 WD40 domain targeting. *Oncogenesis* **2020**, *9*, 1–12. [CrossRef]
30. Shuda, M.; Feng, H.; Kwun, H.J.; Rosen, S.T.; Gjoerup, O.; Moore, P.S.; Chang, Y. T antigen mutations are a human tumor-specific signature for Merkel cell polyomavirus. *Proc. Natl. Acad. Sci. USA* **2008**, *105*, 16272–16277. [CrossRef]
31. Shuda, M.; Guastafierro, A.; Geng, X.; Shuda, Y.; Ostrowski, S.M.; Lukianov, S.; Jenkins, F.J.; Honda, K.; Maricich, S.M.; Moore, P.S.; et al. Merkel cell polyomavirus small T antigen induces cancer and embryonic Merkel cell proliferation in a transgenic mouse model. *PLoS ONE* **2015**, *10*, e0142329. [CrossRef] [PubMed]
32. Spurgeon, M.E.; Cheng, J.; Bronson, R.T.; Lambert, P.F.; DeCaprio, J.A. Tumorigenic activity of Merkel cell polyomavirus T antigens expressed in the stratified epithelium of mice. *Cancer Res.* **2015**, *75*, 1068–1079. [CrossRef] [PubMed]
33. Verhaegen, M.E.; Mangelberger, D.; Harms, P.W.; Eberl, M.; Wilbert, D.M.; Meireles, J.; Bichakjian, C.K.; Saunders, T.L.; Wong, S.Y.; Dlugosz, A.A. Merkel cell polyomavirus small T antigen initiates Merkel cell carcinoma-like tumor development in mice. *Cancer Res.* **2017**, *77*, 3151–3157. [CrossRef] [PubMed]
34. Verhaegen, M.E.; Mangelberger, D.; Harms, P.W.; Vozheiko, T.D.; Weick, J.W.; Wilbert, D.M.; Saunders, T.L.; Ermilov, A.N.; Bichakjian, C.K.; Johnson, T.M.; et al. Merkel cell polyomavirus small T antigen is oncogenic in transgenic mice. *J. Investig. Dermatol.* **2015**, *135*, 1415–1424. [CrossRef] [PubMed]
35. Schmitt, M.; Wieland, U.; Kreuter, A.; Pawlita, M. C-terminal deletions of Merkel cell polyomavirus large T-antigen, a highly specific surrogate marker for virally induced malignancy. *Int. J. Cancer* **2012**, *131*, 2863–2868. [CrossRef]
36. Hesbacher, S.; Pfitzer, L.; Wiedorfer, K.; Angermeyer, S.; Borst, A.; Haferkamp, S.; Scholz, C.-J.; Wobser, M.; Schrama, D.; Houben, R. RB1 is the crucial target of the Merkel cell polyomavirus Large T antigen in Merkel cell carcinoma cells. *Oncotarget* **2016**, *7*, 32956–32968. [CrossRef]
37. Rodig, S.J.; Cheng, J.; Wardzala, J.; Dorosario, A.; Scanlon, J.J.; Laga, A.C.; Martinez-Fernandez, A.; Barletta, J.A.; Bellizzi, A.M.; Sadasivam, S.; et al. Improved detection suggests all Merkel cell carcinomas harbor Merkel polyomavirus. *J. Clin. Investig.* **2012**, *122*, 4645–4653. [CrossRef]
38. Shuda, M.; Kwun, H.J.; Feng, H.; Chang, Y.; Moore, P.S. Human Merkel cell polyomavirus small T antigen is an oncoprotein targeting the 4E-BP1 translation regulator. *J. Clin. Investig.* **2011**, *121*, 3623–3634. [CrossRef]
39. Laude, H.C.; Jonchère, B.; Maubec, E.; Carlotti, A.; Marinho, E.; Couturaud, B.; Peter, M.; Sastre-Garau, X.; Avril, M.-F.; Dupin, N.; et al. Distinct Merkel cell polyomavirus molecular features in tumour and non tumour specimens from patients with Merkel cell carcinoma. *PLoS Pathog.* **2010**, *6*, e1001076. [CrossRef]
40. Bhatia, K.; Goedert, J.J.; Modali, R.; Preiss, L.; Ayers, L.W. Merkel cell carcinoma subgroups by Merkel cell polyomavirus DNA relative abundance and oncogene expression. *Int. J. Cancer* **2009**, *126*, 2240–2246. [CrossRef]
41. Houben, R.; Shuda, M.; Weinkam, R.; Schrama, D.; Feng, H.; Chang, Y.; Moore, P.S.; Becker, J.C. Merkel cell polyomavirus-infected Merkel cell carcinoma cells require expression of viral T antigens. *J. Virol.* **2010**, *84*, 7064–7072. [CrossRef] [PubMed]
42. Katano, H.; Ito, H.; Suzuki, Y.; Nakamura, T.; Sato, Y.; Tsuji, T.; Matsuo, K.; Nakagawa, H.; Sata, T. Detection of Merkel cell polyomavirus in Merkel cell carcinoma and Kaposi's sarcoma. *J. Med. Virol.* **2009**, *81*, 1951–1958. [CrossRef] [PubMed]
43. Shuda, M.; Arora, R.; Kwun, H.J.; Feng, H.; Sarid, R.; Fernández-Figueras, M.-T.; Tolstov, Y.; Gjoerup, O.; Mansukhani, M.M.; Swerdlow, S.H.; et al. Human Merkel cell polyomavirus infection I. MCV T antigen expression in Merkel cell carcinoma, lymphoid tissues and lymphoid tumors. *Int. J. Cancer* **2009**, *125*, 1243–1249. [CrossRef] [PubMed]
44. Busam, K.J.; Jungbluth, A.A.; Rekthman, N.; Coit, D.; Pulitzer, M.; Bini, J.; Arora, R.; Hanson, N.C.; Tassello, J.A.; Frosina, D.; et al. Merkel cell polyomavirus expression in Merkel cell carcinomas and its absence in combined tumors and pulmonary neuroendocrine carcinomas. *Am. J. Surg. Pathol.* **2009**, *33*, 1378–1385. [CrossRef] [PubMed]
45. Schrama, D.; Peitsch, W.K.; Zapatka, M.; Kneitz, H.; Houben, R.; Eib, S.; Haferkamp, S.; Moore, P.S.; Shuda, M.; Thompson, J.F.; et al. Merkel cell polyomavirus status is not associated with clinical course of Merkel cell carcinoma. *J. Investig. Dermatol.* **2011**, *131*, 1631–1638. [CrossRef]
46. Shuda, M.; Chang, Y.; Moore, P.S. Merkel cell polyomavirus-Positive Merkel cell carcinoma requires viral small T-antigen for cell proliferation. *J. Investig. Dermatol.* **2014**, *134*, 1479–1481. [CrossRef]
47. Houben, R.; Adam, C.; Baeurle, A.; Hesbacher, S.; Grimm, J.; Angermeyer, S.; Henzel, K.; Hauser, S.; Elling, R.; Bröcker, E.-B.; et al. An intact retinoblastoma protein-binding site in Merkel cell polyomavirus large T antigen is required for promoting growth of Merkel cell carcinoma cells. *Int. J. Cancer* **2011**, *130*, 847–856. [CrossRef]
48. Schrama, D.; Hesbacher, S.; Angermeyer, S.; Schlosser, A.; Haferkamp, S.; Aue, A.; Adam, C.; Weber, A.; Schmidt, M.; Houben, R. Serine 220 phosphorylation of the Merkel cell polyomavirus large T antigen crucially supports growth of Merkel cell carcinoma cells. *Int. J. Cancer* **2015**, *138*, 1153–1162. [CrossRef]

49. Fan, K.; Gravemeyer, J.; Ritter, C.; Rasheed, K.; Gambichler, T.; Moens, U.; Shuda, M.; Schrama, D.; Becker, J.C. MCPyV Large T antigen-induced atonal homolog 1 is a lineage-dependency oncogene in Merkel cell carcinoma. *J. Investig. Dermatol.* **2020**, *140*, 56–65.e3. [CrossRef]
50. Harold, A.; Amako, Y.; Hachisuka, J.; Bai, Y.; Li, M.Y.; Kubat, L.; Gravemeyer, J.; Franks, J.; Gibbs, J.R.; Park, H.J.; et al. Conversion of Sox2-dependent Merkel cell carcinoma to a differentiated neuron-like phenotype by T antigen inhibition. *Proc. Natl. Acad. Sci. USA* **2019**, *116*, 20104–20114.
51. Arora, R.; Shuda, M.; Guastafierro, A.; Feng, H.; Toptan, T.; Tolstov, Y.; Normolle, D.; Vollmer, L.L.; Vogt, A.; Dömling, A.; et al. Survivin is a therapeutic target in Merkel cell carcinoma. *Sci. Transl. Med.* **2012**, *4*, 133ra56. [CrossRef]
52. Kervarrec, T.; Samimi, M.; Hesbacher, S.; Berthon, P.; Wobser, M.; Sallot, A.; Sarma, B.; Schweinitzer, S.; Gandon, T.; Destrieux, C.; et al. Merkel cell polyomavirus T antigens induce Merkel cell-like differentiation in GLI1-expressing epithelial cells. *Cancers* **2020**, *12*, 1989. [CrossRef]
53. Pietropaolo, V.; Prezioso, C.; Moens, U. Merkel cell polyomavirus and Merkel cell carcinoma. *Cancers* **2020**, *12*, 1774. [CrossRef] [PubMed]
54. Abel, E.L.; Angel, J.M.; Kiguchi, K.; DiGiovanni, J. Multi-stage chemical carcinogenesis in mouse skin: Fundamentals and applications. *Nat. Protoc.* **2009**, *4*, 1350–1362. [CrossRef] [PubMed]
55. Boutwell, R.K. Model systems for defining initiation, promotion, and progression of skin neoplasms. *Prog. Clin. Boil. Res.* **1989**, *298*, 3–15.
56. Maufort, J.P.; Williams, S.M.G.; Pitot, H.C.; Lambert, P. Human Papillomavirus 16 E5 oncogene contributes to two stages of skin carcinogenesis. *Cancer Res.* **2007**, *67*, 6106–6112. [CrossRef]
57. Song, S.; Liem, A.; Miller, J.A.; Lambert, P. Human Papillomavirus types 16 E6 and E7 contribute differently to carcinogenesis. *Virology* **2000**, *267*, 141–150. [CrossRef]
58. Starrett, G.J.; Marcelus, C.; Cantalupo, P.G.; Katz, J.P.; Cheng, J.; Akagi, K.; Thakuria, M.; Rabinowits, G.; Wang, L.C.; Symer, D.E.; et al. Merkel cell polyomavirus exhibits dominant control of the tumor genome and transcriptome in virus-associated Merkel cell carcinoma. *mBio* **2017**, *8*, e02079-16. [CrossRef]
59. Starrett, G.J.; Thakuria, M.; Chen, T.; Marcelus, C.; Cheng, J.; Nomburg, J.; Thorner, A.R.; Slevin, M.K.; Powers, W.; Burns, R.T.; et al. Clinical and molecular characterization of virus-positive and virus-negative Merkel cell carcinoma. *Genome Med.* **2020**, *12*, 1–22. [CrossRef]
60. Li, J.; Wang, X.; Diaz, J.; Tsang, S.H.; Buck, C.B.; You, J. Merkel cell polyomavirus large T antigen disrupts host genomic integrity and inhibits cellular proliferation. *J. Virol.* **2013**, *87*, 9173–9188. [CrossRef]
61. Schrama, D.; Sarosi, E.; Adam, C.; Ritter, C.; Kaemmerer, U.; Klopocki, E.; König, E.; Utikal, J.; Becker, J.; Houben, R. Characterization of six Merkel cell polyomavirus-positive Merkel cell carcinoma cell lines: Integration pattern suggest that large T antigen truncating events occur before or during integration. *Int. J. Cancer* **2019**, *145*, 1020–1032. [CrossRef]
62. Kwun, H.J.; Wendzicki, J.A.; Shuda, Y.; Moore, P.S.; Chang, Y. Merkel cell polyomavirus small T antigen induces genome instability by E3 ubiquitin ligase targeting. *Oncogene* **2017**, *36*, 6784–6792. [CrossRef] [PubMed]
63. Cheng, J.; Park, D.E.; Berrios, C.; White, E.A.; Arora, R.; Yoon, R.; Branigan, T.; Xiao, T.; Westerling, T.; Federation, A.J.; et al. Merkel cell polyomavirus recruits MYCL to the EP400 complex to promote oncogenesis. *PLoS Pathog.* **2017**, *13*, e1006668. [CrossRef] [PubMed]
64. Park, D.E.; Cheng, J.; Berrios, C.; Montero, J.; Cortés-Cros, M.; Ferretti, S.; Arora, R.; Tillgren, M.L.; Gokhale, P.C.; DeCaprio, J.A. Dual inhibition of MDM2 and MDM4 in virus-positive Merkel cell carcinoma enhances the p53 response. *Proc. Natl. Acad. Sci. USA* **2018**, *116*, 1027–1032.
65. Goh, G.; Walradt, T.; Markarov, V.; Blom, A.; Riaz, N.; Doumani, R.; Stafstrom, K.; Moshiri, A.; Yelistratova, L.; Levinsohn, J.; et al. Mutational landscape of MCPyV-positive and MCPyV-negative Merkel cell carcinomas with implications for immunotherapy. *Oncotarget* **2015**, *7*, 3403–3415. [CrossRef] [PubMed]
66. González-Vela, M.D.C.; Curiel-Olmo, S.; Derdak, S.; Beltran, S.; Santibañez, M.; Martínez, N.; Castillo-Trujillo, A.; Gut, M.; Sánchez-Pacheco, R.; Almaraz, C.; et al. Shared oncogenic pathways implicated in both virus-positive and UV-induced Merkel cell carcinomas. *J. Investig. Dermatol.* **2017**, *137*, 197–206. [CrossRef] [PubMed]
67. Harms, P.W.; Vats, P.; Verhaegen, M.E.; Robinson, D.R.; Wu, Y.-M.; Dhanasekaran, S.M.; Palanisamy, N.; Siddiqui, J.; Cao, X.; Su, F.; et al. The distinctive mutational spectra of polyomavirus-negative Merkel cell carcinoma. *Cancer Res.* **2015**, *75*, 3720–3727. [CrossRef]
68. Cohen, P.R.; Tomson, B.N.; Elkin, S.K.; Marchlik, E.; Carter, J.L.; Kurzrock, R. Genomic portfolio of Merkel cell carcinoma as determined by comprehensive genomic profiling: Implications for targeted therapeutics. *Oncotarget* **2016**, *7*, 23454–23467. [CrossRef]
69. Hafner, C.; Houben, R.; Baeurle, A.; Ritter, C.; Schrama, D.; Landthaler, M.; Becker, J.C. Activation of the PI3K/AKT pathway in Merkel cell carcinoma. *PLoS ONE* **2012**, *7*, e31255. [CrossRef]
70. Nardi, V.; Song, Y.C.; Santamaria-Barria, J.A.; Cosper, A.K.; Lam, Q.; Faber, A.C.; Boland, G.M.; Yeap, B.Y.; Bergethon, K.; Scialabba, V.L.; et al. Activation of PI3K signaling in Merkel cell carcinoma. *Clin. Cancer Res.* **2012**, *18*, 1227–1236. [CrossRef]
71. Wong, S.Q.; Waldeck, K.; Vergara, I.A.; Schröder, J.; Madore, J.; Wilmott, J.S.; Colebatch, A.; De Paoli-Iseppi, R.; Li, J.; Lupat, R.; et al. UV-associated mutations underlie the etiology of MCV-negative Merkel cell carcinomas. *Cancer Res.* **2015**, *75*, 5228–5234. [CrossRef]

72. Quintanilla, M.; Brown, K.; Ramsden, M.; Balmain, A. Carcinogen-specific mutation and amplification of Ha-ras during mouse skin carcinogenesis. *Nat. Cell Biol.* **1986**, *322*, 78–80. [CrossRef]
73. Shin, M.-K.; Payne, S.N.; Bilger, A.; Matkowskyj, K.A.; Carchman, E.; Meyer, D.S.; Bentires-Alj, M.; Deming, D.A.; Lambert, P. Activating mutations in Pik3ca contribute to anal carcinogenesis in the presence or absence of HPV-16 oncogenes. *Clin. Cancer Res.* **2019**, *25*, 1889–1900. [CrossRef] [PubMed]
74. Stelzer, M.K.; Pitot, H.C.; Liem, A.; Schweizer, J.; Mahoney, C.; Lambert, P. A Mouse model for human anal cancer. *Cancer Prev. Res.* **2010**, *3*, 1534–1541. [CrossRef] [PubMed]
75. Kemp, C.J. Multistep skin cancer in mice as a model to study the evolution of cancer cells. *Semin. Cancer Biol.* **2005**, *15*, 460–473. [CrossRef] [PubMed]
76. Kervarrec, T.; Samimi, M.; Guyétant, S.; Sarma, B.; Chéret, J.; Blanchard, E.; Berthon, P.; Schrama, D.; Houben, R.; Touzé, A. Histogenesis of Merkel cell carcinoma: A comprehensive review. *Front. Oncol.* **2019**, *9*, 451. [CrossRef] [PubMed]
77. Van Keymeulen, A.; Mascre, G.; Youseff, K.K.; Harel, I.; Michaux, C.; De Geest, N.; Szpalski, C.; Achouri, Y.; Bloch, W.; Hassan, B.A.; et al. Epidermal progenitors give rise to Merkel cells during embryonic development and adult homeostasis. *J. Cell Biol.* **2009**, *187*, 91–100. [CrossRef]
78. Hausen, A.Z.; Rennspiess, D.; Winnepenninckx, V.; Speel, E.-J.M.; Kurz, A.K. Early B-cell differentiation in Merkel cell carcinomas: Clues to cellular ancestry. *Cancer Res.* **2013**, *73*, 4982–4987. [CrossRef]
79. Morrison, K.M.; Miesegaes, G.R.; Lumpkin, E.A.; Maricich, S.M. Mammalian Merkel cells are descended from the epidermal lineage. *Dev. Biol.* **2009**, *336*, 76–83. [CrossRef]
80. Ostrowski, S.M.; Wright, M.C.; Bolock, A.M.; Geng, X.; Maricich, S.M. Ectopic Atoh1 expression drives Merkel cell production in embryonic, postnatal and adult mouse epidermis. *Development* **2015**, *142*, 2533–2544. [CrossRef]
81. Kervarrec, T.; Aljundi, M.; Appenzeller, S.; Samimi, M.; Maubec, E.; Cribier, B.; Deschamps, L.; Sarma, B.; Sarosi, E.-M.; Berthon, P.; et al. Polyomavirus-positive Merkel cell carcinoma derived from a Trichoblastoma suggests an epithelial origin of this Merkel cell carcinoma. *J. Investig. Dermatol.* **2020**, *140*, 976–985. [CrossRef]
82. Dowlatshahi, M.; Huang, V.; Gehad, A.E.; Jiang, Y.; Calarese, A.; Teague, J.E.; Dorosario, A.A.; Cheng, J.; Nghiem, P.; Schanbacher, C.F.; et al. Tumor-specific T cells in human Merkel cell carcinomas: A possible role for Tregs and T-cell exhaustion in reducing T-cell responses. *J. Investig. Dermatol.* **2013**, *133*, 1879–1889. [CrossRef]
83. Herber, R.; Liem, A.; Pitot, H.; Lambert, P. Squamous epithelial hyperplasia and carcinoma in mice transgenic for the human papillomavirus type 16 E7 oncogene. *J. Virol.* **1996**, *70*, 1873–1881. [CrossRef] [PubMed]
84. Song, S.; Pitot, H.C.; Lambert, P. The human Papillomavirus type 16 E6 gene alone is sufficient to induce carcinomas in transgenic animals. *J. Virol.* **1999**, *73*, 5887–5893. [CrossRef] [PubMed]

cancers

MDPI

Article

Merkel Cell Polyomavirus T Antigens Induce Merkel Cell-Like Differentiation in GLI1-Expressing Epithelial Cells

Thibault Kervarrec [1,2,3,*], Mahtab Samimi [2,4], Sonja Hesbacher [3], Patricia Berthon [2], Marion Wobser [3], Aurélie Sallot [5], Bhavishya Sarma [3], Sophie Schweinitzer [3], Théo Gandon [2], Christophe Destrieux [6], Côme Pasqualin [7], Serge Guyétant [1,2], Antoine Touzé [2], Roland Houben [3,†] and David Schrama [3,†]

1 Department of Pathology, Université de Tours, CHU de Tours, Avenue de la République, 37170 Chambray-les-Tours, France; serge.guyetant@univ-tours.fr

2 "Biologie des Infections à Polyomavirus" Team, UMR INRA ISP 1282, Université de Tours, 31 Avenue Monge, 37200 Tours, France; mahtab.samimi@univ-tours.fr (M.S.); patricia.berthon@inra.fr (P.B.); theo.gandon@etu.univ-tours.fr (T.G.); antoine.touze@univ-tours.fr (A.T.)

3 Department of Dermatology, Venereology and Allergology, University Hospital Würzburg, Josef-Schneider-Straße 2, 97080 Würzburg, Germany; hesbacher_s@ukw.de (S.H.); wobser_m@ukw.de (M.W.); Sarma_B@ukw.de (B.S.); sophie.schweinitzer@gmx.de (S.S.); houben_r@ukw.de (R.H.); Schrama_d@ukw.de (D.S.)

4 Dermatology Department, Université de Tours, CHU de Tours, Avenue de la République, 37170 Chambray-les-Tours, France

5 Plastic Surgery Department, Université de Tours, CHU de Tours, Avenue de la République, 37170 Chambray-les-Tours, France; aurelie.sallot@hotmail.fr

6 Neurosurgery Department, UMR 1253, i Brain, Université De Tours, CHU de Tours, Boulevard Tonnelé, 37044 Tours, France; christophe.destrieux@univ-tours.fr

7 CNRS ERL 7368, Signalisation et Transports Ioniques Membranaires, Equipe Transferts Ioniques et Rythmicité Cardiaque, Groupe Physiologie des Cellules Cardiaques et Vasculaires, Université de Tours, 31 Avenue Monge, 37200 Tours, France; come.pasqualin@univ-tours.fr

* Correspondence: thibaultkervarrec@yahoo.fr

† Equal contribution.

Received: 19 June 2020; Accepted: 13 July 2020; Published: 21 July 2020

Abstract: Merkel cell carcinoma (MCC) is an aggressive skin cancer frequently caused by the Merkel cell polyomavirus (MCPyV). It is still under discussion, in which cells viral integration and MCC development occurs. Recently, we demonstrated that a virus-positive MCC derived from a trichoblastoma, an epithelial neoplasia bearing Merkel cell (MC) differentiation potential. Accordingly, we hypothesized that MC progenitors may represent an origin of MCPyV-positive MCC. To sustain this hypothesis, phenotypic comparison of trichoblastomas and physiologic human MC progenitors was conducted revealing GLI family zinc finger 1 (GLI1), Keratin 17 (KRT 17), and SRY-box transcription factor 9 (SOX9) expressions in both subsets. Furthermore, GLI1 expression in keratinocytes induced transcription of the MC marker SOX2 supporting a role of GLI1 in human MC differentiation. To assess a possible contribution of the MCPyV T antigens (TA) to the development of an MC-like phenotype, human keratinocytes were transduced with TA. While this led only to induction of KRT8, an early MC marker, combined GLI1 and TA expression gave rise to a more advanced MC phenotype with SOX2, KRT8, and KRT20 expression. Finally, we demonstrated MCPyV-large T antigens' capacity to inhibit the degradation of the MC master regulator Atonal bHLH transcription factor 1 (ATOH1). In conclusion, our report suggests that MCPyV TA contribute to the acquisition of an MC-like phenotype in epithelial cells.

Keywords: Merkel cell carcinoma; histogenesis; polyomavirus; ATOH1; GLI1; sonic hedgehog; hair follicle

1. Introduction

Merkel cell carcinoma (MCC) is an aggressive cutaneous neoplasm with a five-year overall survival rate of 40% [1]. Morphologically, MCC tumor cells display small cell carcinoma features and express both neuroendocrine and epithelial markers. In 2008, Feng et al. detected the sequence of a hitherto unknown polyomavirus integrated in the genomes of MCC tumor cells [2]. Subsequent studies revealed that approximately 80% of MCC cases are Merkel cell polyomavirus (MCPyV)-positive, and expression of the two viral T antigens (TA) (small T (sT) and large T antigens (LT)) are considered as the main drivers for carcinogenesis and growth of such tumors [2]. Interestingly, while several candidates, such as epithelial cells, fibroblasts, neuronal progenitors, or B cells, have been proposed, the nature of the cells giving rise to MCC following infection remains unknown [3–6].

Based on close phenotypic similarities, the eponymous Merkel cell (MC) was initially regarded as the most probable cell of origin of MCC. MCs can be found either in the appendages of the skin or in the basal layer of the epidermis. They function as mechanoreceptors capable of transmitting tactile stimuli onto Aβ-afferent nerve endings [7]. In mice and humans, MCs can be distinguished immunohistochemically from other intra-epidermal cells by positivity for the SRY-box transcription factor 2 (SOX2) and cytokeratins (KRT) 8, 18, and 20, which sequentially appear during MC differentiation and are also expressed by MCC [8–12].

For a long time, it was a matter of debate whether MCs develop from the neural crest or from the epidermal lineage [13]. Based on genetic mouse models, it is now widely accepted that MCs derive from epidermal progenitors in mammals [12,14,15] and that the transcription factor atonal homolog 1 (Atoh1) is the master regulator of this differentiation process [12,16,17]. While ectopic Atoh1 expression can induce MC differentiation throughout the epidermis of transgenic mice [3], physiological MC development preferentially occurs in hair follicles and in specialized structures named "touch domes" where the epithelial progenitors of MCs are located [18,19]. A critical step for MC differentiation in mice hairy skin is that these progenitors come into contact with dermal nerves leading to activation of the sonic hedgehog pathway (SHH) and subsequent GLI family zinc finger 1 (Gli1) expression [18,19]. Further markers characterizing these Gli1-expressing progenitors in mice are Krt17 [18], Sox9 [20], and CD200 [21], while only one study has shown KRT17 expression in human "touch dome" keratinocytes [22]. Notably, a high tumorigenic potential has been demonstrated for this cell population in transgenic models [23]. Therefore, these MCs' epithelial progenitors, which remain poorly characterized in humans, are one potential candidate for MCC origin [24]. In contrast, due to lack of proliferative activity [25] and insensitiveness to oncogenic stimuli including ectopic TA expression [26], differentiated MCs are regarded as unlikely to be transformable [4].

Besides MCC, a second tumor entity known as trichoblastoma (TB) harbors cells with an MC phenotype. In this regard, TB as a benign epithelial skin tumor displaying hair follicle differentiation [27] is mainly composed of germinative basaloid cells, but is also characterized by sparse intra-tumoral MC cells. The latter probably reflects a preserved potential of TB cells to act as epithelial progenitors and, therefore, to differentiate into MCs [28–30]. Applying massive parallel sequencing on a combined tumor consisting of MCC and TB components, we recently demonstrated that MCPyV integration in a TB cell gave rise to an MCPyV-positive MCC [31] indicating that an MCPyV-positive MCC can arise from an epithelial cell. Moreover, the phenotypical similarities between TB and physiologic hair follicles, where MC progenitors are preferentially located, further support epithelial progenitors with intrinsic MC differentiation potential as possible ancestry for MCPyV-induced MCC [31]. In the present study, we first expanded characterization of such MC progenitors in humans and then aimed to evaluate how the viral T antigens might contribute to the development of an MC-like phenotype in this population using GLI1-expressing keratinocytes as a model system.

2. Results

2.1. MCs Are Often Located in Appendage Structures in Human Skin

MC development has mainly been characterized in mouse models [9,12,14]. Hence, in a first set of experiments, we used immunohistochemistry to compare the MC differentiation process under physiological conditions as well as in the tumor setting in humans. We started with characterizing the MC lineage by assessing physiological density and location of MCs in a set of 15 samples from three human autopsy skin specimens (Figure 1A,B, Figure S1, Table S1). Mean MC density, regardless of the location, was 50 cells/mm^2 of epidermis, and head and neck as well as acral skin were enriched in MCs compared to the other sites (density = 55 and 104 MCs/mm^2, respectively). Moreover, MCs were often located in appendage structures (72% of all observed MCs), i.e., either hair follicles or sweat glands, as depicted in Figure 1B, Figure S1. Of note, contrary to previous reports, some dermal MCs were observed (Figure S1).

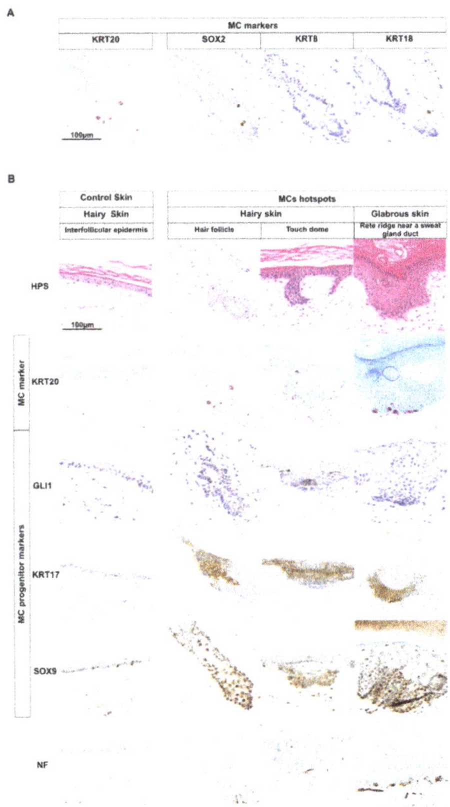

Figure 1. Merkel cells and possible Merkel cell progenitors in human skin. (**A**) Keratin 20 (KRT20), SRY-box transcription factor 2 (SOX2), KRT8, and KRT18 staining was used to identify Merkel cells (MCs) (bar = 100 µm) (only one hotspot investigated for illustration purpose). Merged analysis is available in Figure S1B. (**B**) Identification of potential MC progenitors in human skin: Three MC hotspots as well as interfollicular epidermis for comparison are depicted (bar = 100 µm) (15 hotspots investigated in total). Immunohistochemical staining revealed expression of KRT17 and SOX9 in the epidermal cells surrounding differentiated MCs suggesting that these cells are MC progenitors. Nuclear GLI family zinc finger 1 (GLI1) was detected only close to MC hotspots in hairy, but not in acral skin. Of note, neurofilament (NF)-expressing dermal nerves were observed in contact with the MCs.

2.2. Cells with an MC Progenitor Phenotype Characterized by GLI1 Expression are Found in Close Proximity of MCs in Human Hairy Skin

Since MC epithelial progenitors can be expected to be found preferentially in regions enriched for MCs, we focused on the following on MC hotspots [18,22]. Such areas enriched in KRT20-positive MCs were mostly observed in hair follicles (52% of cases) or in junctions between eccrine sweat ducts and the overlying epidermis (36%). In the latter case, MCs were surrounded by clusters of verticalized basal keratinocytes resembling structures reported as "touch domes" [22] (Figure 1B, Figure S1, Table S1). Slides of MC hotspots were subsequently stained for the epithelial progenitor markers GLI1, SOX9, and KRT17, revealing that epidermal cells surrounding MCs—in contrast to the rest of the epidermis—were characterized by nuclear GLI1 expression and positivity for the stem cell markers KRT17 and SOX9 (Figure 1B, Figure S1). In mice, Gli1-expressing keratinocytes in the hair follicle have been identified as MC progenitors [17–20]. Hence, our results demonstrate that also in human hairy skin an equivalent GLI1-positive population is preferentially located in the hair follicle.

2.3. GLI1 Expression in Keratinocytes Induces MC Lineage Markers

To evaluate a role of GLI1 expression in the establishment of the MC lineage in human epithelial cells, we used primary normal human epidermal keratinocytes (NHEK) as model system (Figure S2). These cells were transduced with a lentiviral vector encoding GLI1. Gene expression analysis after 14 days revealed an increase of the MC lineage markers *SOX2* (110-fold compared to the empty vector control, $p = 0.002$) and *KRT8* (4-fold, $p = 0.05$) in those cells (Figure 2A). Moreover, in GLI1-transduced cells *KRT17* and *SOX9* messenger RNA (mRNA) levels were found to be slightly elevated (2-fold), which, however, did not reach statistical significance. On protein level, we observed increased expression levels of SOX2 upon GLI1 expression by immunocytochemistry and immunoblot (Figure 2B, Figure S3A,B). Additional immunostainings suggested enhanced KRT17 and SOX9 expression in GLI1-transduced NHEK, while no expression of the additional MC markers KRT8 or KRT20 was observed (Figure 2B, Figure S3). The discrepancy between induction of mRNA and lack of KRT8 protein in immunostaining upon GLI1 expression might be explained by protein levels below the detection limit of the antibody used. Nevertheless, together, these results suggest that GLI1, the executor of the sonic hedgehog pathway, is capable of initiating the first step of MC differentiation via SOX2 induction [6,9].

2.4. MC-Progenitor and MC Markers Are Expressed in Trichoblastoma and Merkel Cell Carcinoma

Next, we assessed how the markers defining the MC differentiation status are distributed in the two tumor entities harboring MC-like cells, i.e., TB and MCC. In five out of six MC containing interpretable TBs, we detected sparse SOX2-positive intra-tumoral cells. As typical for trichoblastoma, these expected "MCs" represented only a minority of cells dispersed within a vast majority of germinative tumor cells displaying a MC progenitor phenotype, and may be explained by germinative TB cells undergoing MC differentiation [30,32]. In line with this view and in line with the necessity of active hedgehog pathway signaling for potential MC differentiation in human epithelial cells [9,18], widespread nuclear GLI1 expression in the germinative cells was detectable in seven out of eight TB specimens (Table 1, Table S2, Figure S4A). Furthermore, diffuse expression of the GLI1 target genes, SOX9 and KRT17, was observed in germinative cells of all TB cases (Table 1, Table S2, Figure S4A). In conclusion, these results further substantiate known similarities between MCs' epithelial progenitors and TB cells. In light of our previous report of an MCPyV-positive MCC arising from a TB cell [31], these observations further suggest such MC epithelial progenitors as a potential origin of MCPyV-induced MCC.

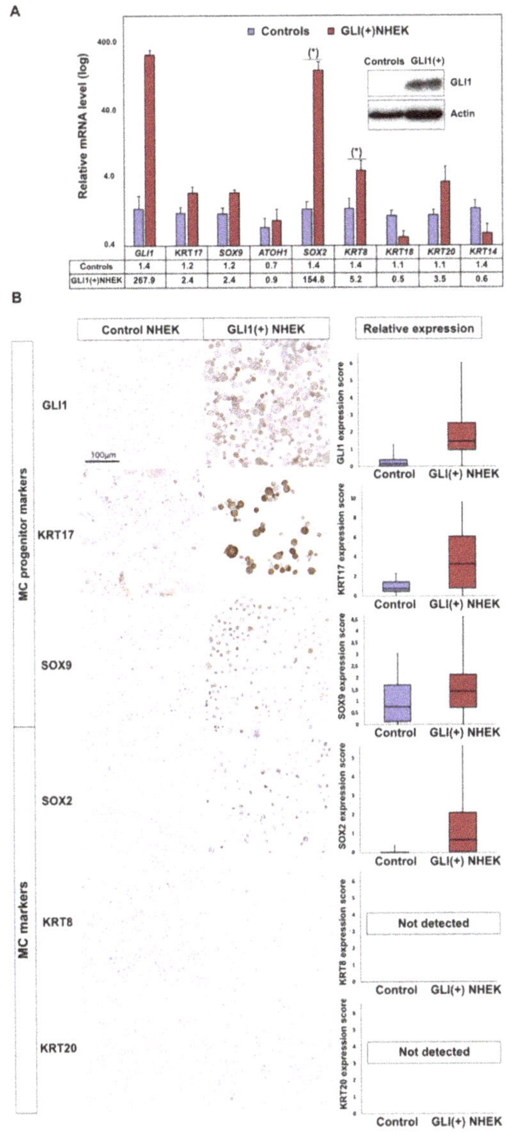

Figure 2. Ectopic GLI1 expression in primary human epidermal keratinocytes induces several MC lineage markers: Normal human epidermal keratinocytes (NHEK) were infected with a lentiviral vector coding for GLI1 and puromycin resistance. Following antibiotic selection, cells were harvested after 14 days of cultivation. (**A**) Immunoblot analysis was performed to confirm GLI1 expression (insert), and isolated RNA was subjected to complementary DNA (cDNA) synthesis and real-time PCR. Relative messenger RNA (mRNA) expression levels of the indicated Merkel cell lineage markers are given as mean (+ standard error of the mean (SEM)) of four independent experiments (* p value < 0.05, paired t test) (mean CT value of the controls was used as reference). (**B**) Expression of GLI1, the MC progenitor (KRT17, SOX9) and the MC markers (SOX2, KRT8, and KRT20) was assessed by immunohistochemistry and relative protein expression quantification was performed on at least 1000 cells/condition using ImageJ software. Results are displayed as box and whiskers diagram with median, Q1, and Q3, as well as first and 99th percentile. These results were confirmed by two additional independent experiments (immunostaining and immunoblot) as shown in Figure S3. Uncropped membranes and Western blot signal quantifications are available in Figures S8 and S9, respectively.

Table 1. Expression of Merkel cell progenitor markers in trichoblastoma (n = 8) and Merkel cell carcinoma (n = 103).

MC Progenitor Markers	TB (n = 8 Cases)	MCC (n = 103 Cases)
GLI1		
Negative	1 (13%)	60 (67%)
Positive (nuclear)	7 (87%)	29 (33%)
No data available	0	14
KRT17		
Negative	0	94 (100%)
Positive (cytoplasmic)	8 (100%)	0
No data available	0	9
SOX9		
Negative	0	7 (8%)
Dot-like (cytoplasmic)	0	59 (64%)
Patchy (nuclear)	0	26 (28%)
Diffuse (nuclear)	8 (100%)	0
No data available	0	11
MC markers	**TB**	**MCC**
SOX2		
Negative	1 (17%)	2 (2%)
Positive (nuclear)	5 (83%)	94 (98%)
No data available	2	7
KRT20		
Negative	0	8
Diffuse (cytoplasmic)	8 (100%)	2
Mixed (cytoplasmic)	0	66
Dot-like pattern (cytoplasmic)	0	19
No data available	0	8

KRT: Cytokeratin; GLI1: GLI family zinc finger 1; MC: Merkel cell; MCC: Merkel cell carcinoma; SOX2: SRY-box transcription factor 2; SOX9: SRY-box 9, TB: Trichoblastoma. Representative photos of SOX9 expression patterns are available in Figure S3. Results are given as numbers and percentage of interpretable cases.

While in TB a mixture of cells with either epithelial progenitor or MC phenotype is present, almost all MCC tumor cells display a phenotype of mature MC. Indeed, in a previous study we observed 100, 99, and 92% of MCC cases with widespread positivity for the MC markers KRT8, 18, and 20, respectively [33,34]. Accordingly, in the present work, diffuse and strong nuclear positivity for SOX2 was detected in almost all analyzed MCC tumors (98%). While the MC progenitor marker KRT17 was not detectable (Table 1, Figure S4A), GLI1 and SOX9 nuclear expression, representing the active forms of these transcription factors, were detected in 33% and 28% of cases, respectively (Table 1, Table S2). Moreover, such findings were more frequently observed in MCPyV-negative than in MCPyV-positive cases (GLI1: 52 versus 24%, $p < 0.03$; SOX9 nuclear positivity: 81 versus 10%, $p < 10^{-9}$, respectively) (Figure S4B,C, Table S3), suggesting that MCPyV presence is associated with a more mature MC phenotype.

2.5. *T Antigens Can Trigger Early MC Differentiation Marker Expression in Epidermal Cells*

On the supposition that MCC arises upon integration of MCPyV in a cell of the MC lineage, the virus might either hit an already determined MC cell or might trigger or promote the acquisition of the MC phenotype in an epithelial progenitor. To investigate a possible contribution of the MCPyV TAs to the development of an MC phenotype, sT and truncated LT were ectopically expressed in NHEK (Figure 3A). Notably, while cells could not be immortalized by the viral proteins, significant morphologic changes with reduction of cell size were observed upon TA expression (Figure 3A). Gene expression analysis after two weeks revealed an increase of mRNAs coding for early MC differentiation markers (*KRT8* $p = 0.02$ and *KRT18* $p = 0.02$), while the keratinocyte marker *KRT14* was slightly reduced upon TA expression ($p = 0.09$) (Figure 2B). Induction of KRT8 upon TA expression in NHEKs was confirmed by immunoblot and immunocytochemical staining, while no expression of SOX2 or KRT20 was observed in three independent experiments (Figure 3C,D, Figure S5). Interestingly, in situ KRT8 staining of TA-expressing NHEK demonstrated that expression of this marker was restricted to a subpopulation of cells with small-medium size and round shape (Figure 3C).

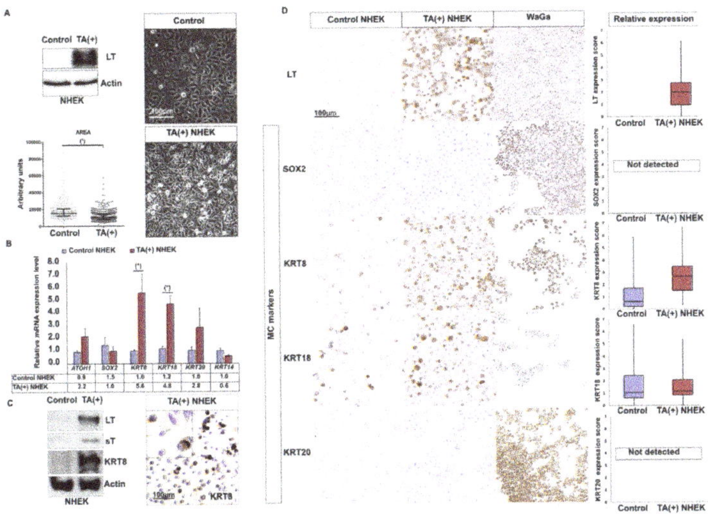

Figure 3. T antigens induce expression of some early MC differentiation markers in primary human keratinocytes. **A:** NHEKs were infected with a lentiviral vector coding for small T (sT) and truncated Large T (LT) as well as a puromycin resistance. Following antibiotic selection, cells were analyzed after 14 days of cultivation. (**A**) Immunoblot analysis confirmed LT expression, and microscopic inspection revealed a less-flattened phenotype and cultures reaching much higher densities. Under microscopic examination such cells harbored reduced cytoplasmic size compared to the controls, as confirmed using imageJ software (bar = 100 µm) (* p value < 0.05, Mann–Whitney U test, $n = 3$ independent experiments). (**B**) Relative mRNA levels of the indicated Merkel cell differentiation markers (* p value < 0.05, paired t test, $n = 4$ independent experiments), (**C**) Immunoblot demonstrated T antigens (TA)-induced KRT8 protein expression and immunohistochemistry additionally revealed KRT8 expression is restricted to a subpopulation of small- to medium-sized round cells. Furthermore, occasionally "dot like" staining was observed (white arrows). (**D**) Immunohistochemical assessment of the indicated MC markers in TA-expressing NHEK, control NHEK and the MCC cell line WaGa (bar = 100 µm). KRT8 induction by T antigens was confirmed in two additional independent experiments, which are depicted in Figure S4. For relative quantification of protein expression levels, at least 1000 cells/condition were evaluated using ImageJ software. Results are displayed as box and whiskers diagram with median, Q1, and Q3 as well as first and 99th percentile. Uncropped membranes and Western blot signal quantifications are available in Figures S8 and S9, respectively.

2.6. *T Antigens Induce Late MC Markers in GLI1-Expressing NHEK*

To model TA expression in GLI1-expressing epithelial progenitor cells, we infected NHEKs with a bicistronic lentiviral construct coding for GLI1 and MCPyV-TA. After two weeks, morphological analysis of these cells in comparison to control cells infected with an empty vector revealed induction of a subpopulation of non-adherent, living cells forming clusters similar to the one observed for MCC cell lines (Figure 4A). Moreover, immunocytochemical staining revealed expression of the MC markers KRT8, SOX2, and, to a lesser extent, KRT20 (Figure 4B,C, Figure S5). Given that NHEKs represent only a limited model for MC progenitor cells, these findings—even though the detection of KRT20 was restricted to only a few cells—indicate that the interplay of GLI1 and MCPyV TA bears the potential of enforcing MC differentiation.

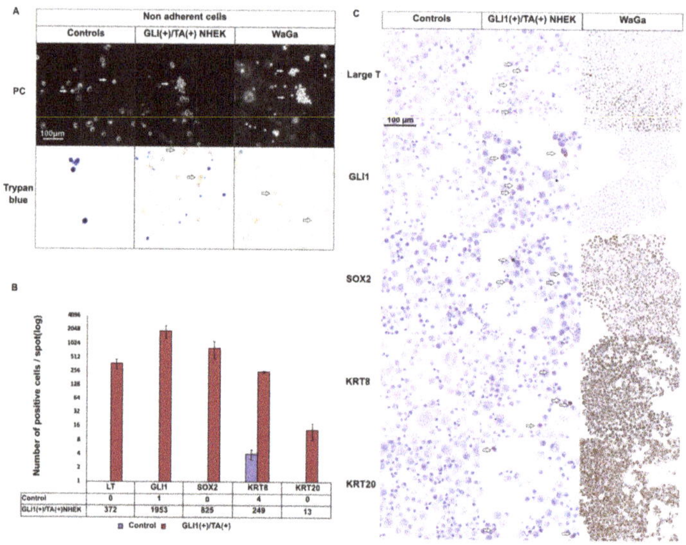

Figure 4. Induction of late MC markers by combined expression of GLI1 and Merkel cell Polyomavirus (MCPyV) T antigens (TA) in primary keratinocytes. NHEKs were infected with a bicistronic lentiviral vector coding for GLI1 as well as sT and truncated LT. Under control of a second promoter, a pure resistance was expressed. Following antibiotic selection, cells were analyzed after 14 days of cultivation. (**A**) GLI1/TA combined ectopic expression was associated with formation of floating clusters of living cells in normal human epidermal keratinocytes (NHEK), while these findings were not observed in controls or when GLI1 and TA were transduced independently (PC: Phase contrast) (Figure S4) ($n = 3$ independent experiments). White arrows indicate the floating cells. (**B**,**C**) Immunohistochemical assessment of Merkel cell markers (SOX2, KRT8, and KRT20) expression levels in GLI1/T antigen-expressing NHEKs and controls. Immunohistochemistry was performed on the respective cells spotted on slides (2×10^5 cells/condition). B. Count of cells expressing the Merkel cell markers in GLI1/T antigens (TA)-expressing NHEK and controls (results are mean ±SEM of three independent experiments). Counting of positive cells was preferred to relative protein level quantification due to the low number of GLI1/TA-expressing cells. C. Representative photos of LT, GLI1, SOX2, KRT8, KRT18, and KRT20 expression in NHEK (controls), GLI1/TA-expressing NHEK, and the WaGa MCC cell line. White arrows indicate cells expressing the respective proteins. The results for two additional independent experiments are shown in Figure S5.

2.7. *T Antigens Prevent ATOH1 Degradation*

In NHEK, MCPyV-TA induced transcription of MC markers without significantly affecting *ATOH1* (Figure 3B), the known master regulator of MC differentiation [12,14]. Indeed, although LT-mediated ATOH1 induction was recently reported [35], we only observed a slight and statistically not significant mRNA increase upon TA expression. Hence, we hypothesized that the TAs might affect ATOH1 protein independent of gene transcription. To test this hypothesis, we transfected U2OS cells either with hemagglutinin (HA)-tagged ATOH1 alone or in combination with MCPyV-TA and analyzed RNA as well as protein levels. To this end, while the *ATOH1* mRNA level was not affected by TA co-expression, ATOH1 protein was increased (Figure 5A). Next, a constant amount of *ATOH1*-encoding plasmid (0.3 µg) and increasing amounts of TA-encoding plasmid (0–1.4 µg) were co-transfected, demonstrating a dose-dependent relation of increasing ATOH1 in the presence of MCPyV-TA (Figure 5B). Then, we asked whether this effect might be due to decreased protein degradation. To test whether protein stability is affected, the co-expression was performed while translation was inhibited in cycloheximide chase assays, allowing to assess ATOH1 protein decay in the presence or absence of TA (Figure 5C). These analyses revealed that TA increased ATOH1 half-life from 2 to 9 h. Interestingly, knockdown of TA expression in the MCC cell lines MKL-1 and WaGa failed to reduce ATOH1 protein levels (Figure S6) suggesting that in established MCC cells ATOH1 does not depend on stabilization by LT.

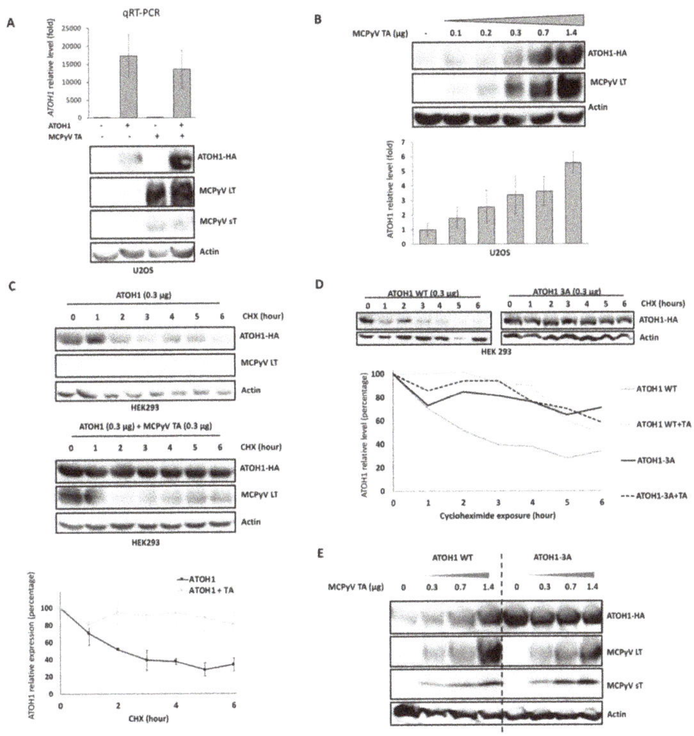

Figure 5. MCPyV T antigens increase the half-life of ATOH1 (**A**) **Hemagglutinin** (HA)-tagged ATOH1- and/or TA-encoding plasmids were transfected either individually or combined into U2OS cells. After two days, real-time PCR and immunoblot analyses were performed. While *ATOH1* mRNA was not affected (mean ± SEM of three independent experiments), ATOH1 protein accumulation in the presence of TA was observed. (**B**) Co-transfection of a constant amount (0.3 µg) of HA-tagged ATOH1 and increasing amounts of TA in U2OS cells followed by immunoblot analysis. ATOH1-HA signals relative to actin were quantified using ImageJ. Mean ± SEM of three independent experiments was displayed. (**C**) Evaluation of ATOH1 half-life in absence or presence of T antigens. Twenty-four hours after transfection, HEK293 cells were exposed to the translation inhibitor cycloheximide (CHX) for variable durations (0–6 h). ATOH1-HA expression was then evaluated by immunoblot analysis and quantified using the Image J Software (mean ± SEM of three independent experiments are depicted). (**D**) A mutant of ATOH1-HA, in which the three serines at positions 331, 337, and 342 were all exchanged to alanines (ATOH1-HA-3A), was generated, and the impact of co-transfected TA on ATOH1-HA wild type and ATOH1-HA-3A expression was analyzed in CHX chase experiments (see C). Quantified signals relative to actin are given in the graphs below. (**E**) Co-transfection of ATOH1-HA-3A with increasing amounts of TA did not affect ATOH1 protein expression level (this was confirmed in a second independent experiment). Uncropped membranes and Western blot signal quantifications are available in Figures S8 and S9, respectively.

In mice, Atoh1 degradation has been shown to be controlled by phosphorylation of three carboxy-terminal serine residues (S331, S337, S341) leading to Atoh1 ubiquitinylation and subsequent targeting to the proteasome [36,37]. Hence, we speculated that TA-dependent stabilization might involve the respective sites in the human protein. Consequently, we generated expression constructs coding for ATOH1 proteins in which the serines were exchanged to alanines, either individually (S331A, S337A, and S342A) or all three combined (ATOH1-3A). Indeed, these modified ATOH1 proteins displayed extended half-lives in cycloheximide chase assays (Figure 5D, Figure S7A). More importantly, however, while T antigens still stabilized ATOH1 proteins harboring single phospho-site mutations (Figure S7B), no additional stabilization could be observed for the triple mutant protein (Figure 5D,E). Therefore, it is likely that the TAs act in the same pathway either by impacting phosphorylation of several serine residues on ATOH1 or by interfering with subsequent phosphorylation-dependent proteasome targeting.

2.8. The MCPyV Unique Region 1 (MUR1) in MCPyV LT Contributes to ATOH1 Stabilization

Irrespective of the fact that the exact mechanism of TA-mediated ATOH1 protein stabilization still requires further investigations, we finally wanted to know which of the two T antigens and which protein subdomains are involved in the process. Hence, we assessed ATOH1 protein levels after co-transfection of ATOH1 with either sT or LT, respectively. These experiments identified LT as the main effector of ATOH1 stability (Figure S7C). To scrutinize which functional domain of large T might be involved in regulating ATOH1 degradation, another series of co-transfections was performed combining ATOH1 with LT mutants devoid of either specific interaction sites or the MCPyV unique region 1 (MUR1) region. Interestingly, mutants, which have been demonstrated to lack any growth-promoting activity, like the heat shock protein 70 (HSC70)-binding mutant D44N [38,39] or the RB transcriptional corepressor 1 (RB1)-binding deficient variants E216K and S220A [38], were still capable of mediating ATOH1 accumulation (Figure S7D). However, co-transfection of ATOH1 with MCPyV-LT$^{\Delta MUR1}$, a LT variant still bearing growth-promoting activity [38], did not result in enhanced protein expression (Figure S7D), suggesting that the MUR1 region of MCPyV-LT is essential for its ATOH1-stabilizing capacity. Since, however, the applied LT antibody (CM2B4) does not recognize LT$^{\Delta MUR1}$, we could not confirm that the protein was de facto expressed in these experiments (Figure S7D). We, therefore, repeated this experiment with V5-tagged versions of LT and LT$^{\Delta MUR1}$. Now, both proteins were detectable and we again observed no stabilization of co-transfected ATOH1 in the case of LT lacking the MUR1 region (Figure S7E). To further confirm the contribution of the MUR1 region, we also tested the truncated large T of AlDo, an MCC cell line expressing a truncated LT with an additional large deletion representing most of MUR1 [40]. Indeed, upon co-expression of AlDo LT, no stabilization but even a reduction of the ATOH1 protein level was observed (Figure S7F).

3. Discussion

Today, the identification of the cell of origin for MCC is still pending. Based on the similarities in phenotype to MCs, the initially described "trabecular carcinoma of the skin" got its name MCC [24]. These phenotypic similarities can result either from transformation of the eponymous cell or inducing phenotypic changes during oncogenesis resulting in a phenotype resembling those cells. Since (1) MCs are regarded as post-mitotic cells with low sensitivity to oncogenic stimuli, (2) they demonstrate different preferred localizations compared to MCCs, (3) lack of infection of MCs by MCPyV, and (4) neuroendocrine tumors tend to derive from epithelial progenitor cells rather than end-differentiated cells [24], a direct transformation of MCs into MCCs is considered as quite unlikely. In this regard, we recently demonstrated that MCPyV integration in a TB gave rise to an MCPyV-positive MCC [31]. Of note, scattered MCs are frequently observed in TB [27,29,30], demonstrating that at least some of the cells possess the potential for MC differentiation, although the molecular determinants of this process are unknown.

Of note, the knowledge on MC development is mainly derived from mouse experiments. In the present study, we confirmed that MC hotspots are mostly located in the hair follicle in human hairy skin. In close vicinity to the MCs, we observed GLI1 and its downstream targets SOX9 [41] and KRT17 [42]-expressing keratinocytes. Similarly, we could confirm nuclear GLI1 positivity and related downstream SOX9 and KRT17 [27] expression in our TB cases suggesting that MC development under human physiological conditions as well as in TB tumors are quite similar and resemble the murine process with GLI1 activation being an early step. Accordingly, upon GLI1-expression in NHEK, we detected an increased expression of SOX9 and KRT17, and—as has been described for other cell lineages [43,44]—a prominent induction of SOX2. Since SOX2 can drive ATOH1 expression by binding to *ATOH1* enhancer [17] or *ATOH1* promoter [6] and thereby promote MC differentiation [45], SOX2 induction appears as a potential mechanism by which GLI1 promotes ATOH1-driven MC development.

Based on our recently reported observation that a MCPyV-positive MCC could arise from a TB, we hypothesized that MCPyV oncoprotein expression is able to induce acquisition of a Merkel cell-like phenotype in epithelial progenitors with intrinsic MC differentiation potential. Indeed, while TA expression in NHEK reduced cell size, triggered KRT8 protein expression, and enhanced *KRT18* mRNA levels, we did not observe expression of KRT20, a marker appearing later during the MC differentiation process [9]. Although Atoh1 alone is able to initiate MC differentiation during embryonic mice development, Sox2 expression is required for Krt20 expression [9]. Accordingly, the two MCC tumors lacking SOX2 expression in our cohort were also KRT20 negative (data not shown). Hence, to test if the lack of KRT20 expression was due to a lack of SHH activation in NHEK, and subsequent lack of SOX2 expression, we generated a MC progenitor model system and assessed TA impact in it, by co-expressing GLI1 and TA in these cells. Although GLI1-expressing NHEKs represent only an artificial and limited model for MC progenitor cells, GLI1 and TA co-expression resulted in cells expressing SOX2 and KRT8, and even to a few cells displaying KRT20 positivity. Of note, similar as to what has been described for ectopic expression of LT in fibroblasts, we detected living cells with suspension growth. In contrast, however, we did not observe a different expression of Merkel cell markers between the adherent and floating cells (Figure S4).

While our results suggest SHH activation is required at some time point in MCC cell development, GLI1 expression was only observed in about 30% of cases in our study, which were mostly MCPyV-negative cases. Accordingly, therapeutic inhibition of SHH pathway using chemical inhibitors failed to reduce MCC tumor cell viability [46]. Therefore, SHH activation might contribute to MCC cell of origin establishment but then be lost during tumor development.

Another important factor in MC development is ATOH1. In this regard, induction of ATOH1 upon large T expression has been recently reported in fibroblasts [35]. In keratinocytes, we observed only a slight, statistically nonsignificant *ATOH1* mRNA level increase upon TA expression. This is in accordance with data obtained in mice where ectopic sT expression in combination with Atoh1 in epidermal cells did initiate a MC-like development [3], but only TA expression did not [3,47]. Thus, cellular context seems to influence the impact of LT expression on *ATOH1*. Indeed, we observed that ATOH1 degradation is impaired in the presence of LT in U2OS and 293 cells while TA knockdown does not affect ATOH1 protein levels in MCC cell lines. This might imply that TA only stabilizes ATOH1 in a specific environment. It is conceivable that, in a hit-and-run type mechanism (although the virus stays integrated in the host genome), LT contributes to initiating MC-like differentiation which later becomes independent of the viral protein. Indeed, T antigens are known to hijack many cellular processes [48], and stabilization of LT by sT via inhibition of the ubiquitin ligase "F-box and WD repeat domain containing 7" (SCF^{Fbw7}) has been proposed [49], although this finding was recently called into question [50]. In mice, phosphorylation of the Atoh1 serine residues S328, S334, and S339 [36,37], equivalent to the amino acids S331, S337, and S342 in human, led to the ubiquitination of the protein by the ubiquitin ligase "HECT, UBA and WWE domain containing E3 ubiquitin protein ligase 1" (HUWE1) and subsequent targeting to the proteasome. Accordingly, human ATOH1 lacking the respective phosphorylation sites presented with an extended half-life in our study. Notably, while LT impaired degradation of wild-type ATOH1, it had no effect on mutant ATOH1. Hence, LT appears to affect the degradation process of ATOH1, either by interfering with the phosphorylation or ubiquitination step. With respect to the latter, although interactions between MCPyV-LT and SCF^{Fbw7} or "beta-transducin repeat-containing protein" (βTrCP) have been reported [49], these ubiquitin ligases appear as unlikely candidates since the (1) LT has been described as their target but not as targeting them, (2) sT, which inhibits both ubiquitin ligases, did not stabilize ATOH1, and (3) none of these ubiquitin ligases was shown to interact with ATOH1 [51]. In contrast, HUWE1 is a ubiquitin ligase that has been identified as ATOH1 binding partner using an unbiased comparative mass spectrometry approach [51]. Therefore, it is possible that the ubiquitin ligase HUWE1 is mediating the ATOH1 stabilization by MCPyV-LT. Moreover, our results suggest that for the ATOH1 stabilization MUR1 in LT is essential. In addition to the several unique functions of MCPyV-sT which have been described [52], this may contribute to the exceptional position of MCPyV among the polyomavirus family in being able to induce a neuroendocrine carcinoma of the skin. Furthermore, these observations suggest that the cell of origin of MCC might already display some degree of ATOH1 expression.

In the present study, we demonstrated that in a specific cellular context, i.e., GLI1-expressing keratinocytes, the expression of MCPyV T antigens can induce a MC-like differentiation. Moreover, the stabilization of ATOH1 by LT might enhance or promote the differentiation of the cell of origin toward an MCC phenotype.

4. Material and Methods

4.1. Human Samples

Healthy cutaneous tissues were obtained from dead people who had signed a body donation procedure for scientific purposes. Skin from five anatomic sites (scalp, face, trunk, finger, lower limb) were collected using a 6-mm-diameter punch in the 24 h following death, and then immediately fixed in formalin and then paraffin embedded. Fifteen TB cases were extracted from the archives of the Dermatology department of Würzburg (Local Würzburg Ethics Committee in Human Research, 196/12). After histological diagnosis confirmation by two pathologists (M.W., T.K.), only cases containing MCs were selected based on KRT20 immunostainings ($n = 8$). MCC cases enrolled in the present work were already included in a tissue microarray used in a previous study [34] (local ethics committee (Tours, France, N° ID RCB2009- A01056-51)). MCPyV status was previously determined using a validated real-time PCR [34].

4.2. Immunohistochemistry

Protein immunochemical detection was performed on formalin-fixed, paraffin-embedded (FFPE) samples (tissue), paraformalin-fixed (cytospin), or living cells. Immunohistochemical staining for KRT20, MCPyV-LT, Neurofilament, and SOX9 were performed using a BenchMark XT Platform, as instructed [34,53]. Immunohistochemical staining for GLI1, KRT8, KRT17, KRT18, and SOX2 as well as all cytospin stainings were performed manually. Microscopic evaluation was performed by a pathologist (T.K.). All details regarding antibodies and dilutions are provided in Supplementary Methods.

4.3. Samples' Management and Interpretation of Immunohistochemical Staining

To determine MC densities, 250 consecutives 5-μm-thick sections were cut from FFPE healthy cutaneous tissues (6-mm-diameter skin punches cut into two equal parts). Every 10th slide, a KRT20 immunohistochemical staining allowing the detection of MC was performed, i.e., one KRT20-stained slide every 50 μm. Unstained slides were preserved for further analyses (MC progenitor markers' evaluation). MC number and location (interfollicular epidermis, hair follicle (infundibulum or isthmus), sebaceous, or sweat glands) were then assessed by a pathologist (T.K.). Since MC are frequently located in the connection area between epidermis and an appendage, i.e., either hair follicles or sweat glands, all MCs located in front of an appendage structure (hair follicle, ostium of a sweat gland or sweat gland duct) were considered to belong to this appendage. Of note, MC hotspots were defined as areas with more than three MCs in one microscopic field at high magnification. Densities of MCs and related hotspots were estimated, taking cut thickness and length of the skin sample into account (estimated evaluated surface = 14.74 mm^2/punch). Unstained slides adjacent to the hotspots were consequently investigated for MC progenitor markers.

4.4. Primary Keratinocytes and Cell Lines

After informed written consent of the patients (*n* = 3), normal human epidermal keratinocytes (NHEK) were extracted, respectively, from abdominal human samples obtained from the plastic surgery of the University Hospital center of Tours (France) using previously described protocols [40,54–56] (Local Ethics Committee in Human Research, Tours, France; no. ID RCB2009-A01056-512016 064). NHEK were cultured in Keratinocyte Serum-Free Medium (K-SFM; Invitrogen Life Technologies), supplemented with epidermal growth factor (5 ng/mL) and bovine pituitary extract (50 µg/mL; all purchased from Invitrogen Life Technologies) at 37 °C, 5% CO_2 in a humidified incubator. HEK293 (RRID: CVCL_0045), HEK293T (RRID: CVCL_0063), U2OS (RRID: CVCL_0045), the MCC cell line WaGa (RRID:CVCL_E998), and MKL-1 (RRID:CVCL_2600) [57] were cultivated in Roswell Park Memorial Institute (RPMI) 1640 supplemented with 10% Foetal calf serum (FCS), 100 U/mL penicillin and 0.1 mg/mL streptomycin. HEK293 (RRID: CVCL_0045) and U2OS (RRID:CVCL_0045) were used for co-transfection experiments. HEK293T (RRID:CVCL_0063), i.e., HEK293 expressing SV40 T antigens, were used for lentivirus production. The MCC cell line WaGa was included as positive control for immunostaining of MC markers.

4.5. Lentiviral Vectors' Generation and Transduction Protocol

The pFLAG-CMV-4-GLI1 plasmid was kindly provided by Dr. J. Vachtenheim (Czech Republic) [58]. *GLI1* was subcloned into pFLAG-CMV backbone (System Biosciences) containing puromycin resistance by classical cloning. Phosphosite mutations (S331A, S337A, S341A) were introduced in *ATOH1* sequence using the Quickchange Lightning mutagenesis kit (Agilent, Frankfurt, Germany) [59]. All TA- and LT-expressing pCDH vectors were previously described [38]. GLI-IRES-TA sequence was cloned into a pCDH backbone. For inducible knockdown of MCPyV-LT, we used the lentiviral single vector TA.shRNA.tet, allowing constitutive green fluorescent protein (GFP) expression and doxycycline (Dox)- inducible expression of an shRNA targeting all transcripts derived from the MCPyV early region [59]. Lentiviral supernatants were produced in HEK293T cells as previously described [60]. Harvested virus supernatant was sterile filtered (0.45 µm) and polybrene (1 µg/mL) was added for infection. Lentiviral transduction of NHEK was performed after seven days of culture. Then, 14–20 h after infection, target cells were washed with medium. NHEK were then subjected to antibiotic selection (puromycin). NHEK were analyzed two weeks after transduction.

4.6. Gene Expression Analyses

Total cellular RNA was isolated by using the peqGOLD total RNA kit (VWR; Darmstadt, Germany) with a subsequent DNaseI digestion step according to the manufacturer's instructions. For cDNA synthesis, the Superscript II RT First Strand Kit (Invitrogen GmbH, Karlsruhe) was used. PCR primer sequences used to detect *ATOH1*, *GLI1*, *KRT8*, *14*, *17*, *18*, *20*, *RPLP0*, *SOX2*, and *SOX9* are given in Supplementary Methods. Thermal profile for the PCR using the Takyon Low Rox Sybr MasterMix (Eurogentec; Cologne, Germany) contained an initial denaturation step at 95 °C for 10 min, followed by 40 cycles of two-step PCR including 15 sec at 95 °C and 60 sec at 60 °C. Quantification was performed in three independent experiments.

4.7. Immunoblot

Cells were lysed in 0.6% SDS, 1 mM Ethylenediaminetetraacetic acid (EDTA), 10 mM Tris- HCl (pH 8.0), 2 mM NaF, 2 mM NaVO3 supplemented with a protease inhibitor cocktail (Roche Diagnostics, Basel, Switzerland). Samples were resolved by SDS-PAGE, transferred to nitrocellulose membrane, blocked for 1 h with Phosphate buffered saline (PBS) containing 0.05% Tween 20 and 5% powdered skim milk, then incubated overnight with anti-HA (ab18181, Abcam, 1:1000), LT (CM2B4, Santa Cruz, 1:200), sT (2T2, Hybridoma obtained from C. Buck laboratory), anti-GLI1 (C68H3, Ozyme, 1:200), anti-SOX2 (EPR3131, Abcam, 1:200), anti-ATOH1 (polyclonal, Proteintech, 1:600), or anti-Actin antibody (A5441, Sigma, 1:1000), washed three times with PBS with 0.05% Tween 20 (PBS/Tween), then incubated for 1 h with a peroxidase-conjugated secondary antibody. Finally, following three washes with PBS/Tween, respective proteins were detected by using a chemiluminescence detection procedure. All primary Western blot membranes' acquisition without cropping and intensity adjustment are available in Figure S8.

4.8. Transient Transfection and ATOH1 Half-Life Evaluation

Transient transfections were done using 2 μg of DNA with polyethylenimine (PEI) and protein expression was analyzed 24 h after transfection. For ATOH1 half-life determination, 24 h after transfection, cells were exposed to cycloheximide (0.3 mg/mL) in a time-course experiment. After harvesting, protein expression was then investigated by immunoblotting, and quantification was performed using ImageJ software.

4.9. Flow Cytometry

Anti-CD200 phycoerythrin (PE)-conjugated (OX-104, BioLegend) and anti-leucine rich repeat containing G protein-coupled receptor 6 (LGR6) Allophycocyanin (APC)-conjugated (Sc-393010, SantaCruz) antibodies were used for NHEK characterization.

4.10. Image Analysis and Expression Score Determination

Cell morphology was analyzed on adherent living cells. After acquisition of five adjacent microscopic fields, cell contouring was performed on 100 cells per conditions (three independent experiments) and cell size was then analyzed using ImageJ software. For protein expression evaluation, 2×10^5 cells were fixed in formalin, spotted on slides, and submitted to immunohistochemical staining. Stained slides were scanned by using NanoZoomer (Hamamatsu, Hamamatsu City, Japan). Computation of the expression score after transduction was performed with a custom software written in ImageJ Macro language. Briefly, color range for each staining was first defined from the whole image data set. Afterwards, cells were segmented in each image. For each cell-related area, the percentage of each type of viral protein staining (low, medium, and high) was computed. H-score was finally calculated for each cell with the following formula:

$$Hscore = \frac{(lowstainingarea \times 1) + (mediumstainingarea \times 2) + (highstainingarea \times 3)}{totalcellarea}$$

Analysis was initially performed on 10 consecutive fields (magnification × 10). In cases in which fewer than 1000 cells per conditions were analyzed, new acquisitions were performed in order to reach this minimal limit of analyzed cells. Results were subsequently expressed as median, quartiles Q1–Q3, and 1st–99th percentiles of the complete cell population analyzed. Protein quantification on immunoblot was performed by ImageJ using the "gel analysis" function.

4.11. Statistical Analysis

Continuous data are described as mean with standard error of mean (SEM), and categorical data with number and as percent. Associations were assessed by two-tailed Fisher exact test for categorical data and Mann–Whitney test for continuous data. Paired *t* test was used for RNA expression analysis without multiple testing correction. The $p < 0.05$ was considered statistically significant. XL-Stat-Life (Addinsoft, Paris, France) was used for statistical analyses.

5. Conclusions

Whether MCC is derived from MC or from another skin lineage is a long-time matter of debate. In this regard, we recently demonstrated that MCPyV integration in a TB gave rise to an MCPyV-positive MCC [31] and, consequently, postulated that MCC tumorgenesis can be initiated in MC epithelial progenitors. In the present work, we confirmed the close similarities between TB tumor cells and epithelial MC progenitors, evident by expression of GLI1 and its related downstream targets, i.e., KRT17 and SOX9, in both settings. While a mixture of cells with either MC progenitor phenotype or already differentiated MCs was observed in TB, almost all MCC tumor cells display a fully differentiated MC phenotype. Consequently, we assessed if TA could contribute to the acquisition of an MC phenotype. In accordance with this hypothesis, ectopic TA expression in NHEK led to induction of early MC markers while concomitant induction of SOX2, KRT8, and KRT20 were only achieved upon co-expression of TA and GLI1. Therefore, our results suggest that TA can induce acquisition of Merkel cell-like phenotype when expressed in epithelial MC progenitors. Accordingly, since large T antigen extends ATOH1 half-life, ATOH1 stabilization by MCPyV oncoproteins might further contribute to the MC-like phenotype observed in MCC.

Supplementary Materials: The following are available online at http://www.mdpi.com/2072-6694/12/7/1989/s1. Figure S1. Further characterization of Merkel cells and related progenitors in human, Figure S2. Characterization of the NHEK, Figure S3. Further characterization of native and GLI1-transduced normal human epidermal keratinocytes, Figure S4. Expression of the Merkel cell progenitor markers in tumors, Figure S5. Further characterization of native, TA, and GLI1/TA-transduced normal human epidermal keratinocytes, Figure S6. Impact of TA knockdown on ATOH1 protein level in the MCC cell lines MKL-1 and WaGa, Figure S7. Further characterization of factors involved in ATOH1 stabilization, Figure S8. Uncropped Western blot membranes' primary acquisitions, Figure S9. Signal quantification of the Western blots, Table S1. MCs density and location depending on the anatomic site, Table S2. Expression of the MC progenitor and MC markers in the trichoblastoma and MCC tumors, Table S3. Expression of GLI1 and SOX9 according to the MCPyV status in MCC tumors.

Author Contributions: Conceptualization, T.K., B.S., C.D., S.G., A.T., R.H., and D.S.; Data curation, T.K., P.B., M.W., T.G., R.H., and D.S.; Formal analysis, T.K., M.W., S.G., R.H., and D.S.; Funding acquisition, T.K., A.T., R.H., and D.S.; Investigation, S.H., B.S., S.S., and C.P.; Methodology, T.K., M.S., S.H., P.B., M.W., A.S., C.P., A.T., R.H., and D.S.; Project administration, T.K., M.S., A.T., R.H., and D.S.; Resources, M.W., A.S., C.D., S.G., R.H., and D.S.; Software, C.P.; Supervision, M.S., S.H., R.H., and D.S.; Validation, T.K., M.S., S.G., R.H., and D.S.; Writing—original draft, T.K.; Writing—review and editing, T.K., M.S., S.H., P.B., M.W., A.S., B.S., S.S., T.G., C.D., C.P., S.G., A.T., R.H., and D.S. All authors have read and agreed to the published version of the manuscript.

Funding: This research was funded by Fondation ARC pour la recherche contre le cancer and la ligue contre le cancer (Comités 16, 18, 28). The study was further supported by the Interdisziplinäres Zentrum für Klinische Forschung Würzburg (IZKF B-343) and by the German Research Foundation (SCHR 1178/3-1).

Acknowledgments: We express our gratitude to the donors involved in the body donation program of the *Association des dons du corps du Centre Ouest*, Tours, who made this study possible by generously donating their bodies for science. We thank J.F. Jégou, J. Vachtenheim, C. Buck, and G. Deluermoz for their help and assistance. In addition, we thank Daniel Sage (EPFL) for the color segmentation imageJ plugin. We thank the foundation ARC and la ligue contre le cancer, the IZKF (B-343) and the German Research Foundation (SCHR 1178/3-1) for funding.

Conflicts of Interest: The authors declare no conflict of interest.

Institutional Review Board: The local Ethics Committee in Human Research of Tours (France) approved the study (no. ID RCB2009-A01056-51).

References

1. Lemos, B.D.; Storer, B.E.; Iyer, J.G.; Phillips, J.L.; Bichakjian, C.K.; Fang, L.C.; Johnson, T.M.; Liegeois-Kwon, N.J.; Otley, C.C.; Paulson, K.G.; et al. Pathologic nodal evaluation improves prognostic accuracy in Merkel cell carcinoma: Analysis of 5823 cases as the basis of the first consensus staging system. *J. Am. Acad. Dermatol.* **2010**, *63*, 751–761. [CrossRef] [PubMed]

2. Feng, H.; Shuda, M.; Chang, Y.; Moore, P.S. Clonal integration of a polyomavirus in human Merkel cell carcinoma. *Science* **2008**, *319*, 1096–1100. [CrossRef] [PubMed]
3. Verhaegen, M.E.; Mangelberger, D.; Harms, P.W.; Eberl, M.; Wilbert, D.M.; Meireles, J.; Bichakjian, C.K.; Saunders, T.L.; Wong, S.Y.; Dlugosz, A.A. Merkel cell polyomavirus small T antigen initiates Merkel cell carcinoma-like tumor development in mice. *Cancer Res.* **2017**, *77*, 3151–3157. [CrossRef] [PubMed]
4. Tilling, T.; Moll, I. Which are the cells of origin in merkel cell carcinoma? *J. Skin Cancer* **2012**, *2012*, 680410. [CrossRef]
5. Becker, J.C.; Zur Hausen, A. Cells of origin in skin cancer. *J. Investig. Dermatol.* **2014**, *134*, 2491–2493. [CrossRef] [PubMed]
6. Harold, A.; Amako, Y.; Hachisuka, J.; Bai, Y.; Li, M.Y.; Kubat, L.; Gravemeyer, J.; Franks, J.; Gibbs, J.R.; Park, H.J.; et al. Conversion of Sox2-dependent Merkel cell carcinoma to a differentiated neuron-like phenotype by T antigen inhibition. *Proc. Natl. Acad. Sci. USA* **2019**. [CrossRef]
7. Ikeda, R.; Cha, M.; Ling, J.; Jia, Z.; Coyle, D.; Gu, J.G. Merkel cells transduce and encode tactile stimuli to drive Aβ-afferent impulses. *Cell* **2014**, *157*, 664–675. [CrossRef]
8. Laga, A.C.; Lai, C.-Y.; Zhan, Q.; Huang, S.J.; Velazquez, E.F.; Yang, Q.; Hsu, M.-Y.; Murphy, G.F. Expression of the embryonic stem cell transcription factor SOX2 in human skin: Relevance to melanocyte and merkel cell biology. *Am. J. Pathol.* **2010**, *176*, 903–913. [CrossRef] [PubMed]
9. Perdigoto, C.N.; Bardot, E.S.; Valdes, V.J.; Santoriello, F.J.; Ezhkova, E. Embryonic maturation of epidermal Merkel cells is controlled by a redundant transcription factor network. *Dev. Camb. Engl.* **2014**, *141*, 4690–4696. [CrossRef] [PubMed]
10. Moll, R.; Moll, I.; Franke, W.W. Identification of Merkel cells in human skin by specific cytokeratin antibodies: Changes of cell density and distribution in fetal and adult plantar epidermis. *Differ. Res. Biol. Divers.* **1984**, *28*, 136–154. [CrossRef]
11. Moll, I.; Paus, R.; Moll, R. Merkel cells in mouse skin: Intermediate filament pattern, localization, and hair cycle-dependent density. *J. Investig. Dermatol.* **1996**, *106*, 281–286. [CrossRef]
12. Van Keymeulen, A.; Mascre, G.; Youseff, K.K.; Harel, I.; Michaux, C.; De Geest, N.; Szpalski, C.; Achouri, Y.; Bloch, W.; Hassan, B.A.; et al. Epidermal progenitors give rise to Merkel cells during embryonic development and adult homeostasis. *J. Cell Biol.* **2009**, *187*, 91–100. [CrossRef]
13. Halata, Z.; Grim, M.; Bauman, K.I. Friedrich Sigmund Merkel and his "Merkel cell", morphology, development, and physiology: Review and new results. *Anat. Rec. A. Discov. Mol. Cell. Evol. Biol.* **2003**, *271*, 225–239. [CrossRef] [PubMed]
14. Morrison, K.M.; Miesegaes, G.R.; Lumpkin, E.A.; Maricich, S.M. Mammalian Merkel cells are descended from the epidermal lineage. *Dev. Biol.* **2009**, *336*, 76–83. [CrossRef] [PubMed]
15. Moll, I.; Lane, A.T.; Franke, W.W.; Moll, R. Intraepidermal formation of Merkel cells in xenografts of human fetal skin. *J. Investig. Dermatol.* **1990**, *94*, 359–364. [CrossRef] [PubMed]
16. Ostrowski, S.M.; Wright, M.C.; Bolock, A.M.; Geng, X.; Maricich, S.M. Ectopic Atoh1 expression drives Merkel cell production in embryonic, postnatal and adult mouse epidermis. *Dev. Camb. Engl.* **2015**, *142*, 2533–2544. [CrossRef]
17. Bardot, E.S.; Valdes, V.J.; Zhang, J.; Perdigoto, C.N.; Nicolis, S.; Hearn, S.A.; Silva, J.M.; Ezhkova, E. Polycomb subunits Ezh1 and Ezh2 regulate the Merkel cell differentiation program in skin stem cells. *EMBO J.* **2013**, *32*, 1990–2000. [CrossRef]
18. Xiao, Y.; Thoresen, D.T.; Williams, J.S.; Wang, C.; Perna, J.; Petrova, R.; Brownell, I. Neural Hedgehog signaling maintains stem cell renewal in the sensory touch dome epithelium. *Proc. Natl. Acad. Sci. USA* **2015**, *112*, 7195–7200. [CrossRef] [PubMed]
19. Xiao, Y.; Thoresen, D.T.; Miao, L.; Williams, J.S.; Wang, C.; Atit, R.P.; Wong, S.Y.; Brownell, I. A Cascade of Wnt, Eda, and Shh Signaling Is Essential for Touch Dome Merkel Cell Development. *PLoS Genet.* **2016**, *12*, e1006150. [CrossRef] [PubMed]
20. Nguyen, M.B.; Cohen, I.; Kumar, V.; Xu, Z.; Bar, C.; Dauber-Decker, K.L.; Tsai, P.-C.; Marangoni, P.; Klein, O.D.; Hsu, Y.-C.; et al. FGF signalling controls the specification of hair placode-derived SOX9 positive progenitors to Merkel cells. *Nat. Commun.* **2018**, *9*, 2333. [CrossRef]
21. Woo, S.-H.; Stumpfova, M.; Jensen, U.B.; Lumpkin, E.A.; Owens, D.M. Identification of epidermal progenitors for the Merkel cell lineage. *Dev. Camb. Engl.* **2010**, *137*, 3965–3971. [CrossRef]

22. Moll, I.; Troyanovsky, S.M.; Moll, R. Special program of differentiation expressed in keratinocytes of human haarscheiben: An analysis of individual cytokeratin polypeptides. *J. Investig. Dermatol.* **1993**, *100*, 69–76. [CrossRef] [PubMed]
23. Peterson, S.C.; Eberl, M.; Vagnozzi, A.N.; Belkadi, A.; Veniaminova, N.A.; Verhaegen, M.E.; Bichakjian, C.K.; Ward, N.L.; Dlugosz, A.A.; Wong, S.Y. Basal cell carcinoma preferentially arises from stem cells within hair follicle and mechanosensory niches. *Cell Stem Cell* **2015**, *16*, 400–412. [CrossRef] [PubMed]
24. Kervarrec, T.; Samimi, M.; Guyétant, S.; Sarma, B.; Chéret, J.; Blanchard, E.; Berthon, P.; Schrama, D.; Houben, R.; Touzé, A. Histogenesis of Merkel Cell Carcinoma: A Comprehensive Review. *Front. Oncol.* **2019**, *9*, 451. [CrossRef] [PubMed]
25. Moll, I.; Zieger, W.; Schmelz, M. Proliferative Merkel cells were not detected in human skin. *Arch. Dermatol. Res.* **1996**, *288*, 184–187. [CrossRef]
26. Shuda, M.; Guastafierro, A.; Geng, X.; Shuda, Y.; Ostrowski, S.M.; Lukianov, S.; Jenkins, F.J.; Honda, K.; Maricich, S.M.; Moore, P.S.; et al. Merkel Cell Polyomavirus Small T Antigen Induces Cancer and Embryonic Merkel Cell Proliferation in a Transgenic Mouse Model. *PLoS ONE* **2015**, *10*, e0142329. [CrossRef]
27. Kurzen, H.; Esposito, L.; Langbein, L.; Hartschuh, W. Cytokeratins as markers of follicular differentiation: An immunohistochemical study of trichoblastoma and basal cell carcinoma. *Am. J. Dermatopathol.* **2001**, *23*, 501–509. [CrossRef]
28. Leblebici, C.; Bambul Sığırcı, B.; Kelten Talu, C.; Koca, S.B.; Huq, G.E. CD10, TDAG51, CK20, AR, INSM1, and Nestin Expression in the Differential Diagnosis of Trichoblastoma and Basal Cell Carcinoma. *Int. J. Surg. Pathol.* **2019**, *27*, 19–27. [CrossRef]
29. McNiff, J.M.; Eisen, R.N.; Glusac, E.J. Immunohistochemical comparison of cutaneous lymphadenoma, trichoblastoma, and basal cell carcinoma: Support for classification of lymphadenoma as a variant of trichoblastoma. *J. Cutan. Pathol.* **1999**, *26*, 119–124. [CrossRef]
30. Collina, G.; Eusebi, V.; Capella, C.; Rosai, J. Merkel cell differentiation in trichoblastoma. *Virchows Arch. Int. J. Pathol.* **1998**, *433*, 291–296. [CrossRef]
31. Kervarrec, T.; Aljundi, M.; Appenzeller, S.; Samimi, M.; Maubec, E.; Cribier, B.; Deschamps, L.; Sarma, B.; Sarosi, E.-M.; Berthon, P.; et al. Polyomavirus-positive Merkel cell carcinoma derived from a trichoblastoma suggests an epithelial origin of this Merkel cell carcinoma. *J. Investig. Dermatol.* **2019**. [CrossRef] [PubMed]
32. Foschini, M.P.; Eusebi, V. Divergent differentiation in endocrine and nonendocrine tumors of the skin. *Semin. Diagn. Pathol.* **2000**, *17*, 162–168. [PubMed]
33. Kervarrec, T.; Tallet, A.; Miquelestorena-Standley, E.; Houben, R.; Schrama, D.; Gambichler, T.; Berthon, P.; Le Corre, Y.; Hainaut-Wierzbicka, E.; Aubin, F.; et al. Morphologic and immunophenotypical features distinguishing Merkel cell polyomavirus-positive and -negative Merkel cell carcinoma. *Mod. Pathol. Off. J. U. S. Can. Acad. Pathol. Inc.* **2019**, *32*, 1605–1616. [CrossRef]
34. Kervarrec, T.; Tallet, A.; Miquelestorena-Standley, E.; Houben, R.; Schrama, D.; Gambichler, T.; Berthon, P.; Le Corre, Y.; Hainaut-Wierzbicka, E.; Aubin, F.; et al. Diagnostic accuracy of a panel of immunohistochemical and molecular markers to distinguish Merkel cell carcinoma from other neuroendocrine carcinomas. *Mod. Pathol. Off. J. U. S. Can. Acad. Pathol. Inc.* **2019**, *32*, 499–510. [CrossRef] [PubMed]
35. Fan, K.; Gravemeyer, J.; Ritter, C.; Rasheed, K.; Gambichler, T.; Moens, U.; Shuda, M.; Schrama, D.; Becker, J.C. MCPyV Large T antigen induced atonal homolog 1 (ATOH1) is a lineage-dependency oncogene in Merkel cell carcinoma. *J. Investig. Dermatol.* **2020**, *140*, 56–65.e3. [CrossRef] [PubMed]
36. Forget, A.; Bihannic, L.; Cigna, S.M.; Lefevre, C.; Remke, M.; Barnat, M.; Dodier, S.; Shirvani, H.; Mercier, A.; Mensah, A.; et al. Shh signaling protects Atoh1 from degradation mediated by the E3 ubiquitin ligase Huwe1 in neural precursors. *Dev. Cell* **2014**, *29*, 649–661. [CrossRef] [PubMed]
37. Cheng, Y.-F.; Tong, M.; Edge, A.S.B. Destabilization of Atoh1 by E3 Ubiquitin Ligase Huwe1 and Casein Kinase 1 Is Essential for Normal Sensory Hair Cell Development. *J. Biol. Chem.* **2016**, *291*, 21096–21109. [CrossRef] [PubMed]
38. Houben, R.; Angermeyer, S.; Haferkamp, S.; Aue, A.; Goebeler, M.; Schrama, D.; Hesbacher, S. Characterization of functional domains in the Merkel cell polyoma virus Large T antigen. *Int. J. Cancer* **2015**, *136*, E290–E300. [CrossRef] [PubMed]
39. Kwun, H.J.; Guastafierro, A.; Shuda, M.; Meinke, G.; Bohm, A.; Moore, P.S.; Chang, Y. The minimum replication origin of merkel cell polyomavirus has a unique large T-antigen loading architecture and requires small T-antigen expression for optimal replication. *J. Virol.* **2009**, *83*, 12118–12128. [CrossRef]

40. Chéret, J.; Bertolini, M.; Ponce, L.; Lehmann, J.; Tsai, T.; Alam, M.; Hatt, H.; Paus, R. Olfactory receptor OR2AT4 regulates human hair growth. *Nat. Commun.* **2018**, *9*, 3624. [CrossRef]
41. Vidal, V.P.I.; Ortonne, N.; Schedl, A. SOX9 expression is a general marker of basal cell carcinoma and adnexal-related neoplasms. *J. Cutan. Pathol.* **2008**, *35*, 373–379. [CrossRef]
42. Mikami, Y.; Fujii, S.; Nagata, K.; Wada, H.; Hasegawa, K.; Abe, M.; Yoshimoto, R.U.; Kawano, S.; Nakamura, S.; Kiyoshima, T. GLI-mediated Keratin 17 expression promotes tumor cell growth through the anti-apoptotic function in oral squamous cell carcinomas. *J. Cancer Res. Clin. Oncol.* **2017**, *143*, 1381–1393. [CrossRef] [PubMed]
43. Jia, Y.; Gu, D.; Wan, J.; Yu, B.; Zhang, X.; Chiorean, E.G.; Wang, Y.; Xie, J. The role of GLI-SOX2 signaling axis for gemcitabine resistance in pancreatic cancer. *Oncogene* **2019**, *38*, 1764–1777. [CrossRef] [PubMed]
44. Santini, R.; Pietrobono, S.; Pandolfi, S.; Montagnani, V.; D'Amico, M.; Penachioni, J.Y.; Vinci, M.C.; Borgognoni, L.; Stecca, B. SOX2 regulates self-renewal and tumorigenicity of human melanoma-initiating cells. *Oncogene* **2014**, *33*, 4697–4708. [CrossRef]
45. Lesko, M.H.; Driskell, R.R.; Kretzschmar, K.; Goldie, S.J.; Watt, F.M. Sox2 modulates the function of two distinct cell lineages in mouse skin. *Dev. Biol.* **2013**, *382*, 15–26. [CrossRef] [PubMed]
46. Carroll, T.M.; Williams, J.S.; Daily, K.; Rogers, T.; Gelb, T.; Coxon, A.; Wang, S.Q.; Crago, A.M.; Busam, K.J.; Brownell, I. Hedgehog Signaling Inhibitors Fail to Reduce Merkel Cell Carcinoma Viability. *J. Investig. Dermatol.* **2017**, *137*, 1187–1190. [CrossRef]
47. Spurgeon, M.E.; Cheng, J.; Bronson, R.T.; Lambert, P.F.; DeCaprio, J.A. Tumorigenic activity of merkel cell polyomavirus T antigens expressed in the stratified epithelium of mice. *Cancer Res.* **2015**, *75*, 1068–1079. [CrossRef]
48. Starrett, G.J.; Marcelus, C.; Cantalupo, P.G.; Katz, J.P.; Cheng, J.; Akagi, K.; Thakuria, M.; Rabinowits, G.; Wang, L.C.; Symer, D.E.; et al. Merkel Cell Polyomavirus Exhibits Dominant Control of the Tumor Genome and Transcriptome in Virus-Associated Merkel Cell Carcinoma. *mBio* **2017**, *8*. [CrossRef]
49. Kwun, H.J.; Shuda, M.; Feng, H.; Camacho, C.J.; Moore, P.S.; Chang, Y. Merkel cell polyomavirus small T antigen controls viral replication and oncoprotein expression by targeting the cellular ubiquitin ligase SCFFbw7. *Cell Host Microbe* **2013**, *14*, 125–135. [CrossRef]
50. Dye, K.N.; Welcker, M.; Clurman, B.E.; Roman, A.; Galloway, D.A. Merkel cell polyomavirus Tumor antigens expressed in Merkel cell carcinoma function independently of the ubiquitin ligases Fbw7 and β-TrCP. *PLoS Pathog.* **2019**, *15*, e1007543. [CrossRef]
51. Cheng, Y.-F. Atoh1 regulation in the cochlea: More than just transcription. *J. Zhejiang Univ. Sci. B* **2019**, *20*, 146–155. [CrossRef] [PubMed]
52. Becker, J.C.; Stang, A.; Hausen, A.Z.; Fischer, N.; DeCaprio, J.A.; Tothill, R.W.; Lyngaa, R.; Hansen, U.K.; Ritter, C.; Nghiem, P.; et al. Epidemiology, biology and therapy of Merkel cell carcinoma: Conclusions from the EU project IMMOMEC. *Cancer Immunol. Immunother. CII* **2018**, *67*, 341–351. [CrossRef]
53. Kervarrec, T.; Samimi, M.; Gaboriaud, P.; Gheit, T.; Beby-Defaux, A.; Houben, R.; Schrama, D.; Fromont, G.; Tommasino, M.; Le Corre, Y.; et al. Detection of the Merkel cell polyomavirus in the neuroendocrine component of combined Merkel cell carcinoma. *Virchows Arch. Int. J. Pathol.* **2018**. [CrossRef]
54. Boniface, K.; Bernard, F.-X.; Garcia, M.; Gurney, A.L.; Lecron, J.-C.; Morel, F. IL-22 inhibits epidermal differentiation and induces proinflammatory gene expression and migration of human keratinocytes. *J. Immunol. Baltim. Md 1950* **2005**, *174*, 3695–3702. [CrossRef] [PubMed]
55. Couderc, E.; Morel, F.; Levillain, P.; Buffière-Morgado, A.; Camus, M.; Paquier, C.; Bodet, C.; Jégou, J.-F.; Pohin, M.; Favot, L.; et al. Interleukin-17A-induced production of acute serum amyloid A by keratinocytes contributes to psoriasis pathogenesis. *PLoS ONE* **2017**, *12*, e0181486. [CrossRef]
56. Langan, E.A.; Philpott, M.P.; Kloepper, J.E.; Paus, R. Human hair follicle organ culture: Theory, application and perspectives. *Exp. Dermatol.* **2015**, *24*, 903–911. [CrossRef] [PubMed]
57. Schrama, D.; Sarosi, E.-M.; Adam, C.; Ritter, C.; Kaemmerer, U.; Klopocki, E.; König, E.-M.; Utikal, J.; Becker, J.C.; Houben, R. Characterization of six Merkel cell polyomavirus-positive Merkel cell carcinoma cell lines: Integration pattern suggest that large T antigen truncating events occur before or during integration. *Int. J. Cancer* **2019**. [CrossRef] [PubMed]
58. Vlčková, K.; Ondrušová, L.; Vachtenheim, J.; Réda, J.; Dundr, P.; Zadinová, M.; Žáková, P.; Poučková, P. Survivin, a novel target of the Hedgehog/GLI signaling pathway in human tumor cells. *Cell Death Dis.* **2016**, *7*, e2048. [CrossRef]

59. Houben, R.; Adam, C.; Baeurle, A.; Hesbacher, S.; Grimm, J.; Angermeyer, S.; Henzel, K.; Hauser, S.; Elling, R.; Bröcker, E.-B.; et al. An intact retinoblastoma protein-binding site in Merkel cell polyomavirus large T antigen is required for promoting growth of Merkel cell carcinoma cells. *Int. J. Cancer* **2012**, *130*, 847–856. [CrossRef]
60. Angermeyer, S.; Hesbacher, S.; Becker, J.C.; Schrama, D.; Houben, R. Merkel cell polyomavirus-positive Merkel cell carcinoma cells do not require expression of the viral small T antigen. *J. Investig. Dermatol.* **2013**, *133*, 2059–2064. [CrossRef]

cancers

MDPI

Article

Merkel Cell Carcinoma Treatment in Finland in 1986–2016—A Real-World Data Study

Helka Sahi [1,*,†], Jenny Their [2,†], Mika Gissler [3,4] and Virve Koljonen [2]

1 Department of Dermatology, Allergology and Venerology, HUS Inflammation Center, University of Helsinki and Helsinki University Hospital, P.O. Box 160, FIN-00029 HUS Helsinki, Finland
2 Department of Plastic Surgery, University of Helsinki and Helsinki University Hospital, FIN-00029 HUS Helsinki, Finland; jenny.their@helsinki.fi (J.T.); virve.koljonen@helsinki.fi (V.K.)
3 Finnish Institute for Health and Welfare, FIN-00271 Helsinki, Finland; mika.gissler@thl.fi
4 Department of Neurobiology, Care Sciences and Society, Karolinska Institutet, 171 65 Solna, Stockholm, Sweden
* Correspondence: helka.sahi@helsinki.fi
† These authors contributed equally and share first authorship.

Received: 20 March 2020; Accepted: 6 May 2020; Published: 13 May 2020

Abstract: Merkel cell carcinoma (MCC) is a rare cutaneous carcinoma that has gained enormous interest since the discovery of Merkel cell polyoma virus, which is a causative oncogenic agent in the majority of MCC tumours. Increased research has focused on effective treatment options with immuno-oncology. In this study, we reviewed the real-world data on different treatments given to MCC patients in Finland in 1986–2016. We used the Finnish Cancer Registry database to find MCC patients and the Hospital Discharge Register and the Cause-of-Death Register to obtain treatment data. We identified 376 MCC patients and 33 different treatment entities and/or combinations of treatment. An increase was noted in the incidence of MCC since 2005. Therefore, the cohort was divided into two groups: the "early" group with time of diagnosis between years 1986 and 2004 and the "late" group with time of diagnosis between 2005 and 2016. The multitude of different treatment combinations is a relatively new phenomenon; before the year 2005, only 11 treatments or treatment combinations were used for MCC patients. Our data show that combining radiation therapy with simple excision provided a survival advantage, which was, however, lost after adjustment for stage or age. Our registry study serves as a baseline treatment efficacy comparison as we move into the age of immunotherapy in MCC. Standardizing the treatment of MCC patients in Finland requires more work on awareness and multidisciplinary co-operation.

Keywords: Merkel cell carcinoma; treatment; radiation therapy; immunotherapy; surgical intervention; survival; multidisciplinary communication

1. Introduction

Merkel cell carcinoma (MCC) is a rare neuroendocrine cutaneous malignancy associated with high mortality, local and regional recurrence and distant metastases [1,2]. The pathogenesis of MCC is driven by the combination of Merkel cell polyoma virus (MCPyV) infection and chronic exposure to UV radiation. The connection between MCC and MCPyV was uncovered in 2008, when a research group led by Patrick Moore discovered the presence of previously unknown polyomavirus DNA in the MCC tumour genome, later proving that viral DNA was present in the majority of MCCs [3]. Similar findings have subsequently been made in numerous studies; however, an Australian study found MCPyV to be present in only 18.3% of MCCs, giving rise to a theory that MCPyV infection is more important in MCC pathogenesis in areas of low UV exposure, whereas its significance is lower in areas of high UV exposure [4–7]. Aside from MCPyV infection and UV radiation exposure, risk factors for

MCC are high age, fair skin and immunosuppression [8]. MCC most often presents in the head and neck region [9–11].

MCC incidence varies globally from 0.1 to 1.6 per 100,000, with the lowest incidence in Europe and the highest incidence in Australia [12,13]. The incidence has been increasing globally, with the exception of most Nordic countries [13]. MCC mainly presents in elderly patients, and rarely in patients under 50 years old, with a median age at time of diagnosis varying between 75 and 82 years [1,14,15]. Globally, MCC incidence is higher in men than in women; however, in Finland the numbers are consistently the opposite [10,13]. MCC is more common in patients with fair skin, and Caucasians present with MCC far more often than Blacks, Hispanics or Asians [13].

The treatment of MCC has seen huge developments since the advent of the sentinel lymph node biopsy (SLNB) over a decade ago. Recommended treatment for primary MCC is wide-margin excision of the primary lesion with SLNB or complete lymph node dissection (CLND) [depending on the lymph node status, followed by adjuvant radiation therapy (RT) on the primary location and/or local lymph nodes [16,17]. In metastatic MCC, chemotherapy, mainly with platinum-based therapeutics, was the only additional treatment available until 2016 [18]. Since 2016, anti-PD1/PD-L1 immunotherapies have become available and have proven effective in treatment of chemotherapy-refractory metastatic MCC [19,20].

MCC is a radiosensitive tumour [21]. Several studies have found that RT as an adjuvant treatment to MCC improves both locoregional control and patient overall and disease-free survival, regardless of excision margin status [14,21,22]. However, RT is not given to all MCC patients, often because the treatment can be strenuous, and the patients are usually elderly and may be suffering from comorbidities. Hypofractionated or single-fraction RT could potentially be beneficial for patients who are not eligible for longer regimens [15,23]. It has also been shown that RT effects are achieved at least partially through immune system response. RT does not only affect the targeted tumour areas, but can have a beneficial abscopal effect on malignant growth elsewhere [23,24].

The aim of this study was to examine in real-world data the different treatments given to MCC patients in Finland during 1986–2016. We also sought to examine whether adjuvant RT has a beneficial effect on patient outcome and survival in MCC in a large real-world treatment cohort of MCC patients. This data set will serve as a baseline treatment efficacy comparison as we move into the age of immunotherapy in MCC.

2. Results

2.1. Patient Demographics and Survival

The specific inclusion criteria resulted in 376 patients with MCC. The annual incidence varied between 1 and 33 patients (Figure 1). Figure 1 illustrates an increase in the incidence of MCC since 2005. Therefore, the cohort was divided into two groups: the "early" group with time of diagnosis between 1986 and 2004 and the "late" group with time of diagnosis between 2005 and 2016.

In Table 1, we illustrate the detailed data of the MCC patients. Mean age at time of diagnosis was 78.7 years. There was a clear female predominance, with a male-to-female ratio of 1:1.8. Most of the tumours, 59.6%, were in the head and neck region, followed by 14.1% in the upper extremities and 12.5% in the lower extremities.

By the closing date of our study on 31 December 2016, two out of three patients (65%) had died either due to MCC (27% of all) or due to other causes (38% of all). The mean overall survival was 4.2 years, and the MCC-specific survival was 1.8 years. Figure 2 shows the comparison of MCC-specific deaths and all deaths stratified by years. In the early and late cohorts, patients' demographic data, tumour location and survival remained similar (Table 1).

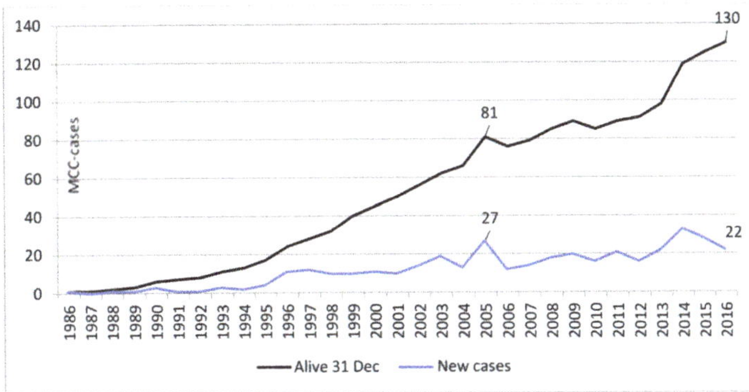

Figure 1. Annual incidence and prevalence of MCC in 1986–2016 in Finland.

Table 1. Demographic data of the 376 MCC patients treated in Finland in 1986–2016.

	Years 1986–2016	Years 1986–2004	Years 2005–2016
	N 376	N 127	N 249
Male/female (%)	132/244 (35/65)	40/87 (31/69)	92/157 (36/63)
Age at diagnosis, years			
Range	27–102	27–100	47–102
Mean (SD)	78.7 (10.6)	76.6 (12.1)	79.8 (9.6)
Tumour location			
440 Skin of lip, NOS	4 (1.1)	- (0.0)	4 (1.6)
441 Eyelid	11 (2.9)	4 (3.1)	7 (2.8)
442 External ear	18 (4.8)	6 (4.7)	12 (4.8)
443 Skin of other and unspecified parts of face	167 (44.4)	57 (44.9)	110 (44)
444 Skin of scalp and neck	24 (6.4)	9 (7.1)	15 (6)
445 Skin of trunk	28 (7.5)	8 (6.3)	20 (8)
446 Skin of upper limb and shoulder	53 (14.1)	17 (13.4)	36 (14)
447 Skin of lower limb and hip	47 (12.5)	16 (12.6)	31 (12.5)
449 Skin, NOS	24 (6.4)	10 (7.9)	14 (5.6)
Stage			
0 Unknown	158 (42.0)	42 (33)	116 (46)
1 Localized	124 (33.0)	64 (50)	60 (24)
2 Non-localised, only regional lymph node metastases	28 (7.5)	9 (7.1)	19 (7.6)
3 Metastasised farther than to regional lymph nodes or invades adjacent tissues	23 (6.1)	4 (3.1)	19 (7.6)
4 Non-localized, no information on extent	40 (10.6)	8 (6.3)	32 (12.9)
5 Non-localized, also distant lymph node metastases	3 (0.8)	- (0.0)	3 (1.2)
Survival years, cut off 31.12.2016			
Alive, n	130 (34.6%)	17 (13.4%)	113 (45.4%)
dead due to this cancer, n	103 (27.4%)	36 (28.3%)	67 (26.9%)
dead due to other cause, n	143 (38.0%)	74 (58.3%)	69 (27.7%)
Mean overall survival years (SD) n	376	127	249
diagnose date to end of surveillance/death	4.2 (4.9)	6.6 (6.7)	2.9 (2.9)
range	0-27	0–27	0–12
Deceased due to this cancer (SD) n	103	36	67
Mean survival, years	1.8 (2.0)	2.4 (2.6)	1.5 (1.4)
range	0–15	0–15	0.0–7
Deceased due to other cause (SD) n	143	74	69
Mean survival, years	4.1 (4.7)	6.1 (5.8)	2.1 (2.2)
range	0–25	0–25	0–11

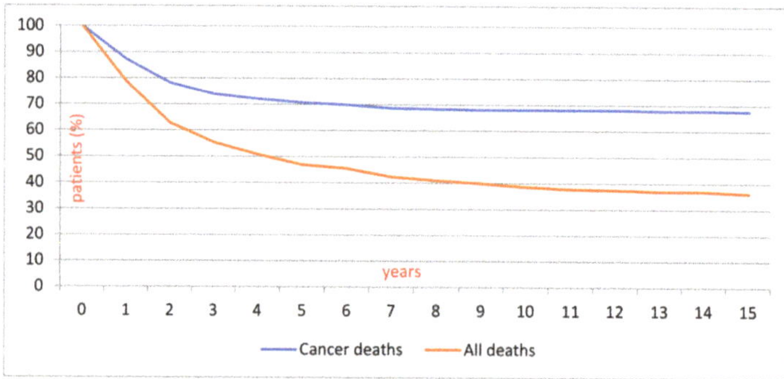

Figure 2. Kaplan–Meier curve depicting MCC specific deaths compared to all MCC patient deaths in 1986–2016.

2.2. Treatment and Outcomes

We identified altogether 33 different treatment entities or combinations of treatments (Table 2). Of the patients, 18% were recorded as having no treatment, 16% as having a single treatment and 66% as having a combination of different treatments.

Table 2. Different treatments and treatment combinations recorded in the study.

		Years 1986–2016	Years 1986–2004	Years 2005–2016
		N 376 (%)	N 127 (%)	N 249 (%)
No Treatment	No treatment	68 (18)	44 (34.6)	24 (9.6)
Single treatment	Re-excision of the primary tumour (Re-ex)	130 (34)	55 (43)	75 (30)
	Sentinel lymph node biopsy (SLNB)	1 (0.2)	-	1 (0.4)
	Complete lymphnode dissection (CLND)	5 (1.3)	2 (1.6)	3 (1.2)
	Pre-operative radiotherapy	1 (0.2)	-	1 (0.4)
	Adjuvant radiotherapy to the primary tumour	4 (1.1)	2 (1.6)	2 (0.8)
	Adjuvant radiotherapy to the regional lymphnodes	1 (0.2)	1 (0.8)	-
	Radiotherapy to metastasis	1 (0.2)	-	1 (0.4)
	Non-specified radiation therapy—recorded as modality	1(0.2)	-	1 (0.4)
	Non-specified radiation therapy	3 (0.8)	-	3 (1.2)
	Palliative radiotherapy	1 (0.2)	-	1 (0.4)
	Radiotherapy of metastases	2 (0.5)	-	2 (0.8)
Multiple treatments	Re-ex + SLNB	37 (9.8)	3 (2.4)	34 (13.7)
	Re-ex + CLND	50 (13.3)	16 (12.6)	34 (13.7)
	Re-ex+ adjuvant radiotherapy to the primary tumour	15 (4)	1 (0.8)	14 (5.6)
	Re-ex + chemotherapy to metastasised malignancy	1 (0.2)	-	1 (0.4)
	Re-ex + non-specified radiation therapy	6 (1.6)	1 (0.8)	5 (2)
	Re-ex + SLNB + CLND	7 (1.9)	1 (0.8)	6 (2.4)
	Re-ex + SLNB + adjuvant radiotherapy to the primary tumour	8 (2.1)	-	8 (3.2)
	Re-ex + CLND + adjuvant radiotherapy to the primary tumour	10 (2.7)	-	10 (4)

Table 2. *Cont.*

	Years 1986–2016	Years 1986–2004	Years 2005–2016
	N 376 (%)	N 127 (%)	N 249 (%)
Re-ex+ adjuvant radiotherapy to the primary tumour + radiotherapy to metastasis	1 (0.2)	-	1 (0.4)
CLND + adjuvant radiotherapy to the regional lymphnodes	1 (0.2)	-	1 (0.4)
CLND + radiotherapy to metastasis	1 (0.2)	-	1 (0.4)
CLND + pre-operative chemotherapy	2 (0.5)	-	2 (0.8)
Adjuvant radiotherapy to the primary tumour + pre-operative chemotherapy	1 (0.2)	-	1 (0.4)
Radiotherapy to metastasis + pre-operative chemotherapy	1 (0.2)	-	1 (0.4)
Re-ex + SLNB + non-specified radiation therapy	2 (0.5)	-	2 (0.8)
Re-ex + SLNB + CLND + adjuvant radiotherapy to the primary tumour	1 (0.2)	-	1 (0.4)
Re-ex + SLNB + CLND + chemotherapy to metastasized malignancy	1 (0.2)	-	1 (0.4)
Re-ex + CLND + non-specified radiation therapy	9 (2.4)	1 (0.8)	8 (3.2)
Re-ex + SLNB + adjuvant radiotherapy to the primary tumour + palliative radiotherapy	2 (0.5)	-	2 (0.8)
Re-ex + CLND + adjuvant radiotherapy to the primary tumour + radiotherapy to metastasis + pre-operative chemotherapy	1 (0.2)	-	1 (0.4)
Re-ex + CLND + adjuvant radiotherapy to the primary tumour + radiotherapy to metastasis + chemotherapy to metastasized malignancy	1 (0.2)	-	1 (0.4)

In Table 3, we grouped the treatments given into three different time periods. The individual procedure codes and numbers of treated patients are listed in more detail in Table S1. The multitude of different treatment combinations is a relatively new phenomenon; before the year 2005, only 11 treatments or treatment combinations were used for MCC patients. We recorded the first SLNB in the year 2005 because the registration of the patients who had their SLNB in 2005 but were diagnosed in 2004 was not recorded in the cohort from 2005 onwards. Even excluding the SLNB, the amount of different treatments in 2005–2016 surpasses that of earlier years.

Table 3. Treatments and their combinations with MCC-specific survival.

Treatment	Years 1986–2016 N 376					Years 1986–2004 N 127					Years 2005–2016 N 249				
	N (%)	Male/Female	Mean Age Years	Death to Cancer %	Survival Years Mean	N (%)	Male/Female	Mean age years	Death to cancer %	Survival, Years Mean	N (%)	Male/Female	Mean Age Years	Death to Cancer %	Survival, Years Mean
No treatment	68 (18)	22/46	76.7	26.5	5.1	44 (34.6)	15/30	75.2	27.3	6.4	24 (9.6)	8/16	79.4	13.0	2.6
Re-excision of the primary tumour (Re-ex)	130 (34)	33/97	80.9	20.0	4.9	55 (43)	13/42	78.0	14.5	8.0	75 (30)	20/55	83.1	29.0	2.7
Sentinel lymph node biopsy (SLNB)	1 (0.2)	0/1	79.2	-	11.5						1 (0.4)	0/1	79.2	0.0	11.5
Complete lymph node dissection (CLND)	5 (1.3)	2/2	79.4	80.0	3.4	2 (1.6)	1/2	73.6	100.0	3.5	3 (1.2)	1/2	83.2	0.0	3.4
Pre-operative radiotherapy	1 (0.2)	0/1	83.9	100.0	1.8						1 (0.4)	0/1	83.9	0.0	1.8
Adjuvant radiotherapy to the primary tumour	4 (1.1)	1/3	82.3	0.0	3.7	2 (1.6)	1/1	84.2	0.0	6.5	2 (0.8)	0/2	80.4	1.0	0.8
Adjuvant radiotherapy to the regional lymph nodes	1 (0.2)	0/1	87.0	0.0	12.6	1 (0.8)	0/1	87.0	0.0	12.6	-	-/-	-	-	-
Radiotherapy to metastasis	1 (0.2)	1/0	79.7	0.0	8.1						1 (0.4)	1/0	79.7	0.0	8.1
Non-specified radiation therapy—recorded as modality	1(0.2)	0/1	86.2	100.0	1.7						1 (0.4)	0/1	86.2	0.0	1.7
Non-specified radiation therapy	3 (0.8)	1/2	89.0	66.7	1.0						3 (1.2)	1/2	89.0	1.0	1.0
Palliative radiotherapy	1 (0.2)	1/0	68.7	100.0	0.1						1 (0.4)	1/0	68.7	0.0	0.1
Radiotherapy of metastases	2 (0.5)	2/0	89.8	0.0	1.2						2 (0.8)	2/0	89.8	1.0	1.2
Re-ex + SLNB	37 (9.8)	18/19	77.8	27.0	3.6	3 (2.4)	2/1	81.1	66.7	4.2	34 (13.7)	16/18	77.5	4.0	3.5
Re-ex + CLND	50 (13.3)	22/28	78.7	34.0	3.7	16 (12.6)	7/9	75.6	56.3	3.9	34 (13.7)	15/19	80.1	9.0	3.6
Re-ex + adjuvant radiotherapy to the primary tumour	15 (4)	7/8	80.7	26.7	2.3	1 (0.8)	0/1	90.9	100.0	1.4	14 (5.6)	7/7	80.0	2.0	2.4
Re-ex + chemotherapy to metastasized malignancy	1 (0.2)	0/1	84.1	0.0	6.8						1 (0.4)	0/1	84.1	0.0	6.8
Re-ex + non-specified radiation therapy	6 (1.6)	1/5	78.7	33.3	3.7	1 (0.8)	1/0	68.7	100.0	0.3	5 (2)	0/5	80.7	2.0	4.3
Re-ex + SLNB + CLND	7 (1.9)	3/4	70.2	14.3	3.9	1 (0.8)	0/1	59.2	0.0	7.2	6 (2.4)	3/3	72.0	2.0	3.4
Re-ex + SLNB + adjuvant radiotherapy to the primary tumour	8 (2.1)	3/5	72.4	12.5	1.7						8 (3.2)	3/5	72.4	1.0	1.7
Re-ex + CLND + adjuvant radiotherapy to the primary tumour	10 (2.7)	2/8	70.2	30.0	2.7						10 (4)	2/8	70.2	0.0	2.7

Table 3. *Cont.*

Treatment	Years 1986–2016					Years 1986–2004					Years 2005–2016				
	N 376					N 127					N 249				
	N (%)	Male/Female	Mean Age Years	Death to Cancer %	Survival Years Mean	N (%)	Male/Female	Mean age years	Death to cancer %	Survival, Years Mean	N (%)	Male/Female	Mean Age Years	Death to Cancer %	Survival, Years Mean
Re-ex + adjuvant radiotherapy to the primary tumour + radiotherapy to metastasis	1 (0.2)	0/1	72.7	0.0	1.2						1 (0.4)	0/1	72.7	0.0	1.2
CLND + adjuvant radiotherapy to the regional lymphnodes	1 (0.2)	1/0	67.2	0.0	11.3						1 (0.4)	1/0	67.2	0.0	11.3
CLND + radiotherapy to metastasis	1 (0.2)	0/1	89.6	100.0	2.5						1 (0.4)	0/1	89.6	0.0	2.5
CLND + pre-operative chemotherapy	2 (0.5)	1/1	84.8	100.0	1.1						2 (0.8)	1/1	84.8	0.0	1.1
Adjuvant radiotherapy to the primary tumour + pre-operative chemotherapy	1 (0.2)	0/1	83.5	100.0	2.2						1 (0.4)	0/1	83.5	0.0	2.2
Radiotherapy to metastasis+ pre-operative chemotherapy	1 (0.2)	1/0	72.6	100.0	1.1						1 (0.4)	1/0	72.6	0.0	1.1
Re-ex + SLNB+ non-specified radiation therapy	2 (0.5)	0/2	83.6	0.0	3.1						2 (0.8)	0/2	83.6	1.0	3.1
Re-ex+SLNB+CLND+ adjuvant radiotherapy to the primary tumour	1 (0.2)	1/0	69.9	0.0	2.6						1 (0.4)	1/0	69.9	0.0	2.6
Re-ex + SLNB + CLND + chemotherapy to metastasized malignanacy	1 (0.2)	1/0	74.4	0.0	1.2						1 (0.4)	1/0	74.4	1.0	1.2
Re-ex + CLND + non-specified radiation therapy	9 (2.4)	5/4	75.3	66.7	2.1	1 (0.8)	1/0	49.8	100.0	2.5	8 (3.2)	4/4	78.5	2.0	2.0
Re-ex + SLNB+ adjuvant radiotherapy to the primary tumour + Palliative radiotherapy	2 (0.5)	1/1	71.6	50.0	0.8						2 (0.8)	1/1	71.6	0.0	0.8
Re-ex + CLND + adjuvant radiotherapy to the primary tumour + radiotherapy to metastasis+ pre-operative chemotherapy	1 (0.2)	1/0	76.0	0.0	1.9						1 (0.4)	1/0	76.0	0.0	1.9
Re-ex + CLND + adjuvant radiotherapy to the primary tumour + chemotherapy to metastasized malignancy	1 (0.2)	1/0	76.7	0.0	1.2						1 (0.4)	1/0	76.7	0.0	1.2

The biggest difference in the frequency of treatments was in the group "No treatment", which declined from 34.6% to only 9.6% during the period from 2005 onwards. The MCC-specific mortality declined as well from 27.3% to 13.0% for the period 2005–2016. Any radiation therapy was given to 4.7% of the patients of the early cohort, and to 26% of the late cohort. The treatment group "Re-excision and adjuvant RT to the primary tumour" increased from 0.8% to 5.6% in the later time period (Table 3). The other treatment combinations and their stratification in separate time periods remained relatively stable.

In 2005–2016, SLNB was performed on 55 patients (22%). In 51 of these cases, re-excision was followed by SLNB, with adjuvant RT to primary tumour in 11 cases and without adjuvant RT to primary tumour in 40 cases. The MCC-specific mortality was four times higher in the group with no adjuvant RT to the primary tumour. Re-excision and SLNB was followed by CLND in only seven cases (Table 3).

Re-excision of the primary tumour was performed on 75 patients (30%) and re-excision combined with adjuvant RT to primary tumour on 14 patients (5.6%). The MCC-specific mortality was 14.5 times higher in the re-excision group that received no adjuvant RT than in the group that received adjuvant RT in addition to re-excision (Table 3). MCC-specific deaths were similar in the early and late periods (Table 1).

2.3. *Effect of Radiation Therapy*

Our data show that adjuvant RT provided a survival advantage to patients receiving simple re-excision ($p < 0.005$, Table 4). However, when standardized by stage or age, the survival advantage was lost. Due to the small number of patients in the re-excision and SLNB group compared with the re-excision and SLNB and CLND group, the difference did not reach statistical significance. MCC patients receiving RT tended to be younger, except in the re-excision and SLNB and CLND group.

Table 4. Comparison of the effect of adjuvant radiation therapy to the primary tumour after excision in patients diagnosed in the year 2005 onwards.

Treatment Categories		N	Male/Female	Mean Age Years	Statistical Difference, Age, *p*	Head and Neck	Trunk	Upper Limb and Shoulder	Lower Limb and Hip	Skin, not specified	Statistical Difference, Location, *p*	Death to Cancer (%)	Statistical Difference, Death to Cancer, *p*
Excision	Re-excision of the primary tumour	75	20/55	83.1	NS	48	8	12	4	3	NS	29 (38.7)	0.005
	Re-ex + adjuvant radiotherapy to the primary tumour	14	7/7	80.0		11	1	0	2	0		0	
Excision and SLNB	Re-ex + SLNB	34	16/18	77.5	NS	18	3	5	5	3	NS	1 (2.9)	NS
	Re-ex + SLNB + adjuvant radiotherapy to the primary tumour	8	3/5	72.4		2	0	1	5	0		0	
Excision and SLNB and CLND	Re-ex + SLNB + CLND	6	3/3	72.0	NA	3	0	2	1	0	NA	1 (16.7)	NA
	Re-ex + SLNB + CLND + adjuvant radiotherapy to the primary tumour	1	1/0	69.9		0	0	1	0	0		0	
Excision and CLND	Re-ex + CLND	34	15/19	80.1	<0.001	22	1	5	3	3	NS	9 (26.5)	NS
	Re-ex + CLND + adjuvant radiotherapy to the primary tumour	10	2/8	70.2		7	1	2	0	0		0	

Re-ex = Re-excision of the primary tumour, SLNB= sentinel lymphnode biopsy, CLND = complete lymphnode dissection.

3. Discussion

We reviewed real-world treatments of 376 patients with MCC from 1986 to 2016 in Finland. We utilized data from several national registries to obtain a representation as realistic as possible of the actual situation. The demographic data of the patients remained fairly stable over the study years. As a peculiarity to Finland [10], we once again recorded a female over-representation among MCC patients; in the years 1986–2004 the male-to-female ratio was 1:2.2 and from 2005 onwards it was 1:1.7. Despite our best efforts, this finding remains unexplained. Otherwise, our patient cohort is similar to that of previously published large series, with a mean age of 78 years at time of diagnosis, 60% of tumours located in the head and neck region and localized stage of disease [11,16,17,25].

Although MCC has been recognized and characterized as its own entity since 1972 [26], it was not until the discovery of Merkel cell polyoma virus in 2008 [3] that a strong interest in MCC arose. In this study, we noted an increase in the number of MCC patients in 2005 (Figure 1). In such a rare cancer as MCC, the effect of chance cannot be ruled out. In the previous literature, the increased incidence has been attributed to advances in immunohistochemistry and morphology code associated with MCC [27,28]. The growing knowledge among pathologists and clinicians translates into more MCC being diagnosed; the first thesis of our group [29] and the first article commissioned by the Finnish Medical Society [30] were published in 2004 and 2005, respectively, which may have increased the knowledge and sensitivity for MCC diagnosis.

The increase in MCC incidence shown in our study coincides with a rise in the number of treatment entities per patient and in the variety of treatment combinations. The NCCN guideline on MCC treatment was first published in 2010, followed by the European guideline in 2016 [16,17,31,32]. The latest national guideline in Finland was published after the advent of avelumab in 2017 [33]. The cornerstones of MCC treatment are re-excision, SLNB and adjuvant RT to primary tumour [16,17]. However, in 2005–2016 only 4.4% of the patients here were treated with this combination. We found that the frequency of any RT given to MCC patients in Finland increased from less than 1% to over 5% after 2005. Compared with the SEER-based results from the United States, where nearly half of the MCC patients were treated with RT [11,34], utilizing RT in MCC treatment is still less common in Finland. Relative to the years 1986–2005, there were no advances in disease-specific survival of MCC patients after 2005, despite the advent of SLNB in 2005 (Table 1). It seems that awareness of the disease or introduction of clinical guidelines has not yet translated into practice and adherence to standardized treatment protocols.

Finland faces certain challenges in the treatment of rare cancers; a small nation of just over 5.5 million people is served by five university-level tertiary centers, leading to dispersion of patients, with the physical distance to treatment facilities potentially being hundreds of kilometers. The Finnish Cancer Registry actively participates in the RARECARE initiative [35], which aims to develop the surveillance and treatment of rare cancers across Europe. In order to reach optimal treatment results while managing the costs in rare diseases, standardizing treatment protocols and establishing strong leader centers with the required expertise are of the utmost importance. Our results point to the need for better interdisciplinary communication and education within tertiary treatment centres to meet the target of standardized treatment protocols. Our findings also clearly suggest that treatment of MCC should be discussed, executed and followed by a multidisciplinary tumour board [36] because the treatment requires a wide range of specialties, including dermatologists, surgeons, radio-oncologists, medical oncologists, pathologists and radiologists.

A significant advantage in MCC-specific survival was seen with the addition of RT to primary tumour location in conjunction with simple excision. The statistical significance was, however, lost when the patients were stratified according to stage or age. Small population sizes also limited the analytical power in other treatment groups. MCC, as other neuroendocrine carcinomas, is responsive to RT and adjuvant RT is advised in the NCCN guidelines in local disease except for small low risk tumours (NCCN). Reports on the survival benefit of adjuvant RT are, however, controversial [11,37–40]. To date, the only randomized controlled study on adjuvant RT was conducted by Jouary et al. [41]

on stage I MCC patients, in which RT to the tumour bed resulted in a significant decrease in local recurrence, but no significant improvement in overall survival. Subgroup analysis is oftentimes hindered in MCC studies by small sample sizes, and as such most of the clinical retrospective studies are weakened by the heterogeneity of comparable patient populations in terms of prior treatment, patient demographics, tumour characteristics and even stage of disease or target of RT. A SEER-based retrospective study postulated that the survival benefits seen in adjuvant RT are, in fact, the result of a selection bias [42].

Previous real-world studies on MCC have focused on metastatic disease [43,44]. Until recently, disseminated MCC was treated with various chemotherapeutic agents with poor results. In 2016, two separate studies proved the efficacy of immunotherapy in metastatic MCC with two molecules, avelumab and pembrolizumab [19,20]. Targeting the PD-L1/PD-1 pathway has revolutionized MCC treatment and survival and bypassed the use of cytotoxic agents in the advanced stages of MCC [17,45]. The role of RT in treatment of metastatic MCC is also undergoing a revolution. Before the advent of immunotherapy, RT was mainly seen as a palliative measure. The search to overcome refractory disease has brought with it a rise in reports on the abscopal effect in metastatic MCC [24,46]. RT modulates the tumour immunoediting process by various and partly unidentified mechanisms, such as altering tumour cell antigen presentation, increasing the infiltration of regulatory T-cells in the tumour microenvironment and staging of T-cell exhaustion with debulking the tumour mass [47,48]. Ongoing clinical trials (Clinical Trials Identifier NCT03071406, A091605) aim to shed light on the efficacy of combination treatment in advanced MCC. These might also provide new insights regarding treatment of early stage MCC.

A surprising notion arising from our large pooled data is that MCC-specific mortality seems to stabilize at seven years after the MCC diagnosis, along with the overall mortality, as seen in the Kaplan–Meier analysis (Figure 2). However, the current guidelines usually advise a five-year follow-up [16,17]. Traditionally, the survival of cancer patients has been presented at five years. Most of the adverse events occur during the first two years [49,50], with a median time to recurrence/relapse varying between 7 and 9 months [51,52].

Some strengths and limitations of the study warrant discussion. Coverage of the Finnish Cancer Registry is nearly 100% [53,54] of all the cancers diagnosed in Finland. Every hospital and pathology/haematology laboratory is required by legislation to submit data to the registry of all cancer patients brought to their attention. However, some individual cases might not be submitted to the Finnish Cancer Registry, and this might be especially true with unusual types of cancers like MCC, which is still poorly recognized. Likewise, the Finnish Hospital Discharge Register, maintained by the Finnish Institute of Health and Welfare, has repeatedly been shown to have completeness and accuracy levels from satisfactory to very good [55]. However, coding of the treatments in the registry is not primarily meant for research purposes or treatments; they are set as clinically indicated. Moreover, apart from data on sex, date of birth and death, we could not retrieve information on patient characteristics, including socioeconomic status or comorbidities. Prognostic tumour characteristics, such as tumour size and MCPyV status, are also beyond the scope of this study. AJCC staging was not employed in the data and staging was not updated beyond four months from diagnosis. More specific clinical findings, such as free margins and the number of positive sentinel lymph nodes, were not recorded.

All in all, our study protocol is best suited to charting the prevailing treatment patterns and general outcomes, but is unable to offer information underlying the treatment decisions such as patient comorbidities or tumour characteristics. MCC patients are elderly, and comorbidities likely explain the scarcity of invasive procedures in MCC treatment patterns. In the future, adding registry data on hospital stays and drug reimbursements could enable cost evaluation, which is needed in MCC with the advent of immunotherapy.

4. Materials and Methods

The study was conducted in accordance with the Declaration of Helsinki, and the protocol was approved by the Ethics Committee of The Helsinki University Hospital (Project identification code HUS/1455/2017). Permissions to identify MCC patients from the Finnish Cancer Registry, their treatment data from the National Hospital Discharge Register and data on death from the Cause-of-Death Register of Finland were obtained from the Finnish Institute for Health and Welfare and from Statistics Finland. Information from the different registers was merged through record linkages based on personal identity codes (PICs). All citizens and permanent residents in Finland have a unique PIC, which was introduced in 1964–1967. As the use of PICs enables the handling of registry data without the risk of patients being identified, patient permissions were not acquired.

In this register linkage study, the data were obtained from the Finnish Cancer Registry on all patients diagnosed with MCC from 1 January 1986 to 31 December 2016. The data included the following:

- Date of diagnosis
- Age at diagnosis
- ICD-O-3 topography
    - 440 Skin of lip, NOS, 441 Eyelid, 442 External ear, 443 Skin of other and unspecified parts of face, 444 Skin of scalp and neck, 445 Skin of trunk, 446 Skin of upper limb and shoulder, 447 Skin of lower limb and hip, 449 Skin, NOS
- Stage
    - 0 Unknown, 1 Localized, 2 Non-localized, only regional lymph node metastases, 3 Metastasized farther than to regional lymph nodes or invades adjacent tissues, 4 Non-localized, no information on extent, 5 Non-localized, also distant lymph node metastases. Stage of disease is recorded in the cancer registry files at four months after diagnosis and is not updated later.

The cohort was linked to the Cause-of-Death Register maintained by Statistics Finland. The closing date for data collection was 31 December 2016. The data included the following:

- Date of death
- Cause of death
- Deceased due to this cancer or due to other causes

The National Hospital Discharge Register maintained by the Finnish Institute of Health and Welfare was queried for the treatments given to these patients after diagnosis was assigned. The Finnish procedure coding is based on the Nordic Classification of Surgical Procedures (NCSP), which was introduced in 1997.

First, we listed all the procedures based on their frequency. In case there was no recorded treatment code and the codes were diagnostic, such as radiologic examinations, the patient was listed as having no treatment after the diagnostic biopsy date. The stratification of the procedure codes was verified by a senior author (VK) who reviewed all treatment codes and their classification case by case. The procedure codes were stratified to eight groups and to further subgroups:

1. Pre-operative RT before the re-excision of the primary tumour
2. Re-excision of primary tumour
3. SLNB
4. CLND, including partial and total parotidectomy
5. Post-operative treatment

   i. adjuvant RT to the primary tumour
   ii. adjuvant RT to the regional lymph nodes
   iii. adjuvant cytostatic therapy

6. Therapy for progressive malignancy

   i. RT of local recidive tumour
   ii. cytostatic therapy of local tumour
   iii. received RT of metastasis

7. Non-specified RT
8. Palliative treatment

The first SLNB was performed in 2005. Thus, the effect of RT on the survival and outcome was analyzed based on the data for 2005–2016. In this sub-cohort, we included patients whose date of diagnosis was in 2005 or later. To compare the effect of adjuvant RT in the treatment of MCC, we statistically compared re-excision, re-excision and SLNB, and CLND with and without adjuvant RT.

Statistical Analysis

All statistical comparisons were done by using the Chi-square test, the test of relative proportion and *t*-test, where appropriate. Kaplan–Meier analysis was performed to determine overall and disease-specific survival (MCC).

All analyses were done using SAS, version 9.3 (SAS Institute Inc., Cary, NC, USA).

5. Conclusions

The treatments patterns of MCC in Finland are highly heterogeneous and rarely follow the international treatment guidelines. RT was rare in all disease stages. Comparison of simple excision to excision combined with adjuvant RT showed an improved survival trend, but no statistical significance was found after stratification according to disease stage. Perhaps unsurprisingly, disease-specific survival has not increased despite advances in diagnostic procedures, SLNB and the advent of treatment guidelines. This registry study serves as a baseline treatment efficacy comparison as we move into the age of immunotherapy in MCC. Importantly, the advent of immunotherapy cannot compensate for the need for proper management at the early stages of disease. Standardizing the treatment of MCC patients in Finland requires more work on awareness and multidisciplinary co-operation.

Supplementary Materials: The following are available online at http://www.mdpi.com/2072-6694/12/5/1224/s1, Table S1: Procedures of MCC patients in 1986–2006 extracted from The National Hospital Discharge Register and listed by the procedure codes according to the Nordic Classification of Surgical Procedures.

Author Contributions: Conceptualization, V.K., H.S., M.G., J.T.; Methodology, H.S.,V.K., M.G.; Validation, M.G.; Formal Analysis, M.G.; Investigation, V.K. and M.G.; Resources, V.K. and M.G.; Data Curation, V.K. and M.G.; Writing—Original Draft Preparation, H.S., J.T. and V.K.; Writing—Review and Editing, H.S., J.T., V.K., M.G.; Visualization, H.S. and V.K.; Supervision, V.K. and M.G.; Project Administration, V.K. All authors have read and agreed to the published version of the manuscript.

Funding: This research received no external funding.

Conflicts of Interest: The authors declare no conflict of interest.

References

1. Becker, J.C. Merkel cell carcinoma. *Ann. Oncol.* **2010**, *21*, vii81–vii85. [CrossRef] [PubMed]
2. Swann, M.H.; Yoon, J. Merkel cell carcinoma. *Semin. Oncol.* **2007**, *34*, 51–56. [CrossRef] [PubMed]
3. Feng, H.; Shuda, M.; Chang, Y.; Moore, P.S. Clonal integration of a polyomavirus in human Merkel cell carcinoma. *Science* **2008**, *319*, 1096–1100. [CrossRef] [PubMed]
4. Erovic, I.; Erovic, B.M. Merkel cell carcinoma: The past, the present, and the future. *J. Ski. Cancer* **2013**, *2013*, 929364. [CrossRef]

5. Sihto, H.; Kukko, H.; Koljonen, V.; Sankila, R.; Bohling, T.; Joensuu, H. Clinical factors associated with Merkel cell polyomavirus infection in Merkel cell carcinoma. *J. Natl. Cancer Inst.* **2009**, *101*, 938–945. [CrossRef]
6. Erovic, B.M.; Al Habeeb, A.; Harris, L.; Goldstein, D.P.; Ghazarian, D.; Irish, J.C. Significant overexpression of the Merkel cell polyomavirus (MCPyV) large T antigen in Merkel cell carcinoma. *Head Neck* **2013**, *35*, 184–189. [CrossRef]
7. Paik, J.Y.; Hall, G.; Clarkson, A.; Lee, L.; Toon, C.; Colebatch, A.; Chou, A.; Gill, A.J. Immunohistochemistry for Merkel cell polyomavirus is highly specific but not sensitive for the diagnosis of Merkel cell carcinoma in the Australian population. *Hum. Pathol.* **2011**, *42*, 1385–1390. [CrossRef]
8. Heath, M.; Jaimes, N.; Lemos, B.; Mostaghimi, A.; Wang, L.C.; Penas, P.F.; Nghiem, P. Clinical characteristics of Merkel cell carcinoma at diagnosis in 195 patients: The AEIOU features. *J. Am. Acad. Dermatol.* **2008**, *58*, 375–381. [CrossRef]
9. Pellitteri, P.K.; Takes, R.P.; Lewis, J.S., Jr.; Devaney, K.O.; Harlor, E.J.; Strojan, P.; Rodrigo, J.P.; Suarez, C.; Rinaldo, A.; Medina, J.E.; et al. Merkel cell carcinoma of the head and neck. *Head Neck* **2012**, *34*, 1346–1354. [CrossRef]
10. Kukko, H.; Bohling, T.; Koljonen, V.; Tukiainen, E.; Haglund, C.; Pokhrel, A.; Sankila, R.; Pukkala, E. Merkel cell carcinoma—A population-based epidemiological study in Finland with a clinical series of 181 cases. *Eur. J. Cancer* **2012**, *48*, 737–742. [CrossRef]
11. Ezaldein, H.H.; Ventura, A.; DeRuyter, N.P.; Yin, E.S.; Giunta, A. Understanding the influence of patient demographics on disease severity, treatment strategy, and survival outcomes in merkel cell carcinoma: A surveillance, epidemiology, and end-results study. *Oncoscience* **2017**, *4*, 106–114. [CrossRef] [PubMed]
12. Becker, J.C.; Stang, A.; Hausen, A.Z.; Fischer, N.; DeCaprio, J.A.; Tothill, R.W.; Lyngaa, R.; Hansen, U.K.; Ritter, C.; Nghiem, P.; et al. Epidemiology, biology and therapy of Merkel cell carcinoma: Conclusions from the EU project IMMOMEC. *Cancer Immunol. Immunother. CII* **2018**, *67*, 341–351. [CrossRef] [PubMed]
13. Stang, A.; Becker, J.C.; Nghiem, P.; Ferlay, J. The association between geographic location and incidence of Merkel cell carcinoma in comparison to melanoma: An international assessment. *Eur. J. Cancer* **2018**, *94*, 47–60. [CrossRef] [PubMed]
14. Strom, T.; Carr, M.; Zager, J.S.; Naghavi, A.; Smith, F.O.; Cruse, C.W.; Messina, J.L.; Russell, J.; Rao, N.G.; Fulp, W.; et al. Radiation Therapy is Associated with Improved Outcomes in Merkel Cell Carcinoma. *Ann. Surg. Oncol.* **2016**, *23*, 3572–3578. [CrossRef]
15. Iyer, J.G.; Parvathaneni, U.; Gooley, T.; Miller, N.J.; Markowitz, E.; Blom, A.; Lewis, C.W.; Doumani, R.F.; Parvathaneni, K.; Anderson, A.; et al. Single-fraction radiation therapy in patients with metastatic Merkel cell carcinoma. *Cancer Med.* **2015**, *4*, 1161–1170. [CrossRef]
16. Lebbe, C.; Becker, J.C.; Grob, J.J.; Malvehy, J.; Del Marmol, V.; Pehamberger, H.; Peris, K.; Saiag, P.; Middleton, M.R.; Bastholt, L.; et al. Diagnosis and treatment of Merkel Cell Carcinoma. European consensus-based interdisciplinary guideline. *Eur. J. Cancer* **2015**, *51*, 2396–2403. [CrossRef]
17. Bichakjian, C.K.; Olencki, T.; Aasi, S.Z.; Alam, M.; Andersen, J.S.; Blitzblau, R.; Bowen, G.M.; Contreras, C.M.; Daniels, G.A.; Decker, R.; et al. Merkel Cell Carcinoma, Version 1.2018, NCCN Clinical Practice Guidelines in Oncology. *J. Natl. Compr. Cancer Netw. JNCCN* **2018**, *16*, 742–774. [CrossRef]
18. Becker, J.C.; Stang, A.; DeCaprio, J.A.; Cerroni, L.; Lebbe, C.; Veness, M.; Nghiem, P. Merkel cell carcinoma. *Nat. Rev. Dis. Primers* **2017**, *3*, 17077. [CrossRef]
19. Kaufman, H.L.; Russell, J.; Hamid, O.; Bhatia, S.; Terheyden, P.; D'Angelo, S.P.; Shih, K.C.; Lebbé, C.; Linette, G.P.; Milella, M.; et al. Avelumab in patients with chemotherapy-refractory metastatic Merkel cell carcinoma: A multicentre, single-group, open-label, phase 2 trial. *Lancet Oncol.* **2016**, *17*, 1374–1385. [CrossRef]
20. Nghiem, P.; Bhatia, S.; Lipson, E.J.; Kudchadkar, R.R.; Miller, N.J.; Annamalai, L.; Berry, S.; Chartash, E.K.; Daud, A.; Fling, S.P.; et al. PD-1 blockade with pembrolizumab in advanced Merkel-cell carcinoma. *N. Engl. J. Med.* **2016**, *374*, 2542–2552. [CrossRef]
21. Garbutcheon-Singh, K.B.; Veness, M.J. The role of radiotherapy in the management of non-melanoma skin cancer. *Australas. J. Dermatol.* **2019**, *60*, 265–272. [CrossRef] [PubMed]
22. Petrelli, F.; Ghidini, A.; Torchio, M.; Prinzi, N.; Trevisan, F.; Dallera, P.; De Stefani, A.; Russo, A.; Vitali, E.; Bruschieri, L.; et al. Adjuvant radiotherapy for Merkel cell carcinoma: A systematic review and meta-analysis. *Radiother. Oncol.* **2019**, *134*, 211–219. [CrossRef] [PubMed]

23. Mullen, T.D.; Nghiem, P.; Bhatia, S.; Parvathaneni, U.; Schwartz, J.L. Merkel Cell Carcinoma Cell Death in Response to Single-Fraction Irradiation: Implications for Immune Therapy. *Int. J. Radiat. Oncol. Biol. Phys.* **2016**, *96*, E714. [CrossRef]
24. Xu, M.J.; Wu, S.; Daud, A.I.; Yu, S.S.; Yom, S.S. In-field and abscopal response after short-course radiation therapy in patients with metastatic Merkel cell carcinoma progressing on PD-1 checkpoint blockade: A case series. *J. Immunother. Cancer* **2018**, *6*, 43. [CrossRef]
25. Schadendorf, D.; Lebbe, C.; Zur Hausen, A.; Avril, M.F.; Hariharan, S.; Bharmal, M.; Becker, J.C. Merkel cell carcinoma: Epidemiology, prognosis, therapy and unmet medical needs. *Eur. J. Cancer* **2017**, *71*, 53–69. [CrossRef]
26. Toker, C. Trabecular carcinoma of the skin. *Arch. Dermatol.* **1972**, *105*, 107–110. [CrossRef]
27. Agelli, M.; Clegg, L.X. Epidemiology of primary Merkel cell carcinoma in the United States. *J. Am. Acad. Dermatol.* **2003**, *49*, 832–841. [CrossRef]
28. Agelli, M.; Clegg, L.X.; Becker, J.C.; Rollison, D.E. The etiology and epidemiology of merkel cell carcinoma. *Curr. Probl. Cancer* **2010**, *34*, 14–37. [CrossRef]
29. Koljonen, V. Prognostic factors in primary merkel cell carcinoma. In *Dissertationes Scholae Doctoralis ad Sanitatem Investigandam Universitatis Helsinkiensis*; Univ. Printing House: Helsinki, Finland, 2004.
30. Koljonen, V.; Tarkkanen, M.; Tukiainen, E.; Bohling, T. Merkel cell carcinoma in elderly. *Duodecim* **2005**, *121*, 2205–2213.
31. Bichakjian, C.K.; Olencki, T.; Alam, M.; Andersen, J.S.; Berg, D.; Bowen, G.M.; Cheney, R.T.; Daniels, G.A.; Glass, L.F.; Grekin, R.C.; et al. Merkel cell carcinoma, version 1.2014. *J. Natl. Compr. Cancer Netw. JNCCN* **2014**, *12*, 410–424. [CrossRef]
32. Miller, S.J.; Alam, M.; Andersen, J.; Berg, D.; Bichakjian, C.K.; Bowen, G.; Cheney, R.T.; Glass, L.F.; Grekin, R.C.; Hallahan, D.E.; et al. NCCN Clinical Practice Guidelines in Oncology: Merkel Cell Carcinoma. *J. Natl. Compr. Cancer Netw. JNCCN* **2009**, *7*, 322–332. [CrossRef] [PubMed]
33. Sahi, H.; Bohling, T.; Koljonen, V. Merkelinsolukarsinooma—Mitä uutta? *Duodecim* **2017**, *133*, 2365–2371.
34. Steuten, L.; Garmo, V.; Phatak, H.; Sullivan, S.D.; Nghiem, P.; Ramsey, S.D. Treatment Patterns, Overall Survival, and Total Healthcare Costs of Advanced Merkel Cell Carcinoma in the USA. *Appl. Health Econ. Health Policy* **2019**, *17*, 733–740. [CrossRef] [PubMed]
35. Gatta, G.; Ciccolallo, L.; Kunkler, I.; Capocaccia, R.; Berrino, F.; Coleman, M.P.; De Angelis, R.; Faivre, J.; Lutz, J.M.; Martinez, C.; et al. Survival from rare cancer in adults: A population-based study. *Lancet Oncol.* **2006**, *7*, 132–140. [CrossRef]
36. Schneider, S.; Thurnher, D.; Erovic, B.M. Merkel cell carcinoma: Interdisciplinary management of a rare disease. *J. Ski. Cancer* **2013**, *2013*, 189342. [CrossRef]
37. Ghadjar, P.; Kaanders, J.H.; Poortmans, P.; Zaucha, R.; Krengli, M.; Lagrange, J.L.; Ozsoy, O.; Nguyen, T.D.; Miralbell, R.; Baize, A.; et al. The essential role of radiotherapy in the treatment of Merkel cell carcinoma: A study from the Rare Cancer Network. *Int. J. Radiat. Oncol. Biol. Phys.* **2011**, *81*, e583–e591. [CrossRef]
38. Howle, J.R.; Hughes, T.M.; Gebski, V.; Veness, M.J. Merkel cell carcinoma: An Australian perspective and the importance of addressing the regional lymph nodes in clinically node-negative patients. *J. Am. Acad. Dermatol.* **2012**, *67*, 33–40. [CrossRef]
39. Han, A.Y.; Patel, P.B.; Anderson, M.; Diaz, M.F.P.; Chin, R.; St John, M.A. Adjuvant radiation therapy improves patient survival in early-stage merkel cell carcinoma: A 15-year single-institution study. *Laryngoscope* **2018**, *128*, 1862–1866. [CrossRef]
40. Fiedler, E.; Vordermark, D. Outcome of Combined Treatment of Surgery and Adjuvant Radiotherapy in Merkel Cell Carcinoma. *Acta Derm. Venereol.* **2018**, *98*, 699–703. [CrossRef]
41. Jouary, T.; Leyral, C.; Dreno, B.; Doussau, A.; Sassolas, B.; Beylot-Barry, M.; Renaud-Vilmer, C.; Guillot, B.; Bernard, P.; Lok, C.; et al. Adjuvant prophylactic regional radiotherapy versus observation in stage I Merkel cell carcinoma: A multicentric prospective randomized study. *Ann. Oncol.* **2012**, *23*, 1074–1080. [CrossRef]
42. Kim, J.A.; Choi, A.H. Effect of radiation therapy on survival in patients with resected Merkel cell carcinoma: A propensity score surveillance, epidemiology, and end results database analysis. *JAMA Dermatol.* **2013**, *149*, 831–838. [CrossRef] [PubMed]

43. Becker, J.C.; Lorenz, E.; Ugurel, S.; Eigentler, T.K.; Kiecker, F.; Pfohler, C.; Kellner, I.; Meier, F.; Kahler, K.; Mohr, P.; et al. Evaluation of real-world treatment outcomes in patients with distant metastatic Merkel cell carcinoma following second-line chemotherapy in Europe. *Oncotarget* **2017**, *8*, 79731–79741. [CrossRef] [PubMed]
44. Cowey, C.L.; Mahnke, L.; Espirito, J.; Helwig, C.; Oksen, D.; Bharmal, M. Real-world treatment outcomes in patients with metastatic Merkel cell carcinoma treated with chemotherapy in the USA. *Future Oncol.* **2017**, *13*, 1699–1710. [CrossRef] [PubMed]
45. Becker, J.C.; Eigentler, T.; Frerich, B.; Gambichler, T.; Grabbe, S.; Holler, U.; Klumpp, B.; Loquai, C.; Krause-Bergmann, A.; Muller-Richter, U.; et al. S2k guidelines for Merkel cell carcinoma (MCC, neuroendocrine carcinoma of the skin)—Update 2018. *J. Dtsch. Dermatol. Ges.* **2019**, *17*, 562–576. [CrossRef]
46. Bloom, B.C.; Augustyn, A.; Pezzi, T.A.; Menon, H.; Mayo, L.L.; Shah, S.J.; Schwartz, D.L.; Chmura, S.J.; Johnson, F.M.; Welsh, J.W.; et al. Rescue of Immunotherapy-Refractory Metastatic Merkel Cell Carcinoma With Conventionally Fractionated Radiotherapy and Concurrent Pembrolizumab. *Front. Oncol.* **2019**, *9*, 223. [CrossRef]
47. Filatenkov, A.; Baker, J.; Mueller, A.M.; Kenkel, J.; Ahn, G.O.; Dutt, S.; Zhang, N.; Kohrt, H.; Jensen, K.; Dejbakhsh-Jones, S.; et al. Ablative Tumor Radiation Can Change the Tumor Immune Cell Microenvironment to Induce Durable Complete Remissions. *Clin. Cancer Res.* **2015**, *21*, 3727–3739. [CrossRef]
48. Garnett, C.T.; Palena, C.; Chakraborty, M.; Tsang, K.Y.; Schlom, J.; Hodge, J.W. Sublethal irradiation of human tumor cells modulates phenotype resulting in enhanced killing by cytotoxic T lymphocytes. *Cancer Res.* **2004**, *64*, 7985–7994. [CrossRef]
49. Harrington, C.; Kwan, W. Outcomes of Merkel cell carcinoma treated with radiotherapy without radical surgical excision. *Ann. Surg. Oncol.* **2014**, *21*, 3401–3405. [CrossRef]
50. Bajetta, E.; Celio, L.; Platania, M.; Lo Vullo, S.; Patuzzo, R.; Maurichi, A.; Santinami, M. Single-Institution Series of Early-Stage Merkel Cell Carcinoma: Long-Term Outcomes in 95 Patients Managed with Surgery Alone. *Ann. Surg. Oncol.* **2009**, *16*, 2985–2993. [CrossRef]
51. Tello, T.L.; Coggshall, K.; Yom, S.S.; Yu, S.S. Merkel cell carcinoma: An update and review: Current and future therapy. *J. Am. Acad. Dermatol.* **2018**, *78*, 445–454. [CrossRef]
52. Santamaria-Barria, J.A.; Boland, G.M.; Yeap, B.Y.; Nardi, V.; Dias-Santagata, D.; Cusack, J.C., Jr. Merkel cell carcinoma: 30-year experience from a single institution. *Ann. Surg. Oncol.* **2013**, *20*, 1365–1373. [CrossRef] [PubMed]
53. Teppo, L.; Pukkala, E.; Lehtonen, M. Data quality and quality control of a population-based cancer registry. Experience in Finland. *Acta Oncol.* **1994**, *33*, 365–369. [CrossRef] [PubMed]
54. Teppo, L.; Pukkala, E.; Saxen, E. Multiple cancer–an epidemiologic exercise in Finland. *J. Natl. Cancer Inst.* **1985**, *75*, 207–217. [PubMed]
55. Sund, R. Quality of the Finnish Hospital Discharge Register: A systematic review. *Scand. J. Public Health* **2012**, *40*, 505–515. [CrossRef] [PubMed]

cancers

MDPI

Article

Artesunate Affects T Antigen Expression and Survival of Virus-Positive Merkel Cell Carcinoma

Bhavishya Sarma [1], Christoph Willmes [1], Laura Angerer [1], Christian Adam [1], Jürgen C. Becker [2,3], Thibault Kervarrec [4], David Schrama [1,†] and Roland Houben [1,*,†]

1 Department of Dermatology, University Hospital Würzburg, 97082 Würzburg, Germany; Sarma_B@ukw.de (B.S.); christoph.willmes@googlemail.com (C.W.); laura-angerer@gmx.de (L.A.); Adam_C@ukw.de (C.A.); Schrama_d@ukw.de (D.S.)
2 Department of Translational Skin Cancer Research, Dermatology, University Hospital Essen, 45147 Essen, Germany; j.becker@dkfz-heidelberg.de
3 German Cancer Consortium (DKTK), Partner Site Essen and the German Cancer Research Center (DKFZ), 69115 Heidelberg, Germany
4 Department of Pathology, Centre Hospitalier Universitaire De Tours, INRA UMR 1282 BIP, 37200 Tours, France; thibaultkervarrec@yahoo.fr
* Correspondence: houben_r@ukw.de; Tel.: +49-931-2012-6756
† These authors contributed equally to this work.

Received: 19 March 2020; Accepted: 7 April 2020; Published: 9 April 2020

Abstract: Merkel cell carcinoma (MCC) is a rare and highly aggressive skin cancer with frequent viral etiology. Indeed, in about 80% of cases, there is an association with Merkel cell polyomavirus (MCPyV); the expression of viral T antigens is crucial for growth of virus-positive tumor cells. Since artesunate—a drug used to treat malaria—has been reported to possess additional anti-tumor as well as anti-viral activity, we sought to evaluate pre-clinically the effect of artesunate on MCC. We found that artesunate repressed growth and survival of MCPyV-positive MCC cells in vitro. This effect was accompanied by reduced large T antigen (LT) expression. Notably, however, it was even more efficient than shRNA-mediated downregulation of LT expression. Interestingly, in one MCC cell line (WaGa), T antigen knockdown rendered cells less sensitive to artesunate, while for two other MCC cell lines, we could not substantiate such a relation. Mechanistically, artesunate predominantly induces ferroptosis in MCPyV-positive MCC cells since known ferroptosis-inhibitors like DFO, BAF-A1, Fer-1 and β-mercaptoethanol reduced artesunate-induced death. Finally, application of artesunate in xenotransplanted mice demonstrated that growth of established MCC tumors can be significantly suppressed in vivo. In conclusion, our results revealed a highly anti-proliferative effect of the approved and generally well-tolerated anti-malaria compound artesunate on MCPyV-positive MCC cells, suggesting its potential usage for MCC therapy.

Keywords: artesunate; Merkel cell carcinoma; MCC; polyomavirus; ferroptosis

1. Introduction

Merkel cell carcinoma (MCC) is an aggressive neuroendocrine skin cancer with increasing incidence and mortality rates [1]. The most recent analysis reported 0.7 new cases per 100,000 person-years in 2013 in the United States of America with a predicted 14% increase in cases for 2020 [2], and 0.43 MCC related deaths per 100,000 were reported for 2011 in another study [3]. 95% of the patients with MCC are more than 50 years old, and the tumors commonly develop in chronically sun exposed body areas [4]. Since immunosuppression is a further known risk factor for MCC, an infectious etiology had been suspected [5]. Indeed, in 2008, a human polyomavirus named Merkel cell polyomavirus (MCPyV) was found to be integrated into the genome of Merkel cell carcinoma cells [6], and subsequent studies confirmed that approximately 80% of all MCC cases are associated with MCPyV [7]. Importantly, the integration patterns suggest that clonal expansion of the tumor cells occurs after MCPyV integration sustaining the assumption that viral proteins are causal for tumorigenesis [6,8,9]. Moreover, in MCPyV-positive MCC cells, expression of the viral oncoproteins small and Large T-antigen (sT and LT) can be detected, and these proteins are essential for growth of the tumor cells [10,11] qualifying them as potential therapeutic targets.

The five-year overall survival rate for patients with MCC is only about 40%, although the relative survival rate (compared to an age- and sex-matched population) is 54% [12]. Primary MCCs are excised by surgery, and adjuvant radiotherapy of the primary tumor location and the lymph node region is recommended [13]. Until recently, the metastatic disease was treated preferentially with various, not-standardized chemotherapeutic regimens, all of which could not improve survival of the patients significantly [14]. Recently, however, antibodies targeting the immune suppressive protein programmed cell death protein 1 (PD-1) or its ligand PD-L1 have demonstrated high response rates of 56 in first-line and 32% in second-line treatment, respectively for patients with stage IV disease [15,16]. Indeed, the PD-L1 targeting antibody Avelumab was the first treatment for metastatic MCC approved both in the US and European Union [17]. Importantly, data available so far suggest that responses of MCC patients to checkpoint inhibition are frequently long-lasting [18,19]. However, despite this encouraging progress, many patients do not respond and a substantial number of patients develop early secondary resistance [18,20]. Therefore, there is strong need for therapeutic approaches for patients' refractory to immune checkpoint inhibition. Furthermore, in developing countries there is a particular need for alternative MCC treatment options, since the high costs of checkpoint antibodies may limit their usage [21]

Artesunate is a semi-synthetic derivative of artemisinin, the active ingredient of the traditional Chinese medicinal herb *Artemisia annua* [22]. Artesunate is applied as first-line drug for the treatment of malaria which is caused by an infection with protozoa of the genus *Plasmodium* [23]. Although artesunate represents the most effective and safe anti-malarial drug [24,25], its mode of action is only incompletely understood [26]. Interestingly, artesunate has also been demonstrated to be specifically cytotoxic to cancer cells from several tumor entities [27,28]. This cytotoxicity was ascribed to artesunate impacting a multitude of signaling pathways and cell death modes [22]. For the latter, induction of apoptosis [29–31] or ferroptotic cell death [32–34] have been reported most frequently. Importantly, besides these anti-cancer effects, it also exerts anti-viral activities towards a broad range of viruses [35,36]. Therefore, we examined whether MCPyV-associated MCC cells are sensitive to this compound.

Here we demonstrate that artesunate effectively induces cell death of MCPyV-positive MCC cells in vitro mainly through ferroptosis, while apoptosis appears not to be involved. Moreover, in a mouse model, we demonstrate that artesunate can be applied to inhibit MCC tumor growth *in vivo*.

2. Results

2.1. Artesunate Effectively Inhibits Growth of MCPyV-Positive MCC Cell Lines In Vitro

Artesunate has been shown to mediate both anti-viral and anti-tumor activity [28,36]. Due to the viral carcinogenesis of most MCCs, we tested in an initial experiment, the effect of artesunate on a panel of MCPyV-positive classical MCC cell lines and some non-classical MCPyV-negative MCC cell lines. Melanoma cell lines and primary fibroblasts were included as further controls. The drug was used at concentrations of 1 and 10 μM and its effect on cell growth and metabolism was determined by the MTS assay. While growth and survival of primary fibroblasts and melanoma cell lines was largely unaffected at the given concentration, in particular the MTS signals of the MCPyV-positive MCC cell lines WaGa and MKL-1 were largely reduced (Supplementary Figure S1).

2.2. Reduced Large T Antigen Expression in Response to Artesunate

MCPyV-positive MCC cell lines depend on expression of the viral T antigens and in particular LT for growth [37,38]. Therefore, seeking for a potentially virus-related mechanism of growth inhibition induced by artesunate, we analyzed whether it affects LT expression. Indeed, in all five analyzed MCC cell lines, immunoblot analysis revealed decreased LT protein expression upon a three-day incubation with artesunate (Figure 1a; Supplementary Figure S2a).

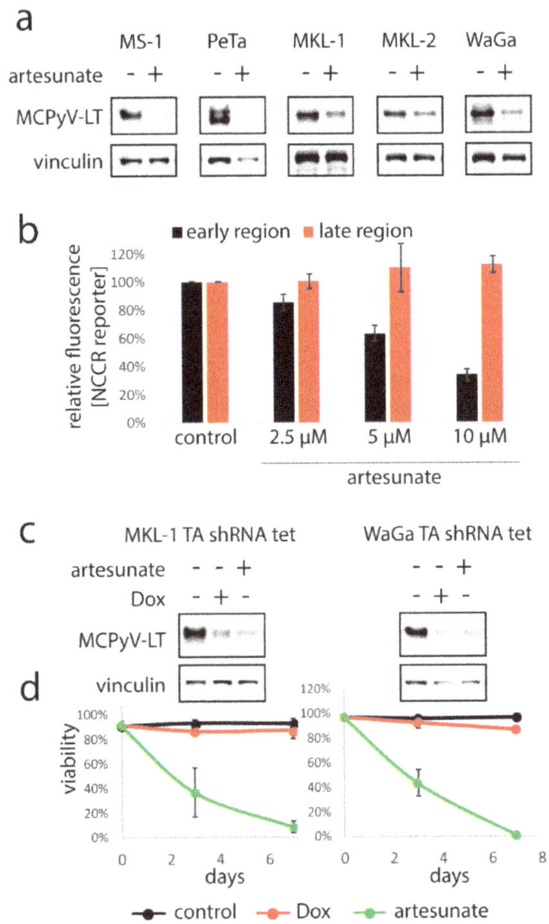

Figure 1. Artesunate-induced repression of MCPyV-LT expression in MCC cells is not crucial for its cytotoxic effects. (**a**) The indicated MCPyV-positive Merkel cell carcinoma (MCC) cell lines were incubated for three days in the absence or presence of artesunate (10 µM for MKL-1, MKL-2 and WaGa and 12.5 µM for MS-1 and PeTa) followed by immunoblot analysis. (**b**) MKL-1 cells stably transduced with a bi-directional non-coding control region (NCCR) reporter construct were treated for five days with the indicated artesunate concentrations followed by flow cytometric analysis. Mean fluorescence for early and late region were recorded, and mean values (± SD) are displayed. (**c**,**d**) MKL-1 and WaGa cells stably transduced with a vector allowing doxycyclin (Dox)-inducible expression of an shRNA targeting MCPyV TA were treated either with Dox (1 µM) or artesunate (10 µM) for 7 days, respectively. (**c**) large T antigen (LT) expression was analyzed by immunoblot. (**d**) Trypan blue exclusion assay was applied to determine viability in the course of time. Mean values (± SD) of at least four independent experiments are depicted.

To investigate whether artesunate affects the promoter driving T antigen expression, we made use of a reporter construct in which the bi-directional MCPyV non-coding control region (NCCR) controls expression of a green and a red fluorescent protein representing the early and late region, respectively. Indeed, MKL-1 cells transduced with the reporter demonstrated a dose dependent reduction of green fluorescence upon treatment with artesunate, while red fluorescence was not affected (Figure 1b; Supplementary Figure S3) suggesting that artesunate may specifically downregulate LT via repression of its NCCR-dependent transcription.

2.3. Artesunate Exerts Stronger Cytotoxic Effects on MCC Cells than TA Knockdown

Next, we asked whether the inhibition of T antigen (TA) expression could be a crucial mediator of the artesunate-induced effects on MCC cells. To answer this question, we compared loss of viability following artesunate treatment with cell death induced upon shRNA-mediated TA knockdown. To this end, MKL-1 and WaGa cells transduced with a lentiviral vector allowing doxycyclin-inducible expression of an shRNA targeting both T antigens were used. Addition of doxycyclin to these cells led to an efficient knockdown evident by reduced LT in immunoblot analysis (Figure 1c; Supplementary Figure S2b), which however was associated with only a minor increase in dead cells as assessed by the trypan blue exclusion assay (Figure 1d). In contrast, incubation with 10 μM artesunate, which was associated with a similar level of LT reduction (Figure 1c; Supplementary Figure S2b), induced massive cell death within 7 days. These results argue against repression of TA expression being the sole mechanism for the observed artesunate-mediated cytotoxicity on MCPyV-positive MCC cell lines.

2.4. Expression of the T Antigens Sensitizes the MCPyV-Positive Cell Line WaGa to Artesunate

As the results so far did not exclude a possible role of MCPyV for the artesunate-induced cytotoxicity, we analyzed next whether TA knockdown in MCPyV-positive MCC cells may affect their artesunate sensitivity. Since many cytotoxic drugs are less effective against non-proliferating cells [39], we used MKL-1, MKL-2 and WaGa cells which in addition to the inducible TA shRNA, constitutively express a Retinoblastoma protein 1 (RB1) shRNA rescuing the growth arrest induced by LT knockdown [38]. Control cells without doxycycline treatment and cells incubated for four days with doxycycline to repress T-antigen expression (Figure 2a) were then treated with artesunate ranging from 1.6 to 50 μM. Two assays, namely the trypan blue dye exclusion assay and DNA staining using propidium iodide were used to analyze cell viability of both groups. Interestingly, both assays demonstrated that WaGa cells with repressed T antigen showed increased cell viability upon artesunate treatment compared to the respective controls without the knockdown (Figure 2b). Therefore, T antigen expression seems to sensitize WaGa cells to artesunate induced cell death. For MKL-1 and MKL-2 cells, however, T antigen knockdown did not alter their sensitivity towards artesunate (Figure 2b).

In addition, artesunate induced cell death was preceded by a G2/M arrest (Supplementary Figure S4), while TA knockdown has been demonstrated to cause an arrest in G1 [11], further sustaining the conclusion that artesunate has important impacts on MCC cells in addition to T antigen repression.

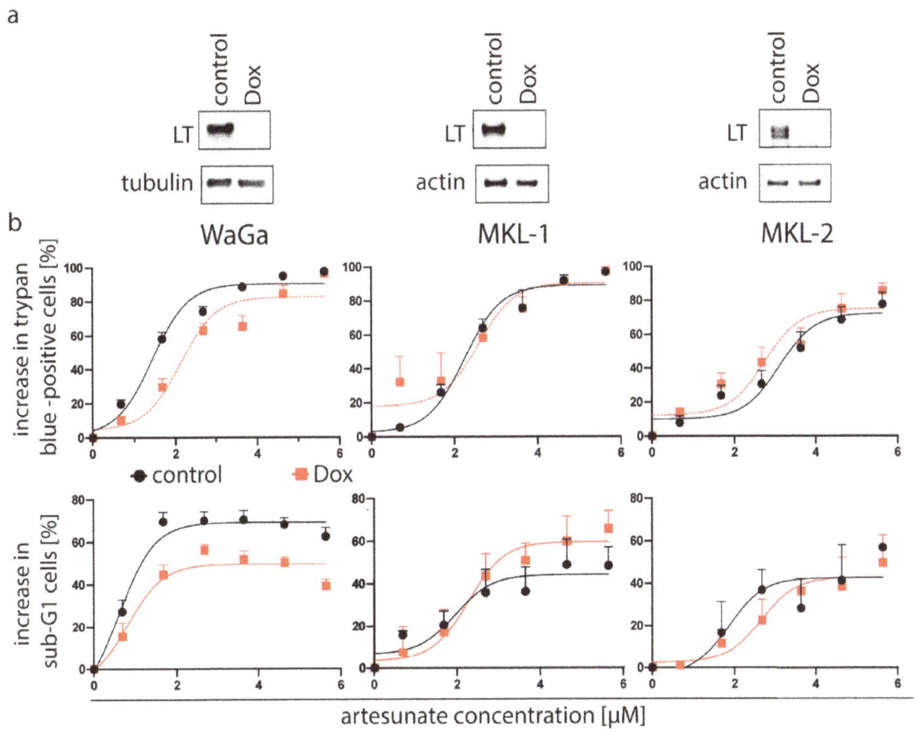

Figure 2. T antigen knockdown is associated with decreased artesunate sensitivity of WaGa but not MKL-1 and MKL-2 cells. We used the indicated cell lines which were stably transduced with a vector allowing doxycyclin (Dox)-inducible expression of a T antigen (TA) shRNA as well as with a vector constitutively expressing an RB1 shRNA. (**a**) Following 5 days in the presence or absence of Dox (1 µM) TA knockdown was evaluated by immunoblot analysis. (**b**) Then artesunate dose-response curves were recorded for control and Dox-treated cells applying the trypan blue exclusion assay as well as determination of the Sub-G1 population following propidium iodide staining of fixed cells. Displayed are mean values (+ SE) of at least three independent experiments.

2.5. No Signs of Apoptotic Cell Death Are Induced by Artesunate in Most MCPyV-Positive MCC Cell Lines

To further scrutinize artesunate's cytotoxicity towards MCPyV-positive MCC cells, we recorded dose response curves for five MCC cell lines applying two different cell death assays. Interestingly, we observed for four of the five cell lines, a significant difference between cell death induction as assayed by trypan blue exclusion compared to the appearance of a sub-G1 population in particular at higher artesunate concentrations (Figure 3a; Supplementary Figure S5). Indeed, cells with DNA less than 2N were less frequent than cells that had lost membrane integrity. This suggests that artesunate-induced death is not preceded by DNA fragmentation, a well-known characteristic of apoptosis [40]. Hence, apoptosis, a frequently described result of artesunate treatment in cancer cells [29–31,41], seems not to represent a crucial mechanism in these MCPyV-positive MCC cell lines. Only for MKL-2, no difference could be observed between the two dose response curves suggesting a possible contribution of apoptotic cell death.

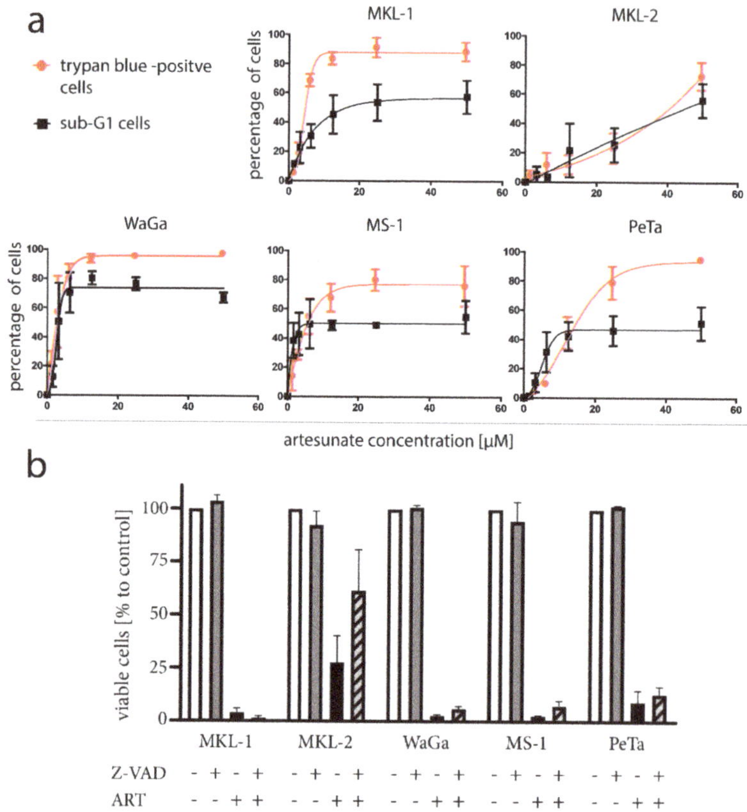

Figure 3. No signs of apoptotic cell death in most artesunate-treated MCC cells. (**a**) The indicated cell lines were treated for three days with increasing concentrations of artesunate. Then cell death was measured by the trypan blue exclusion assay. Additionally, cells were fixed and stained with propidium iodide to determine the increase in cells with a DNA content of less than 2N (sub-G1) (**b**) Cells were treated with 50 μM artesunate (ART) in the presence and absence of 20 μM of the caspase inhibitor Z-VAD. Viability was assessed by the trypan blue exclusion assay. Statistical testing applying ANOVA did not reveal significant differences.

To further evaluate these findings, we applied the pan caspase inhibitor benzyloxycarbonyl -ValAla-Asp (OMe) fluoromethylketone (Z-VAD-FMK), which bears the capability to suppress caspase-dependent apoptosis [42]. Although for MKL-2 an increase of viable cells in the presence of Z-VAD-FMK was observed, a significant rescue from artesunate induced cell death could not be detected for any of the five MCC cell lines (Figure 3b; differences tested with ANOVA and subsequent post hoc tests comparing values to those of artesunate-treated cells).

Finally, we investigated morphologic changes associated with artesunate treatment of MCPyV-positive MCC cell lines since apoptosis is characterized by characteristic features like cell shrinkage, membrane blebbing and formation of apoptotic bodies [43,44]. However, none of these characteristics were detectable when we analyzed the two non-spheroidal cell lines WaGa and PeTa by time lapse microscopy. Indeed, upon artesunate treatment, the opposite of shrinkage, i.e., cell swelling, was observed before death occurred (Supplementary Figure S6).

In conclusion, several observations suggest that at least in most artesunate-treated MCPyV-positive cell lines, apoptosis is not induced, and the morphologic feature of cell swelling hints to either necroptosis or ferroptosis provoked by artesunate [44,45].

2.6. Ferroptosis as a Key Player in Artesunate-Induced Cytotoxicity in MCPyV-Positive Cells

Previous studies had revealed the capability of artesunate to induce ferroptosis, an iron-dependent cell death mode characterized by lipid peroxidation [32–34]. Therefore, we next applied several specific inhibitors to test for ferroptotic features of artesunate-treated MCC cells. In this regard, rescue from cell death by the radical-trapping antioxidant ferrostatin-1 (Fer-1) which blocks lipid peroxidation [46] is regarded as one of the features defining ferroptosis [47]. Indeed, in all investigated MCC cell lines artesunate-induced cell death was significantly reduced by Fer-1. In addition, inhibition of artesunate-triggered viability loss by the iron-chelator deferoxamine (DFO) confirmed a ferroptotic process (Figure 4a).

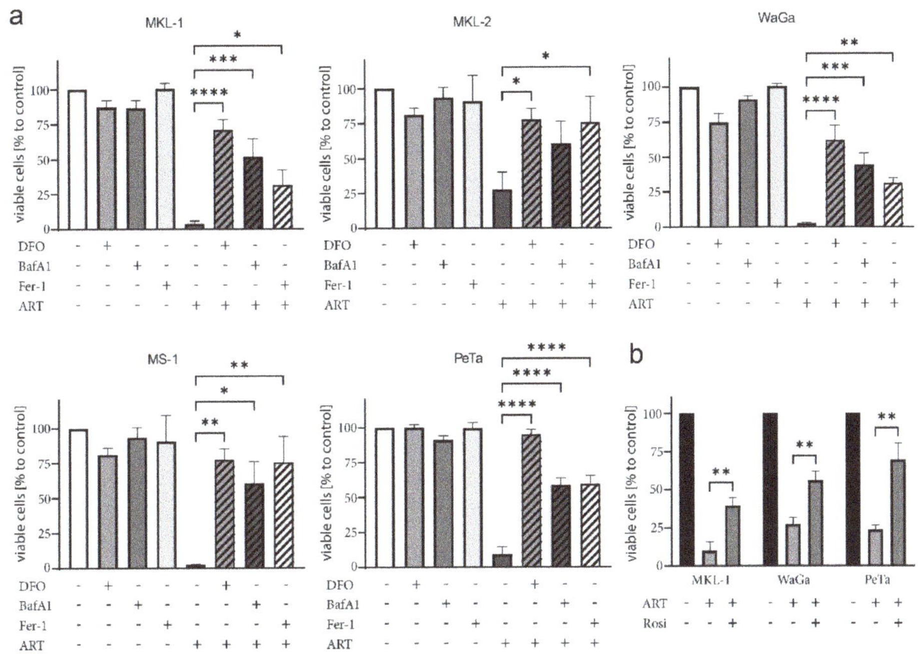

Figure 4. Ferroptosis inhibitors rescue MCPyV-positive MCC cells from artesunate-induced cell death. The indicated MCPyV-positive MCC cell lines were cultured in the absence or presence of 50 µM artesunate (ART). Additionally, either 10 µM of the radical-trapping antioxidant ferrostatin-1 (Fer-1), 100 µM of the iron-chelator deferoxamine (DFO), 50 nM of the autophagy inhibitor bafilomycin-A1 (BAF-A1) (**a**) or 25 µM of the ACSL4 inhibitor rosiglitazone (Rosi) (**b**) were included in the culture medium. After two days of co-treatment, viability was assessed by the trypan blue exclusion assay. Mean values (± SD) of at least three independent experiments are displayed. The effect of multiple treatment and inhibitor combinations was tested by ANOVA followed by post-hoc test comparing the effect always against the one observed for artesunate treatment. (* $p < 0.05$; ** $p < 0.01$; *** $p < 0.001$; **** $p < 0.0001$).

Furthermore, the effect of the vacuolar ATPase inhibitor bafilomycin-A1 (BAF-A1) in combination with artesunate was investigated. Multifaceted outcomes, like apoptosis induction or inhibition of autophagy, have been described for BAF-A1 [48,49]. However, BAF-A1 has also been observed to suppress ferroptosis, giving rise to one of the arguments linking autophagy to the ferroptotic process [47,50,51]. Such a link appears to exist also in MCC cell lines since among the tested inhibitors, BAF-A1 most efficiently suppressed artesunate-induced cell death in the MCPyV-positive MCC cell lines (Figure 4a).

A further reported step essential for ferroptosis is the inhibition of cystine import, which is necessary for antioxidant production [52,53]. In line with the notion that artesunate-induced cell death requires reduced cystine import, β-mercaptoethanol, which promotes cystine uptake [54], repressed cell death in artesunate-treated MCC cells (Supplementary Figure S7).

Finally, we tested rosiglitazone (Rosi), an inhibitor of the Acyl-CoA synthetase long-chain family member 4 (ACSL4). This enzyme has been demonstrated to be involved in ferroptosis execution by converting long-chain poly-unsaturated fatty acids (PUFAs) to their corresponding fatty acyl-CoA variants [55,56]. Indeed, Rosi exerted a protective effect on all three tested artesunate-treated MCC cell lines (Figure 4b).

These results suggest that artesunate kills MCPyV-positive MCC cells by dysregulating lipid metabolism and autophagy resulting in ferroptosis.

2.7. Artesunate Inhibits Tumor Growth In Vivo

To evaluate whether artesunate can affect growth of MCPyV-positive tumors in a living organism, we used xenotransplantation mouse models based on subcutaneous transplantation of the cell lines MKL-1 or WaGa [57]. Following injection of the tumor cells, the animals were monitored until they developed visible and palpable tumors measuring approximately 150 mm^3. Subsequently, 100 mg/kg body weight artesunate was administered intraperitoneally while control mice received the same volume of vehicle control. Artesunate treatment significantly reduced tumor growth of both MKL-1 and WaGa tumors (Figure 5).

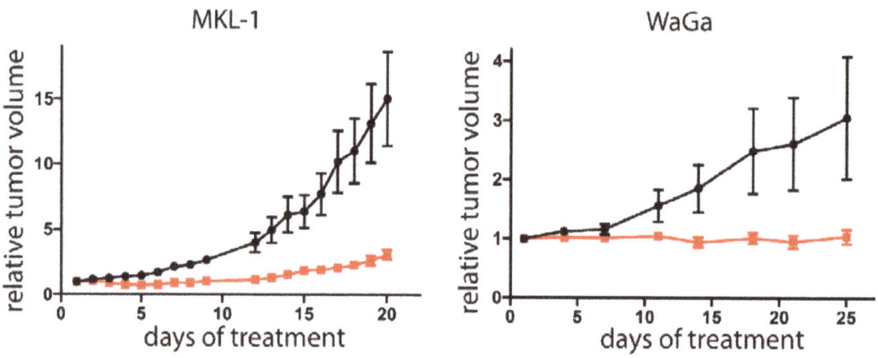

Figure 5. Tumor growth is restricted in artesunate-treated mice. Immunodeficient NOD/Scid mice received subcutaneous injection of either MKL-1 or WaGa cells. When tumors reached a size of 100 mm^3, the mice were randomly assigned to control group (n = 6 for WaGa and n = 5 for MKL-1, since in one animal no tumor growth was observed) or treatment group (n = 6). Each mouse from the treatment group was subjected to daily intraperitoneal injections with 100 mg/kg artesunate. The control group received injection of an equal volume of solvent (2% DMSO in PBS). The experiment was terminated once individual tumors of the control group reached the maximum tolerable size. Depicted are the means (± SEM). Statistical analyses of area under the curves for the two models were $p < 0.001$ for MKL-1 and 0.0305 for WaGa (unpaired *t*-test).

3. Discussion

The term drug repositioning (also called drug repurposing) describes the use of established drugs for new therapeutic purposes. Drug repositioning is a well-established process approved by regulatory agencies that allows fast identification of new treatment options, usually associated with less costs and lower risks for patients compared to the development of new drugs [58]. While some compounds (e.g., thalidomide, zoledronic acid, celecoxib) have already been successfully repositioned for cancer treatment, other drugs like, e.g., artesunate are currently in the process for possible repositioning [58].

Artesunate is a derivative of artemisinin, an extract from the plant *Artemisia annua* Linne [22]. Notably, the discovery that artemisinin-class substances can be applied as potent therapeutics for malaria patients, was awarded with the Nobel Prize in 2015 [59]. Indeed, artesunate exerts superior antimalarial effects in clinical application and is characterized by an excellent safety profile [60]. Furthermore, in recent years, several additional activities beyond anti-malarial activity have been observed [22,61]. In this respect, pre-clinical studies on artesunate have demonstrated anti-tumor activity against many different cancers including colon caancer [29], lung adenocarcinoma [31], pancreatic cancer [33], breast cancer [62] and different hematological malignancies [30,32,63]. The present study adds MCPyV-positive MCC to this list as we demonstrate the capability of artesunate to restrict growth of virus-positive MCC cells in vitro as well as in xenotransplantation mouse models *in vivo*.

The question whether presence of the viral proteins in these cells affects their artesunate sensitivity could not be fully answered. In line with different reported anti-viral effects of artesunate [35,36] including impairment of the polyomavirus life cycle [64,65], we observed repression of T antigen expression in artesunate-treated MCPyV-positive MCC cells. This was different compared to human papilloma virus infected cervical cancer cells in which expression of the viral oncogenes was not affected by the related compound dihydroartemisinin [66]. However, although T antigens are essential for growth of MCPyV-positive MCC cells [11], the cytotoxicity of artesunate towards these cells seems not to depend on viral-protein repression. Indeed, in this respect, artesunate was more potent than T antigen knockdown. It was only in one MCC cell line (WaGa), that sensitivity towards artesunate was reduced upon knockdown of T antigen expression.

Irrespective of a possible contribution of the T antigens to artesunate-induced cell death of MCPyV-positive MCC cells, a set of inhibitor experiments suggest that artesunate induces ferroptosis and not apoptosis in these cells. Among the multitude of different modes of regulated cell death, an important distinction is their dependence on caspases. In this regard, apoptosis and pyroptosis require activation of these proteases while necroptosis, ferroptosis, parthanatos, alkaliptosis and oxeiptosis are caspase-independent [45]. In the case of artesunate-treated MCPyV-positive MCC cells, the pan caspase inhibitor Z-VAD-FMK did not significantly reduce cell death. In contrast, inhibitors targeting different steps of the ferroptotic pathway were effective in rescuing artesunate-triggered killing of virtually all five investigated MCC cell lines.

Ferroptosis is a mode of programmed cell death that is characterized by an iron-dependent accumulation of lipid peroxides [48]. Interestingly, ferroptosis is considered to be pro-inflammatory and immunogenic, due to release of damage-associated molecular patterns (DAMPs) [67,68]. Hence, besides the direct effects on the tumor cells, artesunate may also support anti-tumor immune responses. However, direct evidence for this possibility is still scarce and further investigations on this topic are necessary [67]. Notwithstanding, following preclinical evaluation of the anti-tumoral activity of artesunate in recent years, we have now reached a phase of human trials for the treatment of cancer patients with artesunate. In this respect, several phase 1 and phase two studies (colorectal cancer, hepatocellular carcinoma, breast cancer and several intraepithelial neoplasias) are ongoing (www.clinicaltrials.gov), and for a few trials results have already been published. These reports highlighted the favorable tolerability of artesunate [69–72], and some even found first hints for clinical activity [69,70]. It may be interesting to see how artesunate, maybe even in combination with immune checkpoint inhibitors, performs in cancer trials in the future. Certainly, MCPyV-positive MCC patients may be included in such studies.

4. Materials and Methods

4.1. Ethics Statement

Animal experiments were performed according to the legal requirements and approved by the Regierung von Unterfranken (RUF 55.2.2 -22532.2 -925-18).

4.2. Cloning and Usage of an NCCR Reporter Construct

To allow assessment of the transcriptional activity of the MCPyV noncoding control region (NCCR) by flow cytometry, we cloned a lentiviral reporter construct in which we placed a green and a red fluorescent protein 3′ and 5′ of the NCCR (Supplementary Figure S8). We included in addition to the mere NCCR, also the sequences coding for the N-terminus of sT and VP2 in the construct to prevent losing potential regulatory elements extending into the respective coding region. To this end, mNeongreen and mCherry coding sequences were cloned in frame with the first 78 codons of sT and the first 64 codons of VP2, respectively (Supplementary Figure S8a). This cassette was inserted into the multiple cloning site of pLVX-Puro (Clontech) yielding the construct pLVX NCCR mNeongreen mRuby3 (the map is available upon request).

Lentivirus particles were generated as described [39] and used for infection of MKL-1 cells. Red and green fluorescence was analyzed on a CytoFLEX flow cytometer (Beckman Coulter).

4.3. Cell Culture

MCPyV-positive MCC cell lines MKL-1 [73], MKL-2 [74], MS-1 [75], WaGa and PeTa (both described in [9]) were cultivated in RPMI-1640 medium (Sigma Aldrich) supplemented with 10% FBS (Biochrom GmbH), 100 U/mL penicillin and 0.1 mg/mL streptomycin (Sigma Aldrich).

MKL-1, MKL-2 and WaGa cells with constitutive expression of an shRNA targeting RB1 and doxycyclin-inducible expression of an shRNA targeting both T antigens (TA shRNA tet), have been described previously [38].

4.4. Immunoblotting

Cells were lysed using the ELB lysis buffer containing 150 mM NaCl, 50 mM Hepes pH 7.5, 5 mM EDTA, 0.1% NP-40, 20 mM β-glycerophosphate, 0.5 mM sodium orthovanadate and a protease inhibitor (Roche). The immunoblotting procedure was performed as described [38]. The antibodies used in this study were directed against MCPyV-LT (CM2B4; Santa Cruz Biotechnologies), β-tubulin (TUB 2.1; Sigma-Aldrich, Ottobrunn, Germany) and vinculin (hVIN-1; Sigma-Aldrich). Uncropped blots are given in Supplementary Figures S9 and S10)

4.5. MTS Assay

Cell lines were seeded in sextuplicate per condition in 96-well plates. Following 5 days of incubation with 0. 1 and 10 μM artesunate (Sigma Aldrich), the MTS proliferation assay (Promega, Mannheim, Germany) was applied according to the manufacturer's instructions.

4.6. DNA Staining

Cells were fixed with ice-cold 90% ethanol followed by a one-hour treatment with propidium iodide mix (PBS + 1% FCS + 0.1 mg/mL propidium iodide + 0.1 mg/mL RNAse A). Analysis was then performed by flow cytometry.

4.7. Trypan Blue Exclusion Assay

Cells were stained with 0.4% trypan blue in PBS (Sigma Aldrich), and the number of living, dye-excluding cells as well as the dead blue-stained cells were counted using a hemocytometer.

4.8. Time Lapse Microscopy

WaGa and PeTa cells were seeded at a density of 1×10^5 cells per well in µ-Slides (Ibidi) and treated with 50 µM artesunate. Morphologic changes in the course of time were recorded using a Nikon Eclipse Ti microscope.

4.9. Animal Experiments

Five-week-old female NOD.CB17/*Prkdc*scid mice (Charles River) were used for the xenotransplantation experiments. They were housed under specific pathogen-free conditions. Each mouse was injected subcutaneously with a suspension of 5×10^6 MKL-1 or WaGa tumor cells mixed with an equal volume of Matrigel (Corning) in a total volume of 100 µL. The tumor size was measured daily using a vernier calipers and the volume was calculated using the formula ($V = \pi/6 \times a^2 \times b$ (a: length; b: height). Once the tumor size reached approximately 150 mm^3, the mice were divided into the control group (n = 6 for WaGa and n = 5 for MKL-1, since in one animal, no tumor growth was observed) and treatment group (n = 6). Each mouse from the treatment group was subjected to daily intraperitoneal injections with 100 mg/kg of artesunate, which was dissolved in DMSO and then brought to a total volume of 200 µL with PBS prior to injection. Similarly, the control group was injected with the same volume of DMSO in 200 µL of PBS (2% DMSO). The experiments were terminated once the tumors of the control group reached the maximum tolerable size.

4.10. Statistical Analysis

Statistical analyses were completed with Prism 5.03 (GraphPad Software, Inc;. San Diego, CA 92108, USA). Since cell volume distribution did not pass normality test; the volumes were compared by non-parametric Mann–Whitney test. The effect of multiple treatment and inhibitor combinations was tested by ANOVA followed by post-hoc test comparing the effect always against ones observed for artesunate treatment. Adjusted p values following Dunnett multiple comparison testing were given. For tumor growth curves, first area under the curve of tumor sizes (baseline set to zero) normalized to the size at the start of treatment, were determined. These values were compared by unpaired *t*-test.

5. Conclusions

Artesunate induced ferroptosis in MCPyV-positive MCC cells in vitro and restricted growth of MCC xenograft tumors in vivo. These results suggest that the established antimalarial therapeutic may be applied to treat patients with MCPyV-positive MCC.

Supplementary Materials: The following are available online at http://www.mdpi.com/2072-6694/12/4/919/s1, Figure S1: tMCC cell lines are more sensitive towards artesunate than melanoma cell lines or primary fibroblasts, Figure S2: Densitometric analyses of all immunoblots presented in the publication, Figure S3: Artesunate represses NCCR driven early region transcription (representative histograms of the NCCR Reporter-Assay corresponding to the bar graph in Figure 1b), Figure S4: Artesunate induces G2/M arrest in MCPyV-positive MCC cells, Figure S5: Propidium iodide staining of artesunate treated MCPyV-positive MCC cells (representative histograms corresponding to Figure 3a), Figure S6: Artesunate induces cell swelling of MCPyV-positive MCC cells, Figure S7: β-Mercaptoethanol, an activator of cystine uptake, represses artesunate-induced cell death, Figure S8: An MCPyV NCCR reporter construct, Figure S9: Uncropped blots from Figure 1a,c, Figure S10: Uncropped blots from Figure 2.

Author Contributions: Conceptualization: B.S., C.W., J.C.B., D.S. and R.H.; investigation: B.S., C.W., L.A., C.A., T.K. and R.H.; supervision: D.S. and R.H.; writing—original draft: B.S. and R.H.; writing—review & editing: C.W., J.C.B., T.K. and D.S. All authors have read and agreed to the published version of the manuscript.

Funding: This research was funded by grants from the German Cancer Aid (70112438), German Research Foundation (HO 5280/2-2) and by the IZKF Würzburg (B-343).

Conflicts of Interest: J.C.B. receives speaker's bureau honoraria from Amgen, Pfizer, Merck Serono and Sanofi, is a paid consultant/advisory board member for eTheRNA, Merck Serono, Pfizer, 4SC and Sanofi. His group receives research grants from Bristol-Myers Squibb, Merck Serono and Alcedis. The other authors have no conflict of interest to declare.

References

1. Hodgson, N.C. Merkel cell carcinoma: Changing incidence trends. *J. Surg. Oncol.* **2005**, *89*, 1–4. [CrossRef]
2. Paulson, K.G.; Park, S.Y.; Vandeven, N.A.; Lachance, K.; Thomas, H.; Chapuis, A.G.; Harms, K.L.; Thompson, J.A.; Bhatia, S.; Stang, A.; et al. Merkel Cell Carcinoma: Current United States Incidence and Projected Increases based on Changing Demographics. *J. Am. Acad. Derm.* **2017**. [CrossRef]
3. Fitzgerald, T.L.; Dennis, S.; Kachare, S.D.; Vohra, N.A.; Wong, J.H.; Zervos, E.E. Dramatic Increase in the Incidence and Mortality from Merkel Cell Carcinoma in the United States. *Am. Surg.* **2015**, *81*, 802–806.
4. Miller, R.W.; Rabkin, C.S. Merkel cell carcinoma and melanoma: Etiological similarities and differences. *Cancer Epidemiol. Biomark. Prev.* **1999**, *8*, 153–158.
5. Buell, J.F.; Trofe, J.; Hanaway, M.J.; Beebe, T.M.; Gross, T.G.; Alloway, R.R.; First, M.R.; Woodle, E.S. Immunosuppression and merkel cell cancer. *Transplant. Proc.* **2002**, *34*, 1780–1781. [CrossRef]
6. Feng, H.; Shuda, M.; Chang, Y.; Moore, P.S. Clonal Integration of a Polyomavirus in Human Merkel Cell Carcinoma. *Science* **2008**. [CrossRef]
7. Tothill, R.; Estall, V.; Rischin, D. Merkel cell carcinoma: Emerging biology, current approaches, and future directions. *Am. Soc. Clin. Oncol. Educ. Book. Am. Soc. Clin. Oncol. Meet.* **2015**. [CrossRef]
8. Sastre-Garau, X.; Peter, M.; Avril, M.F.; Laude, H.; Couturier, J.; Rozenberg, F.; Almeida, A.; Boitier, F.; Carlotti, A.; Couturaud, B.; et al. Merkel cell carcinoma of the skin: Pathological and molecular evidence for a causative role of MCV in oncogenesis. *J. Pathol.* **2009**, *218*, 48–56. [CrossRef]
9. Schrama, D.; Sarosi, E.M.; Adam, C.; Ritter, C.; Kaemmerer, U.; Klopocki, E.; Konig, E.M.; Utikal, J.; Becker, J.C.; Houben, R. Characterization of six Merkel cell polyomavirus-positive Merkel cell carcinoma cell lines: Integration pattern suggest that large T antigen truncating events occur before or during integration. *Int. J. Cancer J. Int. Du Cancer* **2019**, *145*, 1020–1032. [CrossRef]
10. Shuda, M.; Kwun, H.J.; Feng, H.; Chang, Y.; Moore, P.S. Human Merkel cell polyomavirus small T antigen is an oncoprotein targeting the 4E-BP1 translation regulator. *J. Clin. Investig.* **2011**, *121*, 3623–3634. [CrossRef]
11. Houben, R.; Shuda, M.; Weinkam, R.; Schrama, D.; Feng, H.; Chang, Y.; Moore, P.S.; Becker, J.C. Merkel cell polyomavirus-infected Merkel cell carcinoma cells require expression of viral T antigens. *J. Virol.* **2010**, *84*, 7064–7072. [CrossRef] [PubMed]
12. Lemos, B.D.; Storer, B.E.; Iyer, J.G.; Phillips, J.L.; Bichakjian, C.K.; Fang, L.C.; Johnson, T.M.; Liegeois-Kwon, N.J.; Otley, C.C.; Paulson, K.G.; et al. Pathologic nodal evaluation improves prognostic accuracy in Merkel cell carcinoma: Analysis of 5823 cases as the basis of the first consensus staging system. *J. Am. Acad. Derm.* **2010**, *63*, 751–761. [CrossRef] [PubMed]
13. Poulsen, M. Merkel cell carcinoma of skin: Diagnosis and management strategies. *Drugs Aging* **2005**, *22*, 219–229. [CrossRef] [PubMed]
14. Eng, T.Y.; Boersma, M.G.; Fuller, C.D.; Goytia, V.; Jones, W.E., III; Joyner, M.; Nguyen, D.D. A comprehensive review of the treatment of Merkel cell carcinoma. *Am. J. Clin. Oncol.* **2007**, *30*, 624–636. [CrossRef]
15. Nghiem, P.T.; Bhatia, S.; Lipson, E.J.; Kudchadkar, R.R.; Miller, N.J.; Annamalai, L.; Berry, S.; Chartash, E.K.; Daud, A.; Fling, S.P.; et al. PD-1 Blockade with Pembrolizumab in Advanced Merkel-Cell Carcinoma. *N. Engl. J. Med.* **2016**, *374*, 2542–2552. [CrossRef]
16. Kaufman, H.L.; Russell, J.; Hamid, O.; Bhatia, S.; Terheyden, P.; D'Angelo, S.P.; Shih, K.C.; Lebbe, C.; Linette, G.P.; Milella, M.; et al. Avelumab in patients with chemotherapy-refractory metastatic Merkel cell carcinoma: A multicentre, single-group, open-label, phase 2 trial. *Lancet Oncol.* **2016**. [CrossRef]
17. Kim, E.S. Avelumab: First Global Approval. *Drugs* **2017**, *77*, 929–937. [CrossRef]
18. Nghiem, P.; Bhatia, S.; Lipson, E.J.; Sharfman, W.H.; Kudchadkar, R.R.; Brohl, A.S.; Friedlander, P.A.; Daud, A.; Kluger, H.M.; Reddy, S.A.; et al. Durable Tumor Regression and Overall Survival in Patients With Advanced Merkel Cell Carcinoma Receiving Pembrolizumab as First-Line Therapy. *J. Clin. Oncol. Off. J. Am. Soc. Clin. Oncol.* **2019**, *37*, 693–702. [CrossRef]
19. D'Angelo, S.P.; Russell, J.; Lebbé, C.; Chmielowski, B.; Gambichler, T.; Grob, J.-J.; Kiecker, F.; Rabinowits, G.; Terheyden, P.; Zwiener, I.; et al. Efficacy and Safety of First-line Avelumab Treatment in Patients With Stage IV Metastatic Merkel Cell Carcinoma: A Preplanned Interim Analysis of a Clinical Trial. *JAMA Oncol.* **2018**, *4*, e180077. [CrossRef]

20. Kaufman, H.L.; Russell, J.S.; Hamid, O.; Bhatia, S.; Terheyden, P.; D'Angelo, S.P.; Shih, K.C.; Lebbé, C.; Milella, M.; Brownell, I.; et al. Updated efficacy of avelumab in patients with previously treated metastatic Merkel cell carcinoma after ≥1 year of follow-up: JAVELIN Merkel 200, a phase 2 clinical trial. *J. Immunother. Cancer* **2018**, *6*, 7. [CrossRef]
21. Marin-Acevedo, J.A.; Dholaria, B.; Soyano, A.E.; Knutson, K.L.; Chumsri, S.; Lou, Y. Next generation of immune checkpoint therapy in cancer: New developments and challenges. *J. Hematol. Oncol.* **2018**, *11*, 39. [CrossRef] [PubMed]
22. Efferth, T. From ancient herb to modern drug: Artemisia annua and artemisinin for cancer therapy. *Semin Cancer Biol.* **2017**, *46*, 65–83. [CrossRef] [PubMed]
23. Burrows, J.N.; Chibale, K.; Wells, T.N. The state of the art in anti-malarial drug discovery and development. *Curr. Top. Med. Chem.* **2011**, *11*, 1226–1254. [CrossRef] [PubMed]
24. Wang, J.; Zhang, C.J.; Chia, W.N.; Loh, C.C.; Li, Z.; Lee, Y.M.; He, Y.; Yuan, L.X.; Lim, T.K.; Liu, M.; et al. Haem-activated promiscuous targeting of artemisinin in Plasmodium falciparum. *Nat. Commun.* **2015**, *6*, 10111. [CrossRef]
25. Yu, L.; Chen, J.F.; Shuai, X.; Xu, Y.; Ding, Y.; Zhang, J.; Yang, W.; Liang, X.; Su, D.; Yan, C. Artesunate protects pancreatic beta cells against cytokine-induced damage via SIRT1 inhibiting NF-kappaB activation. *J. Endocrinol. Investig.* **2016**, *39*, 83–91. [CrossRef]
26. Heller, L.E.; Roepe, P.D. Artemisinin-Based Antimalarial Drug Therapy: Molecular Pharmacology and Evolving Resistance. *Trop. Med. Infect. Dis.* **2019**, *4*. [CrossRef]
27. Efferth, T.; Sauerbrey, A.; Olbrich, A.; Gebhart, E.; Rauch, P.; Weber, H.O.; Hengstler, J.G.; Halatsch, M.E.; Volm, M.; Tew, K.D.; et al. Molecular modes of action of artesunate in tumor cell lines. *Mol. Pharm.* **2003**, *64*, 382–394. [CrossRef]
28. Efferth, T.; Dunstan, H.; Sauerbrey, A.; Miyachi, H.; Chitambar, C.R. The anti-malarial artesunate is also active against cancer. *Int. J. Oncol.* **2001**, *18*, 767–773. [CrossRef]
29. Jiang, F.; Zhou, J.-Y.; Zhang, D.; Liu, M.-H.; Chen, Y.-G. Artesunate induces apoptosis and autophagy in HCT116 colon cancer cells, and autophagy inhibition enhances the artesunate-induced apoptosis. *Int. J. Mol. Med.* **2018**, *42*, 1295–1304. [CrossRef]
30. Våtsveen, T.K.; Myhre, M.R.; Steen, C.B.; Wälchli, S.; Lingjærde, O.C.; Bai, B.; Dillard, P.; Theodossiou, T.A.; Holien, T.; Sundan, A.; et al. Artesunate shows potent anti-tumor activity in B-cell lymphoma. *J. Hematol. Oncol.* **2018**, *11*, 23. [CrossRef]
31. Zhou, C.; Pan, W.; Wang, X.P.; Chen, T.S. Artesunate induces apoptosis via a Bak-mediated caspase-independent intrinsic pathway in human lung adenocarcinoma cells. *J. Cell. Physiol.* **2012**, *227*, 3778–3786. [CrossRef] [PubMed]
32. Wang, N.; Zeng, G.-Z.; Yin, J.-L.; Bian, Z.-X. Artesunate activates the ATF4-CHOP-CHAC1 pathway and affects ferroptosis in Burkitt's Lymphoma. *Biochem. Biophys. Res. Commun.* **2019**, *519*, 533–539. [CrossRef] [PubMed]
33. Eling, N.; Reuter, L.; Hazin, J.; Hamacher-Brady, A.; Brady, N.R. Identification of artesunate as a specific activator of ferroptosis in pancreatic cancer cells. *Oncoscience* **2015**, *2*, 517–532. [CrossRef] [PubMed]
34. Roh, J.-L.; Kim, E.H.; Jang, H.; Shin, D. Nrf2 inhibition reverses the resistance of cisplatin-resistant head and neck cancer cells to artesunate-induced ferroptosis. *Redox Biol.* **2017**, *11*, 254–262. [CrossRef] [PubMed]
35. Youns, M.; Hoheisel, J.D.; Efferth, T. Traditional Chinese medicines (TCMs) for molecular targeted therapies of tumours. *Curr. Drug Discov. Technol.* **2010**, *7*, 37–45. [CrossRef]
36. Efferth, T.; Romero, M.R.; Wolf, D.G.; Stamminger, T.; Marin, J.J.; Marschall, M. The antiviral activities of artemisinin and artesunate. *Clin. Infect. Dis.* **2008**, *47*, 804–811. [CrossRef]
37. Houben, R.; Angermeyer, S.; Haferkamp, S.; Aue, A.; Goebeler, M.; Schrama, D.; Hesbacher, S. Characterization of functional domains in the Merkel cell polyoma virus Large T antigen. *Int. J. Cancer* **2015**, *136*, E290–E300. [CrossRef]
38. Hesbacher, S.; Pfitzer, L.; Wiedorfer, K.; Angermeyer, S.; Borst, A.; Haferkamp, S.; Scholz, C.J.; Wobser, M.; Schrama, D.; Houben, R. RB1 is the crucial target of the Merkel cell polyomavirus Large T antigen in Merkel cell carcinoma cells. *Oncotarget* **2016**, *7*, 32956–32968. [CrossRef]
39. Valeriote, F.; van Putten, L. Proliferation-dependent cytotoxicity of anticancer agents: A review. *Cancer Res.* **1975**, *35*, 2619–2630.
40. Compton, M.M. A biochemical hallmark of apoptosis: Internucleosomal degradation of the genome. *Cancer Metastasis Rev.* **1992**, *11*, 105–119. [CrossRef]

41. Efferth, T.; Giaisi, M.; Merling, A.; Krammer, P.H.; Li-Weber, M. Artesunate induces ROS-mediated apoptosis in doxorubicin-resistant T leukemia cells. *PLoS ONE* **2007**, *2*, e693. [CrossRef] [PubMed]
42. Galluzzi, L.; Vitale, I.; Abrams, J.M.; Alnemri, E.S.; Baehrecke, E.H.; Blagosklonny, M.V.; Dawson, T.M.; Dawson, V.L.; El-Deiry, W.S.; Fulda, S.; et al. Molecular definitions of cell death subroutines: Recommendations of the Nomenclature Committee on Cell Death 2012. *Cell Death Differ.* **2012**, *19*, 107–120. [CrossRef] [PubMed]
43. Bortner, C.D.; Cidlowski, J.A. Uncoupling cell shrinkage from apoptosis reveals that Na+ influx is required for volume loss during programmed cell death. *J. Biol. Chem.* **2003**, *278*, 39176–39184. [CrossRef] [PubMed]
44. Tang, D.; Kang, R.; Berghe, T.V.; Vandenabeele, P.; Kroemer, G. The molecular machinery of regulated cell death. *Cell Res.* **2019**, *29*, 347–364. [CrossRef]
45. Yan, N.; Zhang, J.-J. The Emerging Roles of Ferroptosis in Vascular Cognitive Impairment. *Front. Neurosci.* **2019**, *13*, 811. [CrossRef]
46. Stockwell, B.R.; Friedmann Angeli, J.P.; Bayir, H.; Bush, A.I.; Conrad, M.; Dixon, S.J.; Fulda, S.; Gascón, S.; Hatzios, S.K.; Kagan, V.E.; et al. Ferroptosis: A Regulated Cell Death Nexus Linking Metabolism, Redox Biology, and Disease. *Cell* **2017**, *171*, 273–285. [CrossRef]
47. Dixon, S.J.; Lemberg, K.M.; Lamprecht, M.R.; Skouta, R.; Zaitsev, E.M.; Gleason, C.E.; Patel, D.N.; Bauer, A.J.; Cantley, A.M.; Yang, W.S.; et al. Ferroptosis: An iron-dependent form of nonapoptotic cell death. *Cell* **2012**, *149*, 1060–1072. [CrossRef]
48. Shacka, J.J.; Klocke, B.J.; Roth, K.A. Autophagy, bafilomycin and cell death: The "a-B-cs" of plecomacrolide-induced neuroprotection. *Autophagy* **2006**, *2*, 228–230. [CrossRef]
49. Xie, Z.; Xie, Y.; Xu, Y.; Zhou, H.; Xu, W.; Dong, Q. Bafilomycin A1 inhibits autophagy and induces apoptosis in MG63 osteosarcoma cells. *Mol. Med. Rep.* **2014**, *10*, 1103–1107. [CrossRef]
50. Gao, M.; Monian, P.; Pan, Q.; Zhang, W.; Xiang, J.; Jiang, X. Ferroptosis is an autophagic cell death process. *Cell Res.* **2016**, *26*, 1021–1032. [CrossRef]
51. Zhou, B.; Liu, J.; Kang, R.; Klionsky, D.J.; Kroemer, G.; Tang, D. Ferroptosis is a type of autophagy-dependent cell death. *Semin. Cancer Biol.* **2019**. [CrossRef] [PubMed]
52. Sato, M.; Kusumi, R.; Hamashima, S.; Kobayashi, S.; Sasaki, S.; Komiyama, Y.; Izumikawa, T.; Conrad, M.; Bannai, S.; Sato, H. The ferroptosis inducer erastin irreversibly inhibits system x(c)- and synergizes with cisplatin to increase cisplatin's cytotoxicity in cancer cells. *Sci. Rep.* **2018**, *8*, 968. [CrossRef] [PubMed]
53. Dixon, S.J.; Patel, D.N.; Welsch, M.; Skouta, R.; Lee, E.D.; Hayano, M.; Thomas, A.G.; Gleason, C.E.; Tatonetti, N.P.; Slusher, B.S.; et al. Pharmacological inhibition of cystine-glutamate exchange induces endoplasmic reticulum stress and ferroptosis. *Elife* **2014**, *3*, e02523. [CrossRef]
54. Takahashi, M.; Nagai, T.; Okamura, N.; Takahashi, H.; Okano, A. Promoting effect of beta-mercaptoethanol on in vitro development under oxidative stress and cystine uptake of bovine embryos. *Biol. Reprod.* **2002**, *66*, 562–567. [CrossRef]
55. Doll, S.; Proneth, B.; Tyurina, Y.Y.; Panzilius, E.; Kobayashi, S.; Ingold, I.; Irmler, M.; Beckers, J.; Aichler, M.; Walch, A.; et al. ACSL4 dictates ferroptosis sensitivity by shaping cellular lipid composition. *Nat. Chem. Biol.* **2017**, *13*, 91–98. [CrossRef] [PubMed]
56. Yuan, H.; Li, X.; Zhang, X.; Kang, R.; Tang, D. Identification of ACSL4 as a biomarker and contributor of ferroptosis. *Biochem. Biophys. Res. Commun.* **2016**, *478*, 1338–1343. [CrossRef]
57. Adam, C.; Baeurle, A.; Brodsky, J.L.; Wipf, P.; Schrama, D.; Becker, J.C.; Houben, R. The HSP70 modulator MAL3-101 inhibits Merkel cell carcinoma. *PLoS ONE* **2014**, *9*, e92041. [CrossRef]
58. Serafin, M.B.; Bottega, A.; da Rosa, T.F.; Machado, C.S.; Foletto, V.S.; Coelho, S.S.; da Mota, A.D.; Horner, R. Drug Repositioning in Oncology. *Am. J. Ther.* **2019**. [CrossRef]
59. Tiwari, M.K.; Chaudhary, S. Artemisinin-derived antimalarial endoperoxides from bench-side to bed-side: Chronological advancements and future challenges. *Med. Res. Rev.* **2020**. [CrossRef]
60. Awad Adeel, A. Time to switch from quinine. *Sudan. J. Paediatr.* **2012**, *12*, 6–7.
61. Nass, J.; Efferth, T. The activity of Artemisia spp. and their constituents against Trypanosomiasis. *Phytomed. Int. J. Phytother. Phytopharm.* **2018**, *47*, 184–191. [CrossRef]
62. Chen, K.; Shou, L.M.; Lin, F.; Duan, W.M.; Wu, M.Y.; Xie, X.; Xie, Y.F.; Li, W.; Tao, M. Artesunate induces G2/M cell cycle arrest through autophagy induction in breast cancer cells. *Anticancer Drugs* **2014**, *25*, 652–662. [CrossRef] [PubMed]
63. Li, Y.; Shan, N.N.; Sui, X.H. Research Progress on Artemisinin and Its Derivatives against Hematological Malignancies. *Chin. J. Integr. Med.* **2020**. [CrossRef] [PubMed]

64. Sharma, B.N.; Marschall, M.; Henriksen, S.; Rinaldo, C.H. Antiviral effects of artesunate on polyomavirus BK replication in primary human kidney cells. *Antimicrob. Agents Chemother.* **2014**, *58*, 279–289. [CrossRef]
65. Sharma, B.N.; Marschall, M.; Rinaldo, C.H. Antiviral effects of artesunate on JC polyomavirus replication in COS-7 cells. *Antimicrob. Agents Chemother.* **2014**, *58*, 6724–6734. [CrossRef]
66. Disbrow, G.L.; Baege, A.C.; Kierpiec, K.A.; Yuan, H.; Centeno, J.A.; Thibodeaux, C.A.; Hartmann, D.; Schlegel, R. Dihydroartemisinin is cytotoxic to papillomavirus-expressing epithelial cells in vitro and in vivo. *Cancer Res.* **2005**, *65*, 10854–10861. [CrossRef]
67. Proneth, B.; Conrad, M. Ferroptosis and necroinflammation, a yet poorly explored link. *Cell Death Differ.* **2019**, *26*, 14–24. [CrossRef]
68. Mou, Y.; Wang, J.; Wu, J.; He, D.; Zhang, C.; Duan, C.; Li, B. Ferroptosis, a new form of cell death: Opportunities and challenges in cancer. *J. Hematol. Oncol.* **2019**, *12*, 34. [CrossRef]
69. Krishna, S.; Ganapathi, S.; Ster, I.C.; Saeed, M.E.M.; Cowan, M.; Finlayson, C.; Kovacsevics, H.; Jansen, H.; Kremsner, P.G.; Efferth, T.; et al. A Randomised, Double Blind, Placebo-Controlled Pilot Study of Oral Artesunate Therapy for Colorectal Cancer. *EBioMedicine* **2014**, *2*, 82–90. [CrossRef]
70. Deeken, J.F.; Wang, H.; Hartley, M.; Cheema, A.K.; Smaglo, B.; Hwang, J.J.; He, A.R.; Weiner, L.M.; Marshall, J.L.; Giaccone, G.; et al. A phase I study of intravenous artesunate in patients with advanced solid tumor malignancies. *Cancer Chemother. Pharmacol.* **2018**, *81*, 587–596. [CrossRef]
71. von Hagens, C.; Walter-Sack, I.; Goeckenjan, M.; Storch-Hagenlocher, B.; Sertel, S.; Elsässer, M.; Remppis, B.A.; Munzinger, J.; Edler, L.; Efferth, T.; et al. Long-term add-on therapy (compassionate use) with oral artesunate in patients with metastatic breast cancer after participating in a phase I study (ARTIC M33/2). *Phytomed. Int. J. Phytother. Phytopharm.* **2019**, *54*, 140–148. [CrossRef] [PubMed]
72. König, M.; von Hagens, C.; Hoth, S.; Baumann, I.; Walter-Sack, I.; Edler, L.; Sertel, S. Investigation of ototoxicity of artesunate as add-on therapy in patients with metastatic or locally advanced breast cancer: New audiological results from a prospective, open, uncontrolled, monocentric phase I study. *Cancer Chemother. Pharmacol.* **2016**, *77*, 413–427. [CrossRef] [PubMed]
73. Rosen, S.T.; Gould, V.E.; Salwen, H.R.; Herst, C.V.; Le Beau, M.M.; Lee, I.; Bauer, K.; Marder, R.J.; Andersen, R.; Kies, M.S. Establishment and characterization of a neuroendocrine skin carcinoma cell line. *Lab. Investig.* **1987**, *56*, 302–312. [PubMed]
74. Van Gele, M.; Leonard, J.H.; Van, R.N.; Van, L.H.; Van, B.S.; Cocquyt, V.; Salwen, H.; De, P.A.; Speleman, F. Combined karyotyping, CGH and M-FISH analysis allows detailed characterization of unidentified chromosomal rearrangements in Merkel cell carcinoma. *Int. J. Cancer* **2002**, *101*, 137–145. [CrossRef]
75. Guastafierro, A.; Feng, H.; Thant, M.; Kirkwood, J.M.; Chang, Y.; Moore, P.S.; Shuda, M. Characterization of an early passage Merkel cell polyomavirus-positive Merkel cell carcinoma cell line, MS-1, and its growth in NOD scid gamma mice. *J. Virol. Methods* **2013**, *187*, 6–14. [CrossRef]

cancers

MDPI

Article

Tumor Ulceration, Reduced Infiltration of CD8-Lymphocytes, High Neutrophil-to-CD8-Lymphocyte Ratio and Absence of MC Virus are Negative Prognostic Markers for Patients with Merkel Cell Carcinoma

Simon Naseri [1,*], Torben Steiniche [2], Jeanette Bæhr Georgsen [2], Rune Thomsen [3], Morten Ladekarl [4], Martin Heje [5], Tine Engberg Damsgaard [6] and Marie Louise Bønnelykke-Behrndtz [7]

1 Department of Plastic Surgery, Aalborg University Hospital, 9000 Aalborg, Denmark
2 Department of Pathology, Aarhus University Hospital, 8200 Aarhus, Denmark; torbstei@rm.dk (T.S.); jeanette.georgsen@aarhus.rm.dk (J.B.G.)
3 Department of Biomedicine, Aarhus University, 8200 Aarhus, Denmark; rt@biomed.au.dk
4 Department of Oncology, Clinical Cancer Research Center, Aalborg University Hospital, 9000 Aalborg, Denmark; morten.ladekarl@rn.dk
5 Department of Plastic Surgery, Vejle Hospital, 7100 Vejle, Denmark; martin.heje@dadlnet.dk
6 Department of Plastic Surgery and Burns Treatment, Rigshospitalet, 2100 Copenhagen, Denmark; tinemed@gmail.com
7 Department of Plastic and Reconstructive Surgery, Aarhus University Hospital, 8200 Aarhus, Denmark; louiseboennelykke@gmail.com
* Correspondence: s.nasari@rn.dk

Received: 5 March 2020; Accepted: 4 April 2020; Published: 6 April 2020

Abstract: (1) Background: Merkel cell carcinoma (MCC) is caused by the Merkel cell polyomavirus and UV radiation. Understanding of the underlying biology is limited, but identification of prognostic markers may lead to better prognostic stratification for the patients. (2) Methods: Ninety patients diagnosed with MCC (1996–2012) were included. Virus status was estimated by polymerase chain reaction (qPCR) and immunohistochemistry (IHC). Ulceration status, PD-L1, cd66b neutrophils, cd8 lymphocytes and biomarkers of vascularization (cd34 endothelial cells) and migration (e-cadherin) were estimated by IHC and analyzed with digital pathology. (3) Results: Virus was present in 47% of patient samples and correlated with lower E-cadherin expression ($p = 0.0005$), lower neutrophil-to-CD8 lymphocyte ratio (N:CD8 ratio) ($p = 0.02$) and increased PD-L1 expression ($p = 0.03$). Ulceration was associated with absence of virus ($p = 0.03$), increased neutrophil infiltration ($p < 0.0001$) and reduced CD8 lymphocyte infiltration ($p = 0.04$). In multivariate analysis, presence of virus ($p = 0.01$), ulceration ($p = 0.05$) and increased CD8 lymphocyte infiltration ($p = 0.001$) showed independent prognostic impacts on MCC-specific survival. (3) Conclusions: In this study, we found that a high N:CD8 ratio, ulceration, virus-negative status and absence of CD8 lymphocytes are negative prognostic markers. Accurate prognostic stratification of the patients may be important in the clinical setting for determination of adjuvant treatment.

Keywords: Merkel cell carcinoma; Merkel cell polyoma virus; tumor microenvironment; CD8 lymphocytes; ulceration; E-cadherin

1. Introduction

Merkel cell carcinoma (MCC) is a highly aggressive malignancy of the skin with a five-year overall survival rate of 40% [1]. MCC was first described by Toker in 1972 and has during the past decades shown an up to five-fold increase in incidences in western countries [2–5]. Although the cell of origin of MCC is still debated, the etiology is believed to be UV-radiation (20%) and the recently discovered Merkel cell polyomavirus (MCV) (80%) [6,7]. Despite its poor prognosis, recent clinical trials with immune therapy with checkpoint inhibitors show high response rates, exceeding response rates observed in most other solid tumors. The reason for this might be rooted in the inflammatory microenvironment [8–10]. In most solid tumors, the tumor microenvironment (TME) plays an essential role in both tumor growth and dissemination but also in response to treatment [11]. However, a characterization and understanding of the TME is limited and still largely undescribed in MCC.

Both viral status (MCV-positive or negative) [12] and infiltrating immune cells (e.g., neutrophils and CD8 lymphocytes) [13–15] can be pivotal contributors to either a pro- or anti-TME, which in turn may impact the migratory functions of the tumor cells (e.g., assessed by loss of E-cadherin) [16–18] and response to immune checkpoint inhibitors (generally enhanced in tumors with PD-L1 expression) [19]. In addition, one of the leading prognostic factors in other skin malignancies like melanoma is ulceration [20], which we have previously shown is linked to a tumor-supportive microenvironment [16,17]. We aim to study in MCC the interaction between tumor cell viral status, ulceration and the microenvironment (assessed by PD-L1, E-cadherin, endothelial cells and immune cell stain densities), aiming for a better understanding of these factors that may play an essential role in both the natural and treatment-related biology of MCC.

2. Results

2.1. *Ulceration in MCC Is Associated with Increased Infiltration of Neutrophils and Decreased Infiltration of CD8 Lymphocytes*

Ulceration was present in 29.5 % (n = 23) of primary tumors and absent in 70.5 % (n = 55). The remaining tumors could not be evaluated due to missing epidermal regions in the tumor sections (n = 12). There was no difference in clinical characteristics between ulcerated and nonulcerated MCC (Table S1). Ulcerated tumors were characterized by increased ($p < 0.0001$) stain area fractions of neutrophils (0.02%; 95% CI: 0.00–0.90 vs. 0.06×10^{-3}%; 95% CI: 0.02×10^{-3}–0.18×10^{-3}, Figure S1B,E) and an increased ($p < 0.0001$) neutrophil-to-CD8 lymphocyte ratio (N:CD8) (0.91; 95% CI: 0.12–6.92 vs. 0.33×10^{-3}; 95% CI: 0.09×10^{-3}–1.23×10^{-3}), compared with nonulcerated tumors. In contrast, ulcerated tumors had lower ($p = 0.04$) stain area fractions of CD8 lymphocytes (0.02%; 95% CI: 0.00–0.10 vs. 0.19%; 95% CI: 0.06–0.60), compared with nonulcerated tumors (Figure S1C,F).

2.2. *Ulceration Is Associated with Virus-Negative MCC*

Virus was present in 47% (43/90) of the included MCC patient samples, while 53% (57/90) were virus-negative. Ulceration associated significantly with virus-negative MCC ($p = 0.03$) and was present in 39.5% (17/43) of the virus-negative MCC and only in 17.1% (6/35) of the virus-positive MCC. Ulceration did not associate with tumor size ($p = 0.56$).

2.3. *Virus-Positive MCC Presents Higher Densities of PD-L1, Lower Neutrophil-to-CD8 Lymphocyte Ratio and Lower Density of E-Cadherin*

Virus status was estimated with both qPCR and immunohistochemistry (IHC). Estimated by qPCR, 47% (43/90) of patients were virus-positive. Two additional patients had a positive PCR but were categorized as PCR-negative, as their viral primer/TBP ratio was below the 0.01 cut-off. Estimated by IHC, 40% (36/90) of patients were virus-positive. One additional patient had positive immune staining but was categorized as IHC-negative, as the stained cells were stromal cells. There was a high concordance between IHC and qPCR for virus detection ($p < 0.0001$), with IHC detecting 83.7% of qPCR-positive samples.

Patients with virus-positive MCC were younger (74.7 years vs. 80.8 years; $p = 0.008$), and the primary location of MCC varied significantly between the virus-negative and virus-positive groups ($p = 0.006$). Virus-positive primary tumors were primarily located on the extremities (60.5% vs. 27.6%), and the virus-negative tumors were more often located in the head-and-neck area (61.7% vs. 30.2%), while location on the trunk was rare but equally distributed between the groups (9.3% vs. 10.6%). Factors of the local TME in virus-positive and -negative MCC are illustrated in Table 1. Virus-negative MCC was significantly associated ($p = 0.02$) with an increased N:CD8 ratio (15.93×10^{-3}; 95 % CI: 2.20×10^{-3}–115.16×10^{-3}), compared with virus-positive MCC (0.81×10^{-3}; 95% CI: 0.16×10^{-3}–4.12×10^{-3}). Virus-positive MCC was significantly associated ($p = 0.0005$) with reduced stain area fractions of E-cadherin (0.27×10^{-3} %; 95% CI: 0.04×10^{-3}–2.04×10^{-3}), compared with virus-negative MCC (56.57×10^{-3}; 95 % CI: 6.44×10^{-3}–497.02×10^{-3}, Figure S2D,H). In addition, presence of the virus associated ($p = 0.03$) with an increased stain area fraction of PD-L1 (59.28×10^{-3}%; 95 % CI: 9.46×10^{-3}–371.29×10^{-3}), compared with virus-negative samples (4.36×10^{-3} %; 95 % CI: 0.84×10^{-3}–22.68×10^{-3}) (Figure S2C,G).

Table 1. This stain area fraction (in %) of immune cells and biomarkers in virus-positive and -negative Merkel cell carcinoma (MCC).

Mean Area Marker (%)	Virus-Positive MCC Mean Area Fraction of Marker (95% CI)	Virus-Negative MCC Mean Area Fraction of Marker (95% CI)	p-Value
Lymphocytes (CD8, intratumoral)	0.23 (0.06–0.89)	0.06 (0.02–0.19)	$p = 0.11$
PD-L1 (intratumoral)	59.28×10^{-3} (9.46×10^{-3}–371.29×10^{-3})	4.36×10^{-3} (0.84×10^{-3}–22.68×10^{-3})	$p = 0.03$
Neutrophils (CD66b, intratumoral)	0.19×10^{-3} (0.07×10^{-3}–0.52×10^{-3})	0.89×10^{-3} (0.19×10^{-3}–4.07×10^{-3})	$p = 0.09$
Neutrophil-to-lymphocyte ratio, (CD66b/CD8, intratumoral) *	0.81×10^{-3} (0.16×10^{-3}–4.12×10^{-3})	15.93×10^{-3} (2.20×10^{-3}–115.16×10^{-3})	$p = 0.02$
E-cadherin (intratumoral)	0.27×10^{-3} (0.04×10^{-3}–2.04×10^{-3})	56.57×10^{-3} (6.44×10^{-3}–497.02×10^{-3})	$p = 0.0005$
Endothelia (CD34, intratumoral)	3.74 (0.65–21.39)	4.40 (1.23–15.78)	$p = 0.87$

* No unit.

2.4. *Density of CD8 Lymphocytes and PD-L1 Are Associated*

Increasing stain area fractions of CD8 lymphocytes in the tumor ($p < 0.0001$) and a low N:CD8 ratio ($p = 0.0003$) associated with an increased PD-L1 stain area fraction.

2.5. *Density of CD8 Lymphocytes, Neutrophil-to-CD8 Lymphocyte Ratio, Virus-Positive Status, Ulceration and Nodal Involvement Have Independent Impact on MCC Specific Survival*

In univariate analysis, a significantly reduced MCC-specific survival was seen in patients with an ulcerated primary tumor (HR = 2.49; 95% CI= 1.18–5.25; $p = 0.02$), increased N:CD8 ratio (HR = 1.21; 95% CI= 1.06–1.37; $p = 0.004$) and nodal involvement (HR = 3.17; 95% CI = 1.47–6.81; $p = 0.003$). A significantly improved MCC-specific survival was seen in patients with an increased stain area fraction of CD8 lymphocytes (HR = 0.70; 95% CI= 0.57–0.87; $p = 0.001$) and with a positive viral status (HR = 0.47; 95% CI = 0.22–1.00; $p = 0.05$). No significant difference in MCC-specific survival was seen based on the stain area fraction of PD-L1 expression ($p = 0.21$), E-cadherin ($p = 0.73$), endothelia ($p = 0.74$), neutrophils ($p = 0.32$) or tumor size ($p = 0.35$). The results of the univariate analysis are illustrated in Table 2.

Table 2. Univariate analysis showing MCC-specific survival based on immune cells and biomarkers in the tumor microenvironment.

Characteristics	Number of Patients (*n*)	Univariate Analysis HR (95% CI)	*p*-Value
Presence of virus	90	0.47 (0.22–1.00)	$p = 0.05$
Presence of ulceration	78	2.49 (1.18–5.25)	$p = 0.02$
Lymphocytes (CD8, intratumoral)	90	0.70 (0.57– 0.87)	$p = 0.001$
Neutrophils (CD66b, intratumoral)	89	1.10 (0.91–1.34)	$p = 0.32$
Neutrophil-to-lymphocyte ratio (CD66b/CD8, intratumoral)	89	1.21 (1.06–1.37)	$p = 0.004$
Endothelia (CD34, intratumoral)	89	0.97 (0.82–1.15)	$p = 0.74$
E-cadherin (intratumoral)	89	0.98 (0.88–1.10)	$p = 0.73$
PD-L1 (intratumoral)	38	0.81 (0.59–1.12)	$p = 0.21$

For the multivariate analysis, we chose to adjust for T-size over and under 2 cm and lymph node involvement, as these factors are known and accepted prognostic markers of MCC. Presence of ulceration (HR = 2.22; 95% CI= 0.99–4.98; $p = 0.05$) and an increased N:CD8 ratio (HR = 1.14; 95% CI = 1.00–1.31; $p = 0.04$) had negative independent prognostic impacts on MCC-specific survival. Kaplan-Meier survival curves for ulcerated and nonulcerated MCC are illustrated in Figure S1G. A significantly improved MCC-specific survival was seen in patients with an increased stain area fraction of CD8 lymphocytes (HR = 0.68; 95% conf. 0.54–0.85; $p = 0.001$) and with a positive viral status (HR = 0.32; 95% CI = 0.13–0.78; $p = 0.01$). Kaplan-Meier survival curves for virus-positive and -negative MCC are illustrated in Figure S2I. No significant difference in MCC-specific survival was seen based on the stain area fractions of PD-L1 ($p = 0.29$), neutrophils ($p = 0.87$), endothelia (0.77) or E-cadherin ($p = 0.73$). The results of the multivariate analysis are illustrated in Table 3.

Table 3. Multivariate analysis showing MCC-specific survival based on immune cells and biomarkers in the tumor microenvironment.

Characteristics	Number of Patients (*n*)	Multivariate Analysis HR (95% CI)	*p*-Value
Presence of virus	82	0.32 (0.13–0.78)	$p = 0.01$
Presence of ulceration	70	2.22 (0.99–4.98)	$p = 0.05$
Lymphocytes (CD8, intratumoral)	82	0.68 (0.54–0.85)	$p = 0.001$
Neutrophils (CD66b, intratumoral)	81	1.02 (0.82–1.26)	$p = 0.87$
Neutrophil-to-lymphocyte ratio (CD66b/CD8, intratumoral)	89	1.14 (1.00–1.31)	$p = 0.04$
Endothelia (CD34, intratumoral)	81	1.03 (0.86–1.23)	$p = 0.77$
E-cadherin (intratumoral)	81	0.98 (0.86–1.11)	$p = 0.73$
PD-L1 (intratumoral)	31	0.80 (0.53–1.20)	$p = 0.29$

3. Discussion

The primary aim of this study was to investigate prognostic markers of MCC, an aggressive skin tumor with worse prognosis than melanoma [21]. We collected the majority of primary MCC samples from patients diagnosed between 2007–2012 in Denmark. We aimed to characterize and associate the virus status; ulceration status; factors of the TME (PD-L1 expression, E-cadherin expression and CD34 endothelial cells) and important immune cells in primary MCC and link these factors to disease-specific survival.

Importantly, we found that ulceration is an independent negative prognostic marker for patients with MCC. In melanoma, ulceration is a part of staging and is an established negative prognostic marker [22]; however, only few studies have looked at its role in MCC. Several studies have found no association [23–26], while Bob et al. found correlation between ulceration and poor MCC-specific survival [27]. Important limitations of many of these studies include a low number of ulcerated samples, unclear definition of ulceration or if analysis was performed on primary or metastatic tumors. In this study, ulceration was present in 29.5% (23/55) of primary tumors, with previous reports ranging between 6.7–40% [23–26,28]. Ulceration associated with absence of the virus and a high N:CD8, with the latter suggesting that ulceration may contribute to a tumor-supporting microenvironment by attracting neutrophils to the wound and surrounding tumor cells, in line with what has been previously shown in melanoma [16,29]. Neutrophils, inflammation and UV exposure can suppress the levels and functions of CD8 lymphocytes and induce inflammation and a local immune-suppressive microenvironment [30,31]. An alternative explanation may be that virus-negative tumors are larger and, therefore, more likely to be ulcerated; however, in our cohort, there was no significant difference in tumor size based on viral or ulceration status.

In our study, a virus-positive status estimated by qPCR associated with improved MCC-specific survival, confirming the results of several studies [32,33], although a virus-positive status estimated by IHC did not impact survival significantly (data not shown). In our cohort, 47% (43/90) of primary MCC samples were virus-positive in line with aggregate studies demonstrating 76% (453 of 595 MCCs) virus positivity, although ranges vary between 24% and 100% [32,34,35]. This variance is largely unexplained, as the hypothesis that this may be due to viral degradation in old FPPE patient samples has been rejected by digital transcriptome analysis of frozen virus-negative samples [36,37]. In support of our results, we used the same viral primers as previous published studies, and our bimodal approach of detecting the virus showed high concordance [35].

E-cadherin is an important adhesion molecule, and its loss is among the factors that are downregulated in epithelial-to-mesenchymal transition, allowing tumor cells to migrate [17,38]. In our sample, a reduced E-cadherin area fraction associated with virus-negative patients. This was unexpected, as virus-negative patients more often present with advanced disease, compared with virus-positive patients (66.7% vs. 48.3%) [32]. This is the first time E-cadherin expression has been linked to virus-negative status, and it may be rooted in the controversies regarding the cellular origin of MCC. Recent studies suggest that virus-positive MCC may originate from the epidermal keratinocyte, and virus-negative MCC may originate from the dermal fibroblast [39]. Based on these results, the difference in E-cadherin expression may be an intrinsic trait of each MCC host cell. An alternative explanation may be that the increased E-cadherin stain area fraction is an extrinsic, viral-mediated trait. Virus-mediated downregulation of E-cadherin has been reported for the Epstein-Barr virus in nasopharyngeal carcinoma and for the hepatitis C virus in hepatocellular carcinoma [40,41]. Future experiments with the knockdown of viral proteins may provide additional knowledge to this question.

The positive prognostic impact of CD8 lymphocytes and its association with PD-L1 is well-recognized [15,42,43]. The latter is well-known to occur through a CD8 lymphocyte-mediated induction of the interferon-γ pathway [44]. However, to the best of our knowledge, this is the first time that the N:CD8 ratio in the TME has been examined in MCC. In this current study, with 89 patients included in the analysis, a high N:CD8 ratio in the tumor was an independent prognostic marker of poor MCC-specific survival in both univariate and multivariate analysis. One recently published study examined its role in the peripheral blood of MCC patients, where a high N:CD8 ratio at baseline associated with a poor MCC-specific survival [45]. This may be due to the role of neutrophils in suppressing the antitumor effect of lymphocytes [30].

Our study had several important limitations, including its retrospective design. Ninety included patients in our analysis represent a large number in the scope of MCC research but is a relatively small sample size in statistical analysis. Formalin-fixed paraffin-embedded (FFPE) blocks were obtained from different pathology departments with different protocols from the time of tissue excision to final tissue preparation. We were therefore unable to control for the difference in fixation time, which could potentially affect the IHC. We used a digital image analysis that measures the immune stain area while manual assessments involve counting the number of stained cells, although comparative studies of these two evaluation methods show high concordance [46]. The strict legislation on the acquisition of patient journal materials meant that we could not obtain information on patient treatments. This may be a confounder when evaluating prognostic markers. Tumor size was not a prognostic marker in our cohort. This might be rooted in several factors, including the size and composition of our cohort, and may subsequently limit our findings. Due to the previous reported and accepted prognostic role of tumor size, we found it most correct adjusting for both lymph node involvement and tumor size in the multivariate analyses [1].

4. Materials and Methods

4.1. Patients and Samples

Patients diagnosed with MCC between 1 January 1996 to 31 December 2012 at Aarhus University Hospital and between 1 January 2007 to 31 December 2012 at Aalborg University Hospital, Vejle Hospital, Odense University Hospital, Herlev & Nordsjaelland Hospital, Bispebjerg Hospital and Rigshospitalet were included while searching the Aarhus Pathology Database and the Danish National Pathology Database using the SNOMED code M8247* for Merkel cell tumors. One-hundred and twenty-one (n = 121) patients matched the search criteria. After exclusion, ninety (n = 90) patients were included in the analyses (Figure S3). Clinical endpoints including the time of death and cause of death were obtained from the Danish Register of Causes of Death filed by a local doctor with knowledge of the patient's admissions and disease history. Data on tumor size and pathology-confirmed regional lymph node involvement (fine needle aspiration and sentinel lymph node biopsy) were obtained from the Danish Pathology Database. This project was approved by the regional central Denmark Ethics Committee (Ethics code: 1-10-72-280-16)

4.2. Tumor Specimens

Formalin-fixed paraffin-embedded (FFPE) tissue blocks with primary MCC were evaluated at the Department of Pathology, Aarhus University Hospital. To confirm the diagnosis and presence of tumor tissues, 2-µm-thick sections were cut and stained with haematoxylin and eosin (HE) and evaluated by the departments senior pathologist (TS). Serial sections for further analysis with IHC and macro-dissections for DNA extraction were prepared.

4.3. DNA Extraction and Quantification

A 2-µm-thick section was cut and H&E-stained to mark a representative tumor-only area to guide the macro-dissection. Three sections (10-µm-thick) were cut and macro-dissected of the slide into a sterile tube. Between each patient sample, the microtome, gloves and knife were changed to avoid cross-contamination. DNA extraction was performed on the QIAsymphony SP (QIAGEN, Germany) following the manufacturer's protocol. DNA purity and quantity were estimated on the Implen nanophotometer (Implen GmbH, Germany).

4.4. Real-time Taqman Polymerase Chain Reaction

Real-time quantitative PCR was performed on the Stratagene Mx3000P at the Department of Pathology, Aarhus University Hospital with previously tested Taqman viral primer sets (LT2, LT3, Set6 and Set7) with Onyx Quencher A (Sigma-Aldrich Company, Ltd, St. Louis, MO, USA) [35]. These primers are designed to amplify sequences within nucleotide position 196–1257 in the MCV genome. This region is known to be present in all variations of sequenced MCV-DNA from MCC. The housekeeping gene TATA-binding protein (TBP) was used as a reference (LGC Biosearch Technologies, United Kingdom; forward primer CACCACAGCTCTTCCACTCA; reverse primer GGGGAGGGATACAGTGGAGT; Probe AGACTCTCACAACTGCACCCTTGC). The testing was done with duplicates of each patient sample, negative controls (H_2O, tonsillar tissue) and positive control with a Merkel cell virus-positive cell line (MKL-1, Sigma-Aldrich). qPCR was performed for 40 cycles at 95 °C for 3 s and 60 °C for 20 s.

4.5. Immunohistochemical Staining

IHC was performed on the Ventana Benchmark XT-automated immunohistochemistry platform (Oro Valley, AZ, USA) and the Dako Autostainer Link48 (Santa Clara, CA, USA). From each FFPE, five consecutive sections (3-μm-thick) were cut and prepared for staining of CD8 lymphocytes (Dako, C8/144b, 1:200, OV dab); PD-L1 (Dako, 22C3, RTU, Dab); CD34 endothelia (Ventana, Oro Valley, AZ, USA, QBEnd/10, RTU, OV dab); CD66b neutrophils (BD Bioscience, Franklin Lakes, NJ, USA, G10F5, 1:200, UV red); E-cadherin (Ventana, Oro Valley, AZ, USA, 36, RTU, UV red) and CMB2B4 virus antigen (Santa Cruz, CA, USA, Poly, 1:100, OV dab) (Figure 1A–E). IHC was performed in large batches to reduce batch-to-batch variance between runs. Control tissue with internal negative and positive controls were used for all IHC staining. Control tissue for CMB2B4 virus antigen consisted of an MCV-positive patient sample estimated by qPCR and CMB2B4 staining, while tonsillar tissue was used for the remaining IHC stains.

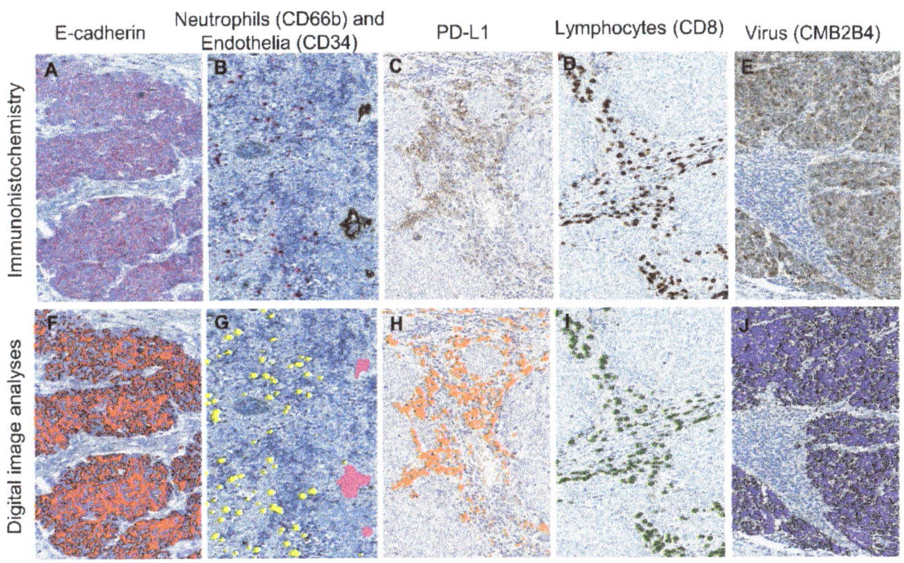

Figure 1. Sections stained with immunohistochemistry (IHC) (top row) analyzed with digital pathology (bottom row). Stained IHC sections of (**A**) E-cadherin, (**B**) CD66b neutrophils & CD34 endothelia, (**C**) PD-L1, (**D**) CD8 lymphocytes and (**E**) CMB2B4 at 20× magnification with comparable illustrations of digital image analysis (**F–J**). (**B**) CD34 endothelia (brown) and CD66b neutrophils (red) are stained on the same section. The digital image analysis software converts the IHC dye into a digital color that is used for the calculation of stain area fraction.

4.6. Digital Pathology

Software from Visiopharm (Visiopharm A/S, Denmark) was used to attain a quantitative estimate of all analyzed factors. Image analysis protocols were developed by training the software to recognize specific colors of the stains used (Figure 1F–J). The results of image analyses of all sections were reviewed by the observer to exclude errors. A region of interest for the automatic evaluation of IHC stains was manually marked. The region of interest included tumor epithelium and adjacent intratumoral stroma. In this region, CD8, PD-L1, cd66b (both intra- and extravascular neutrophils) and CD34 (vascularization) were assessed, whereas E-cadherin and CMB2B4 (virus) were assessed only in the contained tumor epithelium. The IHC-stain area fraction per region of interest in percent was calculated regarding CD8 lymphocytes, PD-L1 and CD34 (vascularization), whereas the stain area fractions of virus-positive cells and E-cadherin were defined as the area of CMB2B-positive and E-cadherin-positive MCC cells, respectively, divided by the area of tumor epithelium. The tumor neutrophil-to-CD8 lymphocyte ratio (N:CD8 ratio) was estimated by the stain area fraction of cd66b divided by the stain area fraction of CD8 in the tumor.

4.7. Ulceration Status

Ulceration was defined as the full-thickness loss of the epidermis overlying MCC tissue in which epidermal loss was associated with a host reaction. The H&E-stained section was used for ulceration estimation, which was consensual based between SN and MLB, verified if in doubt by a senior pathologist (TS).

4.8. Viral Status

IHC: The Allred scoring system combines the intensity of staining (0–3) and proportion of cells stained (0-5), into a 0–8 points score. This method of semiquantative evaluation has previously been used to determine if a sample is considered positive for the MCV antigen, with a threshold set to 2 equating < 1% of cells with weak staining [32,47]. With this threshold in mind, the objective estimate in Visiopharm was set to analyze the stain area fraction of virus-positive cells with 1% as the cut-off.

qPCR: MCV is part of the skin flora and may therefore be present in tissue samples with virus-negative MCC [48]. To match the cut-off of immune staining, samples with less than 1% of cells containing viral DNA were categorized as "PCR-negative" (equating a viral primer/TBP ratio < 0.01). In this study, virus status was based on the qPCR results.

4.9. Statistical Methods

The stain area faction of CD8 lymphocytes, PD-L1, neutrophils, CD34 endothelial cells, E-cadherin and virus antigen expression were log transformed, and the assumption of normal distribution assessed using the residuals. Correlations between the different markers were analyzed using linear regression and estimation of spearman correlation coefficients, and the differences in means between the groups were tested using a *t*-test. Data concerning the viral status (qPCR and immune staining) and ulcerated status was dichotomized and tested with a chi-square test. The study endpoint was disease-specific survival, defined as the time from the date of surgery to date of death from MCC. Statistical analysis of survival was performed using the Cox proportional hazards. Each variable was tested in multivariate analysis adjusted for two variables (tumor size and lymph node involvement) to retain sufficient statistical power with $n > 10$ events per adjusted factor. These variables are known prognostic markers in MCC, included in the 8th AJCC staging system [1]. Survival probabilities were illustrated using the Kaplan–Meier method. Level of significance of 0.05 was used for all analyses.

5. Conclusions

The results of this study show that patients with ulcerated primary tumors, absence of virus, scarce infiltration of CD8 lymphocytes and a high N:CD8 ratio have a significantly worse prognosis. In the clinical setting, we therefore suggest that these factors should be reported, as this may provide a more accurate prognosis and lead to better prognostic stratification for the patients in determination of the resection margin size and in the stratification of patients for adjuvant treatment based on the predicted risk of recurrence and death. Furthermore, estimation of ulceration status is easy, fast and does not require additional staining, while detection of virus, neutrophils and CD8 lymphocytes with IHC is reliable and easy to implement in the clinical labs.

Supplementary Materials: The following are available online at http://www.mdpi.com/2072-6694/12/4/888/s1: Table S1: Characteristics between ulcerated and nonulcerated MCC, Figure S1: Images showing differences in staining and survival between ulcerated and nonulcerated MCC, Figure S2: Images showing differences in staining and survival between virus-positive and -negative MCC, Figure S3: Flowchart of included and excluded patients and samples.

Author Contributions: Writing and laboratory work and statistics and visiopharm, S.N.; pathology confirmation of the diagnosis of MCC and determination of ulceration status and providing facilities for laboratory work and financial support, T.S.; immunohistochemical staining, J.B.G.; qPCR, R.T.; patient database design and review and editing, M.L.; review and editing M.H.; review and editing, T.E.D. and study design and visiopharm and statistics and writing and editing, M.L.B.-B. All authors have read and agreed to the published version of the manuscript.

Funding: This research was funded by KRÆFTENS BEKÆMPELSE (THE DANISH CANCER SOCIETY), grant number R157-A10338, P. A. MESSERSCHMIDT OG HUSTRUS FOND, grant number 028077-0006 jhw/mkol, FABRIKANT FRANDS KØHLER NIELSEN OG HUSTRUS MINDELEGAT, grant number 819167 UBJ/gr and AUGUST FREDERIK WEDELL ERICHSENS LEGAT, grant number 13655.

Acknowledgments: Acknowledgements to the Department of Pathology, Aarhus University Hospital for providing the facilities for the laboratory work and Aarhus University for providing a research year scholarship.

Conflicts of Interest: The authors declare no conflict of interest.

References

1. Harms, K.L.; Healy, M.A.; Nghiem, P.; Sober, A.J.; Johnson, T.M.; Bichakjian, C.K.; Wong, S.L. Analysis of Prognostic Factors from 9387 Merkel Cell Carcinoma Cases Forms the Basis for the New 8th Edition AJCC Staging System. *Ann. Surg. Oncol.* **2016**, *23*, 3564–3571. [CrossRef] [PubMed]
2. Toker, C. Trabecular carcinoma of the skin. *Arch. Dermatol.* **1972**, *105*, 107–110. [CrossRef] [PubMed]
3. Fitzgerald, T.L.; Dennis, S.; Kachare, S.D.; Vohra, N.A.; Wong, J.H.; Zervos, E.E. Dramatic Increase in the Incidence and Mortality from Merkel Cell Carcinoma in the USA. *Am. Surg.* **2015**, *81*, 802–806. [PubMed]
4. Lyhne, D.; Lock-Andersen, J.; Dahlstrom, K.; Drzewiecki, K.T.; Balslev, E.; Muhic, A.; Krarup-Hansen, A. Rising incidence of Merkel cell carcinoma. *J. Plast. Surg. Hand Surg.* **2011**, *45*, 274–280. [CrossRef] [PubMed]
5. Youlden, D.R.; Soyer, H.P.; Youl, P.H.; Fritschi, L.; Baade, P.D. Incidence and survival for Merkel cell carcinoma in Queensland, Australia, 1993–2010. *Jama Dermatol.* **2014**, *150*, 864–872. [CrossRef]
6. Feng, H.; Shuda, M.; Chang, Y.; Moore, P.S. Clonal integration of a polyomavirus in human Merkel cell carcinoma. *Science* **2008**, *319*, 1096–1100. [CrossRef]
7. Wong, S.Q.; Waldeck, K.; Vergara, I.A.; Schroder, J.; Madore, J.; Wilmott, J.S.; Colebatch, A.J.; De Paoli-Iseppi, R.; Li, J.; Lupat, R.; et al. UV-Associated Mutations Underlie the Etiology of MCV-Negative Merkel Cell Carcinomas. *Cancer Res.* **2015**, *75*, 5228–5234. [CrossRef]
8. Kaufman, H.L.; Russell, J.; Hamid, O.; Bhatia, S.; Terheyden, P.; D'Angelo, S.P.; Shih, K.C.; Lebbe, C.; Linette, G.P.; Milella, M.; et al. Avelumab in patients with chemotherapy-refractory metastatic Merkel cell carcinoma: A multicentre, single-group, open-label, phase 2 trial. *Lancet Oncol.* **2016**, *17*, 1374–1385. [CrossRef]
9. Nghiem, P.T.; Bhatia, S.; Lipson, E.J.; Kudchadkar, R.R.; Miller, N.J.; Annamalai, L.; Berry, S.; Chartash, E.K.; Daud, A.; Fling, S.P.; et al. PD-1 Blockade with Pembrolizumab in Advanced Merkel-Cell Carcinoma. *N. Engl. J. Med.* **2016**, *374*, 2542–2552. [CrossRef]
10. Yarchoan, M.; Hopkins, A.; Jaffee, E.M. Tumor Mutational Burden and Response Rate to PD-1 Inhibition. *N. Engl. J. Med.* **2017**, *377*, 2500–2501. [CrossRef]
11. Hanahan, D.; Weinberg, R.A. Hallmarks of cancer: The next generation. *Cell* **2011**, *144*, 646–674. [CrossRef] [PubMed]
12. Lipson, E.J.; Vincent, J.G.; Loyo, M.; Kagohara, L.T.; Luber, B.S.; Wang, H.; Xu, H.; Nayar, S.K.; Wang, T.S.; Sidransky, D.; et al. PD-L1 expression in the Merkel cell carcinoma microenvironment: Association with inflammation, Merkel cell polyomavirus and overall survival. *Cancer Immunol. Res.* **2013**, *1*, 54–63. [CrossRef] [PubMed]
13. Sihto, H.; Bohling, T.; Kavola, H.; Koljonen, V.; Salmi, M.; Jalkanen, S.; Joensuu, H. Tumor infiltrating immune cells and outcome of Merkel cell carcinoma: A population-based study. *Clin. Cancer Res.* **2012**, *18*, 2872–2881. [CrossRef] [PubMed]
14. Miller, N.J.; Church, C.D.; Dong, L.; Crispin, D.; Fitzgibbon, M.P.; Lachance, K.; Jing, L.; Shinohara, M.; Gavvovidis, I.; Willimsky, G.; et al. Tumor-Infiltrating Merkel Cell Polyomavirus-Specific T Cells Are Diverse and Associated with Improved Patient Survival. *Cancer Immunol. Res.* **2017**, *5*, 137–147. [CrossRef]
15. Paulson, K.G.; Iyer, J.G.; Simonson, W.T.; Blom, A.; Thibodeau, R.M.; Schmidt, M.; Pietromonaco, S.; Sokil, M.; Warton, E.M.; Asgari, M.M.; et al. CD8+ lymphocyte intratumoral infiltration as a stage-independent predictor of Merkel cell carcinoma survival: A population-based study. *Am. J. Clin. Pathol.* **2014**, *142*, 452–458. [CrossRef]
16. Antonio, N.; Bonnelykke-Behrndtz, M.L.; Ward, L.C.; Collin, J.; Christensen, I.J.; Steiniche, T.; Schmidt, H.; Feng, Y.; Martin, P. The wound inflammatory response exacerbates growth of pre-neoplastic cells and progression to cancer. *EMBO J.* **2015**, *34*, 2219–2236. [CrossRef]

17. Bonnelykke-Behrndtz, M.L.; Steiniche, T.; Norgaard, P.; Danielsen, A.V.; Damsgaard, T.E.; Christensen, I.J.; Bastholt, L.; Moller, H.J.; Schmidt, H. Loss of E-cadherin as Part of a Migratory Phenotype in Melanoma Is Associated With Ulceration. *Am. J. Dermatopathol.* **2017**, *39*, 672–678. [CrossRef]
18. Bald, T.; Quast, T.; Landsberg, J.; Rogava, M.; Glodde, N.; Lopez-Ramos, D.; Kohlmeyer, J.; Riesenberg, S.; van den Boorn-Konijnenberg, D.; Homig-Holzel, C.; et al. Ultraviolet-radiation-induced inflammation promotes angiotropism and metastasis in melanoma. *Nature* **2014**, *507*, 109–113. [CrossRef]
19. Kaufman, H.L.; Russell, J.S.; Hamid, O.; Bhatia, S.; Terheyden, P.; D'Angelo, S.P.; Shih, K.C.; Lebbe, C.; Milella, M.; Brownell, I.; et al. Updated efficacy of avelumab in patients with previously treated metastatic Merkel cell carcinoma after >/=1 year of follow-up: JAVELIN Merkel 200, a phase 2 clinical trial. *J. Immunother. Cancer* **2018**, *6*, 7. [CrossRef]
20. Balch, C.M.; Gershenwald, J.E.; Soong, S.J.; Thompson, J.F. Update on the melanoma staging system: The importance of sentinel node staging and primary tumor mitotic rate. *J. Surg. Oncol.* **2011**, *104*, 379–385. [CrossRef]
21. Grabowski, J.; Saltzstein, S.L.; Sadler, G.R.; Tahir, Z.; Blair, S. A comparison of merkel cell carcinoma and melanoma: Results from the california cancer registry. *Clin. Med. Oncol.* **2008**, *2*, 327–333. [CrossRef] [PubMed]
22. Gershenwald, J.E.; Scolyer, R.A.; Hess, K.R.; Sondak, V.K.; Long, G.V.; Ross, M.I.; Lazar, A.J.; Faries, M.B.; Kirkwood, J.M.; McArthur, G.A.; et al. Melanoma staging: Evidence-based changes in the American Joint Committee on Cancer eighth edition cancer staging manual. *CA Cancer J. Clin.* **2017**, *67*, 472–492. [CrossRef] [PubMed]
23. Frohm, M.L.; Griffith, K.A.; Harms, K.L.; Hayman, J.A.; Fullen, D.R.; Nelson, C.C.; Wong, S.L.; Schwartz, J.L.; Bichakjian, C.K. Recurrence and Survival in Patients With Merkel Cell Carcinoma Undergoing Surgery Without Adjuvant Radiation Therapy to the Primary Site. *Jama Dermatol.* **2016**, *152*, 1001–1007. [CrossRef] [PubMed]
24. Andea, A.A.; Coit, D.G.; Amin, B.; Busam, K.J. Merkel cell carcinoma: Histologic features and prognosis. *Cancer* **2008**, *113*, 2549–2558. [CrossRef]
25. Mott, R.T.; Smoller, B.R.; Morgan, M.B. Merkel cell carcinoma: A clinicopathologic study with prognostic implications. *J. Cutan. Pathol.* **2004**, *31*, 217–223. [CrossRef]
26. Skelton, H.G.; Smith, K.J.; Hitchcock, C.L.; McCarthy, W.F.; Lupton, G.P.; Graham, J.H. Merkel cell carcinoma: Analysis of clinical, histologic, and immunohistologic features of 132 cases with relation to survival. *J. Am. Acad. Dermatol.* **1997**, *37*, 734–739. [CrossRef]
27. Bob, A.; Nielen, F.; Krediet, J.; Schmitter, J.; Freundt, D.; Terhorst, D.; Rowert-Huber, J.; Kanitakis, J.; Stockfleth, E.; Ulrich, C.; et al. Tumor vascularization and clinicopathologic parameters as prognostic factors in merkel cell carcinoma. *J. Cancer Res. Clin. Oncol.* **2017**, *143*, 1999–2010. [CrossRef]
28. Llombart, B.; Monteagudo, C.; Lopez-Guerrero, J.A.; Carda, C.; Jorda, E.; Sanmartin, O.; Almenar, S.; Molina, I.; Martin, J.M.; Llombart-Bosch, A. Clinicopathological and immunohistochemical analysis of 20 cases of Merkel cell carcinoma in search of prognostic markers. *Histopathology* **2005**, *46*, 622–634. [CrossRef]
29. Nagase, K.; Kimura-Kaku, H.; Inoue, T.; Shinogi, T.; Narisawa, Y. Usefulness of ulceration and hyperkeratosis as clinical predictors of Merkel cell polyomavirus-negative and combined Merkel cell carcinoma: A retrospective study. *J. Dermatol.* **2019**, *46*, 103–109. [CrossRef]
30. Michaeli, J.; Shaul, M.E.; Mishalian, I.; Hovav, A.H.; Levy, L.; Zolotriov, L.; Granot, Z.; Fridlender, Z.G. Tumor-associated neutrophils induce apoptosis of non-activated CD8 T-cells in a TNFalpha and NO-dependent mechanism, promoting a tumor-supportive environment. *Oncoimmunology* **2017**, *6*, e1356965. [CrossRef]
31. Rana, S.; Byrne, S.N.; MacDonald, L.J.; Chan, C.Y.; Halliday, G.M. Ultraviolet B suppresses immunity by inhibiting effector and memory T cells. *Am. J. Pathol.* **2008**, *172*, 993–1004. [CrossRef] [PubMed]
32. Moshiri, A.S.; Doumani, R.; Yelistratova, L.; Blom, A.; Lachance, K.; Shinohara, M.M.; Delaney, M.; Chang, O.; McArdle, S.; Thomas, H.; et al. Polyomavirus-Negative Merkel Cell Carcinoma: A More Aggressive Subtype Based on Analysis of 282 Cases Using Multimodal Tumor Virus Detection. *J. Investig. Dermatol.* **2017**, *137*, 819–827. [CrossRef] [PubMed]
33. Coursaget, P.; Samimi, M.; Nicol, J.T.; Gardair, C.; Touze, A. Human Merkel cell polyomavirus: Virological background and clinical implications. *Apmis* **2013**, *121*, 755–769. [CrossRef] [PubMed]

34. Garneski, K.M.; Warcola, A.H.; Feng, Q.; Kiviat, N.B.; Leonard, J.H.; Nghiem, P. Merkel cell polyomavirus is more frequently present in North American than Australian Merkel cell carcinoma tumors. *J. Investig. Dermatol.* **2009**, *129*, 246–248. [CrossRef] [PubMed]
35. Rodig, S.J.; Cheng, J.; Wardzala, J.; DoRosario, A.; Scanlon, J.J.; Laga, A.C.; Martinez-Fernandez, A.; Barletta, J.A.; Bellizzi, A.M.; Sadasivam, S.; et al. Improved detection suggests all Merkel cell carcinomas harbor Merkel polyomavirus. *J. Clin. Investig.* **2012**, *122*, 4645–4653. [CrossRef]
36. Touzé, A.; Gaitan, J.; Maruani, A.; Le Bidre, E.; Doussinaud, A.; Clavel, C.; Durlach, A.; Aubin, F.; Guyétant, S.; Lorette, G.; et al. Merkel Cell Polyomavirus Strains in Patients with Merkel Cell Carcinoma. *Emerg. Infect. Dis* **2009**, *15*, 960–962.
37. Harms, P.W.; Vats, P.; Verhaegen, M.E.; Robinson, D.R.; Wu, Y.M.; Dhanasekaran, S.M.; Palanisamy, N.; Siddiqui, J.; Cao, X.; Su, F.; et al. The Distinctive Mutational Spectra of Polyomavirus-Negative Merkel Cell Carcinoma. *Cancer Res.* **2015**, *75*, 3720–3727. [CrossRef]
38. Kalluri, R.; Weinberg, R.A. The basics of epithelial-mesenchymal transition. *J. Clin. Investig.* **2009**, *119*, 1420–1428. [CrossRef]
39. Sunshine, J.C.; Jahchan, N.S.; Sage, J.; Choi, J. Are there multiple cells of origin of Merkel cell carcinoma? *Oncogene* **2018**, *37*, 1409–1416. [CrossRef]
40. Krishna, S.M.; Kattoor, J.; Balaram, P. Down regulation of adhesion protein E-cadherin in Epstein-Barr virus infected nasopharyngeal carcinomas. *Cancer Biomark* **2005**, *1*, 271–277. [CrossRef]
41. Arora, P.; Kim, E.O.; Jung, J.K.; Jang, K.L. Hepatitis C virus core protein downregulates E-cadherin expression via activation of DNA methyltransferase 1 and 3b. *Cancer Lett.* **2008**, *261*, 244–252. [CrossRef] [PubMed]
42. Paulson, K.G.; Iyer, J.G.; Tegeder, A.R.; Thibodeau, R.; Schelter, J.; Koba, S.; Schrama, D.; Simonson, W.T.; Lemos, B.D.; Byrd, D.R.; et al. Transcriptome-wide studies of merkel cell carcinoma and validation of intratumoral CD8+ lymphocyte invasion as an independent predictor of survival. *J. Clin. Oncol.* **2011**, *29*, 1539–1546. [CrossRef] [PubMed]
43. Feldmeyer, L.; Hudgens, C.W.; Ray-Lyons, G.; Nagarajan, P.; Aung, P.P.; Curry, J.L.; Torres-Cabala, C.A.; Mino, B.; Rodriguez-Canales, J.; Reuben, A.; et al. Density, Distribution, and Composition of Immune Infiltrates Correlate with Survival in Merkel Cell Carcinoma. *Clin. Cancer Res.* **2016**, *22*, 5553–5563. [CrossRef] [PubMed]
44. Taube, J.M.; Anders, R.A.; Young, G.D.; Xu, H.; Sharma, R.; McMiller, T.L.; Chen, S.; Klein, A.P.; Pardoll, D.M.; Topalian, S.L.; et al. Colocalization of inflammatory response with B7-h1 expression in human melanocytic lesions supports an adaptive resistance mechanism of immune escape. *Sci. Transl. Med.* **2012**, *4*, 127ra137. [CrossRef]
45. Zaragoza, J.; Kervarrec, T.; Touze, A.; Avenel-Audran, M.; Beneton, N.; Esteve, E.; Wierzbicka Hainaut, E.; Aubin, F.; Machet, L.; Samimi, M. A high neutrophil-to-lymphocyte ratio as a potential marker of mortality in patients with Merkel cell carcinoma: A retrospective study. *J. Am. Acad. Dermatol.* **2016**, *75*, 712–721. [CrossRef] [PubMed]
46. Carus, A.; Donskov, F.; Nielsen, P.; Hager, H.; Nedergaard, B.; Steiniche, T.; Ladekarl, M. Strong Prognostic Value of Tumor-infiltrating Neutrophils and Lymphocytes Assessed by Automated Digital Image Analysis in Early Stage Cervical Cancer: A Comparator Study with Observer-assisted Stereological Assessments. *J. OncoPathol.* **2014**, *2*. [CrossRef]
47. Allred, D.C.; Harvey, J.M.; Berardo, M.; Clark, G.M. Prognostic and predictive factors in breast cancer by immunohistochemical analysis. *Mod. Pathol.* **1998**, *11*, 155–168.
48. Schowalter, R.M.; Pastrana, D.V.; Pumphrey, K.A.; Moyer, A.L.; Buck, C.B. Merkel cell polyomavirus and two previously unknown polyomaviruses are chronically shed from human skin. *Cell Host Microbe* **2010**, *7*, 509–515. [CrossRef]

cancers

MDPI

Article

Highly Expressed miR-375 is not an Intracellular Oncogene in Merkel Cell Polyomavirus-Associated Merkel Cell Carcinoma

Kaiji Fan [1,2,3,4], Armin Zebisch [5,6], Kai Horny [1,2,3], David Schrama [7] and Jürgen C. Becker [1,2,3,8,*]

1 Department of Translational Skin Cancer Research, University Hospital Essen, 45141 Essen, Germany; k.fan@dkfz-heidelberg.de (K.F.); k.horny@dkfz-heidelberg.de (K.H.)
2 German Cancer Consortium (DKTK), 45141 Essen, Germany
3 German Cancer Research Center (DKFZ), 69120 Heidelberg, Germany
4 Department of Dermatology, Medical University of Graz, 8010 Graz, Austria
5 Division of Hematology, Medical University of Graz, 8010 Graz, Austria; armin.zebisch@medunigraz.at
6 Otto Loewi Research Center for Vascular Biology, Immunology and Inflammation, Division of Pharmacology, Medical University of Graz, 8010 Graz, Austria
7 Department of Dermatology, University Hospital Würzburg, 97080 Würzburg, Germany; schrama_d@ukw.de
8 Department of Dermatology, University Hospital Essen, 45147 Essen, Germany
* Correspondence: j.becker@dkfz.de; Tel.: +49-201-1836-727

Received: 6 January 2020; Accepted: 24 February 2020; Published: 25 February 2020

Abstract: miR-375 is a highly abundant miRNA in Merkel cell carcinoma (MCC). In other cancers, it acts as either a tumor suppressor or oncogene. While free-circulating miR-375 serves as a surrogate marker for tumor burden in patients with advanced MCC, its function within MCC cells has not been established. Nearly complete miR-375 knockdown in MCC cell lines was achieved using antagomiRs via nucleofection. The cell viability, growth characteristics, and morphology were not altered by this knockdown. miR-375 target genes and related signaling pathways were determined using Encyclopedia of RNA Interactomes (ENCORI) revealing Hippo signaling and epithelial to mesenchymal transition (EMT)-related genes likely to be regulated. Therefore, their expression was analyzed by multiplexed qRT-PCR after miR-375 knockdown, demonstrating only a limited change in expression. In summary, highly effective miR-375 knockdown in classical MCC cell lines did not significantly change the cell viability, morphology, or oncogenic signaling pathways. These observations render miR-375 an unlikely intracellular oncogene in MCC cells, thus suggesting that likely functions of miR-375 for the intercellular communication of MCC should be addressed.

Keywords: miR-375; antagomiRs; Merkel cell carcinoma; Hippo signaling; focal adhesion

1. Introduction

Merkel cell carcinoma (MCC) is an aggressive skin cancer. Risk factors for MCC include an advanced age, ultraviolet (UV) light exposure, and immune suppression [1]. About 80% of MCC tumors are associated with genomic integration of the Merkel cell polyomavirus (MCPyV) bearing truncating tumor-specific large T antigen mutations, while the others are characterized by a UV-induced tumor mutational burden [1]. The pathogenesis of these two types of MCC tumors is surmised to be distinct: MCPyV-positive MCC is associated with MCPyV T antigen-mediated tumor suppressor gene inhibition and/or oncogene induction [1–3], while in MCPyV-negative MCC tumors, the comparable oncogenic observations are caused by UV-induced DNA mutations [1,4–6]. However, the specific molecular alterations caused by either MCPyV or UV-mutations are just starting to emerge [7,8].

Transcription factor Atonal homolog 1 is characterized as a lineage-dependency oncogene in MCC, which induces miR-375 expression [9]. microRNAs (miRNAs) are small, ~21nt single-stranded RNAs, which post-transcriptionally regulate the stability and translation of genes, mainly by binding to the 3′ UTR of mRNAs [10,11]. Each miRNA can bind a specific set of genes, which are referred to as its target genes. The dysregulation of miRNAs has been reported in almost all types of human cancer [10,12]. miRNA expression profiling in MCCs revealed miR-375 as one of the most abundant miRNAs in classical MCC cell lines and tumor tissues [13–16]. Physiologically, miR-375 acts as a pancreatic-islet miRNA essential for β-cell formation and the regulation of insulin secretion [17,18]. Divergent miR-375 expression has been described for multiple cancer types, e.g., reduced expression in gastric [19,20], pancreatic [21], colon [22,23], and liver cancer [24], and high expression in medullary thyroid carcinoma [25], prostate cancer [26], and MCC [13–16]. Therefore, miR-375 was assumed to be an oncogenic miRNA in the latter group.

However, when the function of miR-375 in MCC was studied by different groups, the results were inconsistent. Abraham et al. reported that miR-375 was involved in neuroendocrine differentiation and miR-375 knockdown in classical MCC cell lines (MKL-1 and MS-1) and did not alter their growth properties [13]. Our preliminary results of miR-375 knockdown experiments were consistent with their report for the tested MCC cell lines [9]. In contrast, Kumar et al. reported that miR-375 inhibition in WaGa and MKL-1 cells reduced cell growth and induced apoptosis by targeting lactate dehydrogenase b (*LDHB*) [27]. Recent reports from the same group showed that miR-375, together with other miRNAs, inhibits autophagy, thus protecting MCC cells from autophagy-associated cell death [28]. To resolve these controversies, here, we scrutinize the function of miR-375 in MCC. For this, we established a highly efficient method for miR-375 knockdown in classical MCC cell lines and analyzed the inflected effects, with an emphasis on intracellular signaling.

2. Results

2.1. *Effective Knockdown of miR-375 by Nuclear Transfection Using miR-375 AntagomiRs*

To explore the function of miRNAs, it is essential to achieve largely complete knockdown. Achieving highly effective knockdown in classical MCC cell lines is trivial. Therefore, we tested different transfection methods, i.e., lipofectamine and nucleofection, in the two classical MCC cell lines WaGa and PeTa using miR-375 antagomiRs.

The transfection of miR-375 antagomiRs by lipofectamine reduced miR-375 expression in a dose-dependent manner, but was not sufficient for complete knockdown of the highly expressed miR-375 (Figure 1a,b). Next, we performed nucleofection and optimized the transfection conditions. Program D23 with 25nM miR-375 antagomiRs was determined as the optimal protocol for knockdown, which rendered dramatically reduced miR-375 expression in both WaGa and PeTa cells (Figure 1c,d and Figure S1). All further experiments were carried out using these conditions.

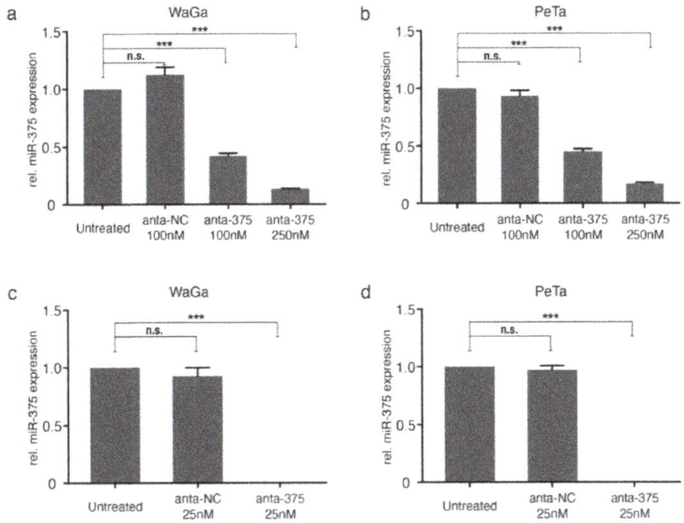

Figure 1. Knockdown of miR-375 in classical Merkel cell carcinoma (MCC) cell lines. Relative miR-375 expression was determined in triplicate by qRT-PCR in WaGa (**a**,**c**) and PeTa (**b**,**d**) cells transfected with miR-375 antagomiRs (anta-375) or a negative control (anta-NC) using lipofectamine (top row; **a**,**b**) or nucleofection (bottom row; **c**,**d**). Quantification cycle threshold (Cq) values were normalized to the small nucleolar RNA RNU6B (U6) and calibrated to the untreated WaGa cells. All experiments were independently repeated three times. Error bars represent SD, *** indicates $p < 0.001$. n.s.: non-significant.

2.2. *miR-375 Knockdown Does Not Impact the Morphology, Proliferative Capacity, or Apoptosis of MCC Cells*

We were able to confirm our previous observation that miR-375 knockdown has no major impact on cell proliferation, survival, growth characteristics, or cell morphology (Figure 2 and Figures S2 and S3). Notably, even the highly effective miR-375 knockdown did not alter the morphologic appearance as cells still showed a neuroendocrine growth pattern as loose spheroids or single cells, which was identical to the growth pattern in cells transfected with unspecific control antagomiRs (Figure 2a,b). Furthermore, neither the metabolic nor proliferative activity was affected by the miR-375 knockdown (Figure 2c,d). While the harsh transfection conditions for the highly efficient miR-375 knockdown inhibited the proliferation of MCC cells per se, we observed around 40% apoptotic cells 24h after nucleofection in both WaGa and PeTa, and no difference was observed in MCC cells transfected with miR-375 antagomiRs or the negative control (Figure 2e,f and Figure S3). Sequential analyses on days 3 and 5 after transfection further supported that miR-375 knockdown had no specific impact on cell survival or metabolic activity (Figure 2c–f and Figure S3).

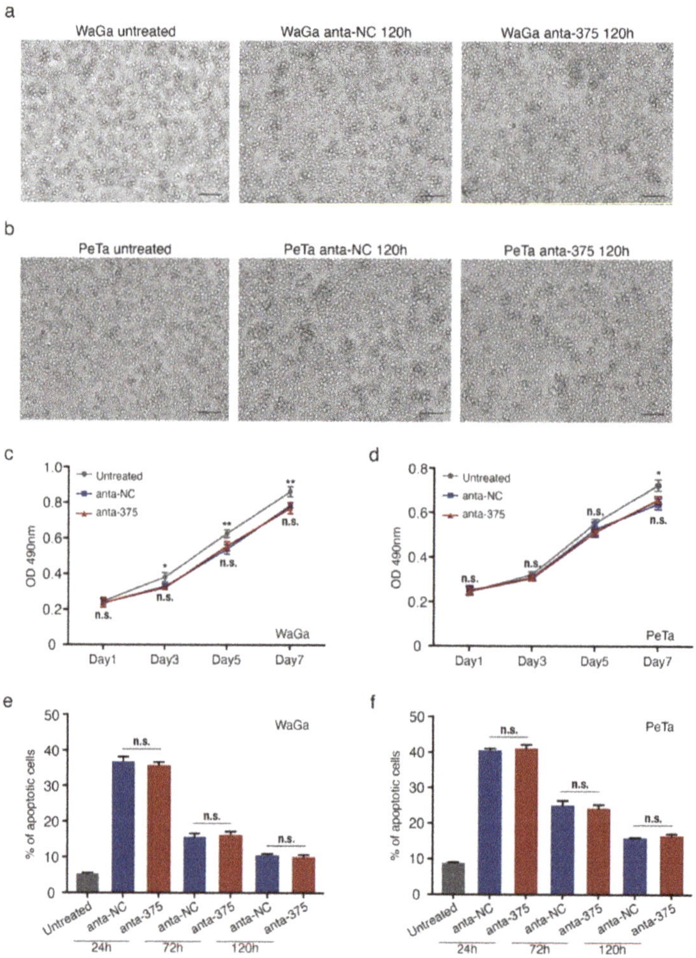

Figure 2. miR-375 knockdown does not alter the cell morphology, viability, and apoptosis of MCC cells. (**a**,**b**) Morphology of WaGa (**a**) and PeTa (**b**) cells, untransfected (untreated) and 120 h after nucleofection with either anta-NC or anta-375. (**c**,**d**) Cell proliferation (metabolic activity) of WaGa (**c**) and PeTa (**d**) cells after nucleofection with miR-375 antagomiRs or a negative control was measured by MTS assays at the indicated time points. Absorbance values at 490 nm are presented. Scale bar: 100 μM. (**e**,**f**) The apoptotic cell rate of untreated or nuclear transfected WaGa (**e**) and PeTa (**f**) cells was determined by flow cytometry using the NucView 488/ MitoView 633 apoptosis assay. Scale bar represents 50μm. All experiments were independently repeated three times, error bars represent SD, * indicates $p < 0.05$, and ** indicates $p < 0.01$. n.s.: non-significant.

2.3. *miR-375 Target Genes are Involved in Hippo- and EMT-Related Signaling Pathways*

To further investigate the role of miR-375 in MCCs, we predicted target genes of this miRNA using the miRNA target prediction tool ENCORI. This tool has the advantage that the results can be filtered for experimentally-validated target genes. Nevertheless, more than 3000 target genes were predicted; thus, the top 500 ranked genes were selected for further analysis (Table S1). Gene Ontology (GO) analysis showed that miR-375 target genes contribute to several signaling pathways, including Golgi transport, cell junction assembly, Hippo signaling, and neuron differentiation (Figure 3a). To test the relevance of these predictions in MCC, we re-analyzed previously published transcriptome microarray data of MCC cell lines [29]. Of this data set, four MCC cell lines were selected according to their miR-375 expression level: WaGa and MKL-1 with high and, MCC13 and MCC26 with low, miR-375 expression [14]. Gene Set Enrichment Analysis (GSEA) confirmed that particularly genes related to the focal adhesion signaling pathway were lower expressed in cell lines with high miR-375 expression (Figure 3b). Moreover, focal adhesion signaling pathways included most of the experimentally-confirmed miR-375 target genes; this notion also applies for miR-375 target genes related to the Hippo signaling pathway. Both pathways regulate epithelial to mesenchymal transition (EMT) [30] (Figure 3c).

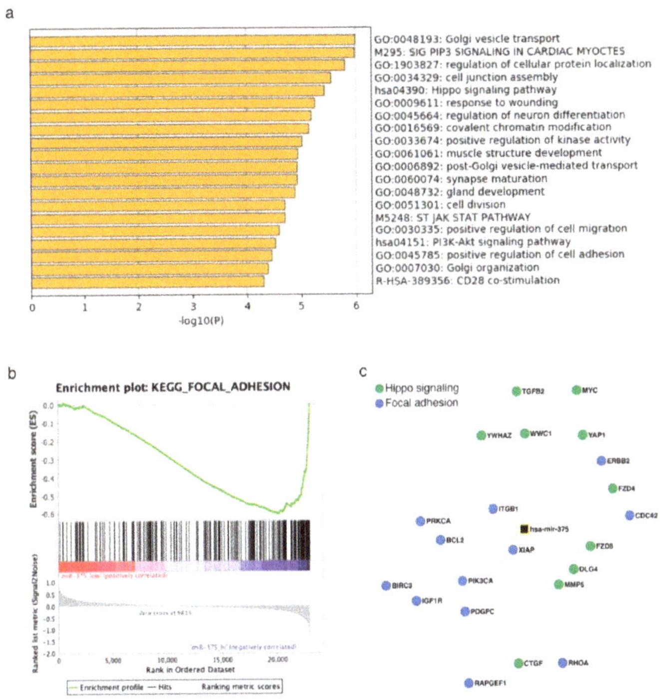

Figure 3. miR-375 target genes are involved in Hippo and epithelial to mesenchymal transition (EMT) signaling pathways in MCC cells. (**a**) Gene ontology analysis was performed in Metascape using the top 500 predicted miR-375 target genes. (**b**) Gene set enrichment analysis was performed using previously published transcriptome microarray data of MCC cell lines with high (WaGa and MKL1) and low (MCC13 and MCC26) miR-375 expression. Enrichment plot of the kegg_focal_adhesion signaling pathway is depicted. (**c**) miR-375 target genes involved in Hippo and focal adhesion signaling pathways.

2.4. Hippo and EMT Signaling Pathway-Related Genes are Marginally Altered by miR-375 Knockdown

Since our in-silico analysis suggested that miR-375 may regulate Hippo- and EMT-related signaling pathways, we tested this hypothesis by miR-375 knockdown experiments, together with qRT-PCR-based expression arrays for Hippo and EMT signaling-related genes. These experiments, however, did not reveal any statistically significant changes in the gene expression of compounds of these two signaling pathways in MCC cell lines upon miR-375 knockdown. Specifically, miR-375 knockdown only resulted in a non-significant (i.e., less than +/- two-fold change in expression) upregulation of eleven genes (11/84, 13.1%) and downregulation of four genes (4/84, 4.8%) related to the Hippo signaling pathway, as well as a non-significant upregulation of eleven genes (11/84, 13.1%) and downregulation of three genes (3/84, 3.5%) with respect to the EMT-signaling pathway (Figure 4, Table S2).

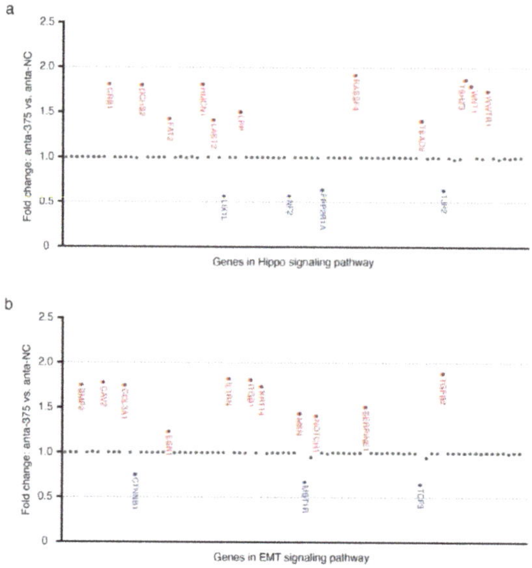

Figure 4. Moderate changes in the expression of Hippo (**a**) and EMT (**b**) signaling pathway-related genes by miR-375 knockdown.

The expression of genes related to Hippo (a) and EMT (b) signaling pathways was determined by a multiplexed qRT-PCR expression array in WaGa cells transfected with anta-375 or anta-NC, normalized to the average Cq values of housekeeping genes (GAPDH, HPRT, and RPLP0) and calculated for the ΔCq of WaGa cells transfected with anta-NC. Gene names colored in red represent genes upregulated upon miR-375 knockdown, while gene names colored in blue represent downregulated genes. Doted lines represent +/- two-fold changes. Experiments were independently repeated twice.

3. Discussion

Despite the fact that miR-375 is highly expressed in classical MCC cell lines and MCC tumors, its function in MCC is not clear. To study the relevance of miR-375 in intracellular signaling in detail, we performed a series of knockdown experiments using specific antagomiRs. Surprisingly, even the nearly complete knockdown of miR-375 expression did not affect the proliferation, growth pattern, or cell morphology. Similarly, the impact of miR-375 knockdown on the expression of Hippo and EMT signaling pathway-related genes, i.e., pathways predicted to be regulated by miR-375, was only marginal. These results, taken together with our previous observations that miR-375 is present in MCC cell line-conditioned medium in sera of preclinical xenotransplantation animal models and in sera of MCC patients [14], suggest that miR-375 may serve intercellular rather than intracellular signaling in MCC. Indeed, miR-375 was recently characterized as an exosomal shuttle miRNA [31,32].

In previous reports, miR-375 knockdown or inhibition in MCC cell lines resulted in different consequences. miR-375 knockdown using antagomiRs did not alter growth properties [13], whereas the inhibition of miR-375 using an miRNA sponge suppressed cell growth and induced cell death via downregulation of the *LDHB* gene [27]. Recently, the same group demonstrated that miR-375 inhibits autophagy to protect MCC cells from cell death [28]. AntagomiRs bind particular miRNAs, causing their degradation, while sponge RNAs compete with target mRNAs. Differences in the specificity and/or effectivity of the used methods are likely to explain some of the conflicting results. The quantification of miRNA expression after miR-375 knockdown by antagomiRs might be helpful to better understand this controversy. Notably, miR-375 is lowly expressed in variant MCC cell lines and the ectopic expression of miR-375 decreased their cell viability and migratory potential [13,27], suggesting that miR-375 might be a tumor suppressor in these cells. However, several reports question if these variant MCC cell lines are indeed representative of MCC tumors [29,33].

The knockdown of abundant miRNAs can be challenging [34]. AntagomiRs have been employed for miRNA silencing in vitro and in vivo via miRNA degradation for years [34,35]. In our study, we introduced the respective antagomiRs with two different transfection conditions, which revealed that nuclear transfection was much more efficient and only this method succeeded in nearly complete knockdown up to five days post-transfection. To be noted, we observed a slight increase in miR-375 expression over time after antagomiRs transfection. Therefore, the described method is very effective for short-term knockdown, but not for long-term inhibition (Figure S2). Besides miRNA antagomiRs, the miRNA sponge is another powerful tool that can be employed to inhibit the miRNA function. Notably, an miRNA sponge was used by Kumar et al. to inhibit the miR-375 function in MCC cells [27]. To achieve long-term miRNA inhibition, viral vectors based on stable miRNA antagomiRs or sponge expression and CRISPR-mediated miRNA knockout are feasible [36–38].

By testing for the expression of Hippo and EMT signaling pathway-related genes after miR-375 knockdown, we observed moderate expression changes of only a few genes. Furthermore, even these changes did not clearly reveal any relevant role of miR-375 in regulating these pathways. Indeed, both EMT-negative (*CAV2* and *IL1RN*) and -positive (*BMP2*, *ITGB1*, and *TGFB2*) regulators were induced upon miR-375 knockdown. Therefore, miR-375 may inhibit or induce EMT in MCC cells. More importantly, none of the changes were greater than two-fold. The gap between the predicted and functional observed effects of miR-375 knockdown is not entirely unexpected. Several reports have provided a possible explanation: long non-coding RNAs, such as *TINCR*, *HNGA1*, and *CircFAT1*, act as an miR-375 sponge [39–41]. Alternatively, other miRNAs in MCC may have redundant functional targets as miR-375 [42].

In summary, we have demonstrated that even the highly efficient, almost complete knockdown of the highly abundant miR-375 in classical MCC cells lines, has no relevant impact on the cell viability, metabolic activity, morphology, or oncogenic signaling pathways targeted by miR-375. These observations render miR-375 unlikely to function as an intracellular oncogene in MCC cells.

4. Materials and Methods

4.1. Cell Culture

The classical, MCPyV-positive MCC cell lines WaGa and PeTa were maintained in RPMI-1640 (PAN Biotech, Aidenbach, Germany) supplemented with 10% fetal bovine serum (Sigma-Aldrich, Munich, Germany) and 1% penicillin/streptomycin (Biochrome, Berlin, Germany), as previously described [43].

4.2. miR-375 Knockdown

For miR-375 knockdown, specific miR-375 inhibitors (Assay ID: MH10327, Catalog: 4464084, Thermo Fisher Scientific, Frankfurt, Germany) or respective controls (Catalog: 4464076, Thermo Fisher Scientific) were used.

For the transfection of MCC cells, two methods were compared. Lipofectamine 3000 reagent (Thermo Fisher Scientific) was used according to the instructions of the manufacturer, i.e., 2×10^6 cells were seeded into a 6-well-plate 24 h before transfection with 100 nM or 250 nM of antagomiRs. Alternatively, the Nucleofector™ 2b Device (Lonza, Basel, Switzerland) with the Cell Line Nucleofector® Kit V (Lonza) was used. D-23 was established as the appropriate program to transfect MCC cells (https://bioscience.lonza.com/lonza_bs/CH/en/nucleofector-technology). A total of 100µL of buffer V was mixed with 10 µL miRNA antagomiRs (25 nM) and 2×10^6 MCC cells before being transferred into an electroporation cuvette. After the pulse, cells were immediately transferred into 6-well-plates containing pre-warmed culture media.

4.3. qRT-PCR for miR-375

Applied Biosystems TaqMan MicroRNA assays (Thermo Fisher Scientific) were performed according to the manufacturer's instructions. Pre-designed TaqMan microRNA assays for miR-375 (ID000564) were used. The quantification cycle threshold (Cq) values of miR-375 were normalized to the small nucleolar RNA RNU6B (ID001093) and the relative expression of the respective comparator was calculated using the 2-ΔΔCq method.

4.4. MTS Assay

Dead cells and cell debris after nucleofection were removed using Ficoll-mediated gradient centrifugation (Biochrom, Berlin, Germany). For MTS assays, 10^4 living cells per well of each group (untreated, anta-NC and anta-375) were seeded into 96-well-plates. CellTiter 96 AQ_{ueous} One Solution (Promega, Walldorf, Germany) was used to determine the relative cell proliferation every other day. In brief, 20 uL of the CellTiter solution was added to each well and incubated for two hours, and the absorbance was then measured using a plate reader at 490 nm.

4.5. Apoptosis Assay

The NucView 488/MitoView 633 apoptosis assay kit (Biotium, Fremont, CA, USA) was used to determine the apoptotic cell rate, according to the manufacturer's instruction. Viable cells were stained red with MitoView 633 (red, APC-A channel), and apoptotic cells were stained green with NucView 488 (green, PE-A channel). Twenty-four hours' post-nucleofection and subsequently every other day, cells were analyzed using a CytoFLEX flow cytometer (Beckman Coulter, Krefeld, Germany).

4.6. Prediction of miR-375 Target Genes, Gene Ontology (GO), and Gene Set Enrichment Analysis (GSEA)

The Encyclopedia of RNA Interactomes (ENCORI, http://starbase.sysu.edu.cn/index.php) provides miRNA–target gene interactions, which are based on miRNA target prediction programs, i.e., TargetScan, miRanda, microT, PITA, miRmap, and PicTar. All miRNA target predictions are supported by published Argonaute-crosslinking and immunoprecipitation (AGO-CLIP) data [44]. Predicted target genes are ranked based on the predicted efficacy of targeting, as calculated using cumulative weighted context++ scores of the sites and related AGO-CLIP scores (clipExpNum, Table S1) [44,45]. The top 500 highest ranking predicted target genes were selected for the following analysis.

Metascape (http://metascape.org) was applied for GO analysis [46]. Metascape incorporates a core set of default ontologies, including GO processes, KEGG pathways, Reactome gene sets, canonical pathways, and CORUM complexes, for enrichment analysis.

GSEA, the desktop application from the MSigDB of Broad Institute (Cambridge, MA, USA), was used for re-analysis of the transcriptome microarray of selected MCC cell lines (http://software.broadinstitute.org/gsea/msigdb/index.jsp) [47]. The transcriptome microarray data set GSE50451 was downloaded from Gene Expression Omnibus. Four MCC cell lines were selected and analyzed in GSEA: WaGa and MKL-1 as miR-375_ high, and MCC13 and MCC26 as miR-375_low.

4.7. Pathway Finder Gene Expression Arrays

The RT2 Profiler PCR arrays (SABioscience via Qiagen, Hilden, Germany) for epithelial to mesenchymal transition (EMT) (PAHS-090Z) and Hippo signaling (PAHS-172Z) were performed according to the manufacturer's instructions. Total RNA was isolated from MCC cell lines three days after nucleofection with miR-375 inhibitors or the respective control. A total of 200ng of total RNA was transcribed into cDNA using the RT2 first strand kit (Qiagen). The relative gene expression was determined using the RT^2 Profiler PCR Array software from Qiagen (https://dataanalysis.qiagen.com/pcr/arrayanalysis.php).

4.8. Statistical Analysis

Statistical analyses were performed using GraphPad Prism 8.0 Software (GraphPad Software Inc., San Diego, CA, USA). Experiments containing more than two groups were analyzed using the Kruskal–Wallis test, and an unpaired nonparametric ANOVA. R studio (version 3.6.0) was used in the statistical analysis as indicated: the ggpubr R package (version 3.2.0) for the dot plot of gene expression in EMT and Hippo signaling. A p-value smaller than 0.05 was considered significant; the respective p-values are indicated in the figures as follows: * $p < 0.05$, ** $p < 0.01$, and *** $p < 0.001$.

5. Conclusions

The highly efficient knockdown of abundant miR-375 achieved by miR-375 antagomiRs with nucleofection did not cause obvious effects on the cell viability, apoptosis, morphology, or oncogenic Hippo- and EMT-related signaling pathways. These observations render miR-375 unlikely to function as an intracellular oncogene in MCC cells.

Supplementary Materials: The following are available online at http://www.mdpi.com/2072-6694/12/3/529/s1, Figure S1: Expression of miR-375 and U6 in MCC cells transfected with antagomiRs, depicted in amplification curves, Figure S2: Relative expression of miR-375 in MCC cells transfected with antagomiRs, Figure S3: Apoptosis of MCC cells transfected with antagomiRs was determined in flow cytometry, Table S1: List of predicted miR-375 target genes in ENCORI, Table S2: Alteration of gene expression (Hippo and EMT related) after miR-375 knockdown in WaGa cells.

Author Contributions: Conceptualization, J.C.B.; methodology, K.F., D.S., and A.Z.; formal analysis, K.F. and K.H.; investigation, K.F. and A.Z.; resources, A.Z. and J.C.B.; data curation, K.F. and K.H.; writing, K.F., K.H., and J.C.B.; visualization, K.F.; supervision, J.C.B. and D.S.; project administration, J.C.B.; funding acquisition, J.C.B. All authors have read and agreed to the published version of the manuscript.

Funding: This research was funded by the European Commission, Grant Agreement number HEALTH-F2-2012-277775, IMMOMEC.

Acknowledgments: We thank Ashwin Sriram, TSCR, DKTK site University of Essen, Essen, Germany and the DKFZ, Heidelberg, Germany for critically reading the final version of the manuscript.

Conflicts of Interest: J.C. Becker is receiving speaker's bureau honoraria from Amgen, Pfizer, MerckSerono, and Sanofi, and is a paid consultant/advisory board member for eTheRNA, MerckSerono, Pfizer, 4SC, REcordati, InProTher, and Sanofi. His group receives research grants from IQVIA, Merck Serono, and Alcedis. None of the activities are related to the submitted work. None of the other authors indicated any potential conflicts of interest.

References

1. Becker, J.C.; Stang, A.; DeCaprio, J.A.; Cerroni, L.; Lebbe, C.; Veness, M.; Nghiem, P. Merkel cell carcinoma. *Nat. Rev. Dis. Primers* **2017**, *3*, 17077. [CrossRef] [PubMed]
2. Arora, R.; Shuda, M.; Guastafierro, A.; Feng, H.C.; Toptan, T.; Tolstov, Y.; Normolle, D.; Vollmer, L.L.; Vogt, A.; Domling, A.; et al. Survivin is a therapeutic target in Merkel cell carcinoma. *Sci. Transl. Med.* **2012**, *4*, 133ra56. [CrossRef]
3. Hesbacher, S.; Pfitzer, L.; Wiedorfer, K.; Angermeyer, S.; Borst, A.; Haferkamp, S.; Scholz, C.J.; Wobser, M.; Schrama, D.; Houben, R. RB1 is the crucial target of the Merkel cell polyomavirus Large T antigen in Merkel cell carcinoma cells. *Oncotarget* **2016**, *7*, 32956–32968. [CrossRef]
4. Wong, S.Q.; Waldeck, K.; Vergara, I.A.; Schroder, J.; Madore, J.; Wilmott, J.S.; Colebatch, A.J.; De Paoli-Iseppi, R.; Li, J.; Lupat, R.; et al. UV-associated mutations underlie the etiology of MCV-negative Merkel cell carcinomas. *Cancer Res.* **2015**, *75*, 5228–5234. [CrossRef] [PubMed]
5. Gonzalez-Vela, M.D.C.; Curiel-Olmo, S.; Derdak, S.; Beltran, S.; Santibanez, M.; Martinez, N.; Castillo-Trujillo, A.; Gut, M.; Sanchez-Pacheco, R.; Almaraz, C.; et al. Shared oncogenic pathways implicated in both virus-positive and UV-Induced Merkel cell carcinomas. *J. Investig. Derm.* **2017**, *137*, 197–206. [CrossRef] [PubMed]
6. Rozenblatt-Rosen, O.; Deo, R.C.; Padi, M.; Adelmant, G.; Calderwood, M.A.; Rolland, T.; Grace, M.; Dricot, A.; Askenazi, M.; Tavares, M.; et al. Interpreting cancer genomes using systematic host network perturbations by tumour virus proteins. *Nature* **2012**, *487*, 491–495. [CrossRef]

7. Veija, T.; Sarhadi, V.K.; Koljonen, V.; Bohling, T.; Knuutila, S. Hotspot mutations in polyomavirus positive and negative Merkel cell carcinomas. *Cancer Genet.* **2016**, *209*, 30–35. [CrossRef]
8. Knepper, T.C.; Montesion, M.; Russell, J.S.; Sokol, E.S.; Frampton, G.M.; Miller, V.A.; Albacker, L.A.; McLeod, H.L.; Eroglu, Z.; Khushalani, N.I.; et al. The genomic landscape of Merkel cell carcinoma and clinicogenomic biomarkers of response to immune checkpoint Inhibitor therapy. *Clin. Cancer Res.* **2019**, *25*, 5961–5971. [CrossRef]
9. Fan, K.; Gravemeyer, J.; Ritter, C.; Rasheed, K.; Gambichler, T.; Moens, U.; Shuda, M.; Schrama, D.; Becker, J.C. MCPyV large T antigen-induced Atonal homolog 1 is a lineage-dependency oncogene in Merkel cell carcinoma. *J. Investig. Derm.* **2020**, *140*, 56–65. [CrossRef]
10. Gebert, L.F.R.; MacRae, I.J. Regulation of microRNA function in animals. *Nat. Rev. Mol. Cell Bio.* **2019**, *20*, 21–37. [CrossRef]
11. Lagos-Quintana, M.; Rauhut, R.; Lendeckel, W.; Tuschl, T. Identification of novel genes coding for small expressed RNAs. *Science* **2001**, *294*, 853–858. [CrossRef] [PubMed]
12. Peng, Y.; Croce, C.M. The role of MicroRNAs in human cancer. *Signal. Transduct. Tar.* **2016**, *1*, 15004. [CrossRef] [PubMed]
13. Abraham, K.J.; Zhang, X.; Vidal, R.; Pare, G.C.; Feilotter, H.E.; Tron, V.A. Roles for miR-375 in neuroendocrine differentiation and tumor suppression via Notch pathway suppression in Merkel cell carcinoma. *Am. J. Pathol.* **2016**, *186*, 1025–1035. [CrossRef]
14. Fan, K.; Ritter, C.; Nghiem, P.; Blom, A.; Verhaegen, M.E.; Dlugosz, A.; Odum, N.; Woetmann, A.; Tothill, R.W.; Hicks, R.J. Circulating cell-free miR-375 as surrogate marker of tumor burden in Merkel cell carcinoma. *Clin. Cancer Res.* **2018**, *24*, 5873–5882. [CrossRef]
15. Renwick, N.; Cekan, P.; Masry, P.A.; McGeary, S.E.; Miller, J.B.; Hafner, M.; Li, Z.; Mihailovic, A.; Morozov, P.; Brown, M.; et al. Multicolor microRNA FISH effectively differentiates tumor types. *J. Clin. Investig.* **2013**, *123*, 2694–2702. [CrossRef]
16. Xie, H.; Lee, L.; Caramuta, S.; Hoog, A.; Browaldh, N.; Bjornhagen, V.; Larsson, C.; Lui, W.O. MicroRNA expression patterns related to Merkel cell polyomavirus infection in human Merkel cell carcinoma. *J. Investig. Derm.* **2014**, *134*, 507–517. [CrossRef]
17. Poy, M.N.; Eliasson, L.; Krutzfeldt, J.; Kuwajima, S.; Ma, X.; Macdonald, P.E.; Pfeffer, S.; Tuschl, T.; Rajewsky, N.; Rorsman, P.; et al. A pancreatic islet-specific microRNA regulates insulin secretion. *Nature* **2004**, *432*, 226–230. [CrossRef]
18. Poy, M.N.; Hausser, J.; Trajkovski, M.; Braun, M.; Collins, S.; Rorsman, P.; Zavolan, M.; Stoffel, M. miR-375 maintains normal pancreatic alpha- and beta-cell mass. *Proc. Natl. Acad. Sci. USA* **2009**, *106*, 5813–5818. [CrossRef]
19. Ding, L.; Xu, Y.; Zhang, W.; Deng, Y.; Si, M.; Du, Y.; Yao, H.; Liu, X.; Ke, Y.; Si, J.; et al. MiR-375 frequently downregulated in gastric cancer inhibits cell proliferation by targeting JAK2. *Cell Res.* **2010**, *20*, 784–793. [CrossRef]
20. Tsukamoto, Y.; Nakada, C.; Noguchi, T.; Tanigawa, M.; Lam, T.N.; Uchida, T.; Hijiya, N.; Matsuura, K.; Fujioka, T.; Seto, M.; et al. MicroRNA-375 is downregulated in gastric carcinomas and regulates cell survival by targeting PDK1 and 14-3-3 zeta. *Cancer Res.* **2010**, *70*, 2339–2349. [CrossRef]
21. Zhou, J.; Song, S.D.; Cen, J.N.; Zhu, D.M.; Li, D.C.; Zhang, Z.X. MicroRNA-375 is downregulated in pancreatic cancer and inhibits cell proliferation in vitro. *Oncol. Res.* **2012**, *20*, 197–203. [CrossRef] [PubMed]
22. Dai, X.; Chiang, Y.; Wang, Z.; Song, Y.; Lu, C.; Gao, P.; Xu, H. Expression levels of microRNA-375 in colorectal carcinoma. *Mol. Med. Rep.* **2012**, *5*, 1299–1304. [PubMed]
23. Faltejskova, P.; Svoboda, M.; Srutova, K.; Mlcochova, J.; Besse, A.; Nekvindova, J.; Radova, L.; Fabian, P.; Slaba, K.; Kiss, I.; et al. Identification and functional screening of microRNAs highly deregulated in colorectal cancer. *J. Cell Mol. Med.* **2012**, *16*, 2655–2666. [CrossRef] [PubMed]
24. He, X.X.; Chang, Y.; Meng, F.Y.; Wang, M.Y.; Xie, Q.H.; Tang, F.; Li, P.Y.; Song, Y.H.; Lin, J.S. MicroRNA-375 targets AEG-1 in hepatocellular carcinoma and suppresses liver cancer cell growth in vitro and in vivo. *Oncogene* **2012**, *31*, 3357–3369. [CrossRef]
25. Hudson, J.; Duncavage, E.; Tamburrino, A.; Salerno, P.; Xi, L.; Raffeld, M.; Moley, J.; Chernock, R.D. Overexpression of miR-10a and miR-375 and downregulation of YAP1 in medullary thyroid carcinoma. *Exp. Mol. Pathol.* **2013**, *95*, 62–67. [CrossRef] [PubMed]

26. Szczyrba, J.; Nolte, E.; Wach, S.; Kremmer, E.; Stohr, R.; Hartmann, A.; Wieland, W.; Wullich, B.; Grasser, F.A. Downregulation of Sec23A protein by miRNA-375 in prostate carcinoma. *Mol. Cancer Res.* **2011**, *9*, 791–800. [CrossRef]
27. Kumar, S.; Xie, H.; Scicluna, P.; Lee, L.; Bjornhagen, V.; Hoog, A.; Larsson, C.; Lui, W.O. miR-375 regulation of LDHB plays distinct roles in Polyomavirus-positive and -negative Merkel cell carcinoma. *Cancers* **2018**, *10*, 443. [CrossRef]
28. Kumar, S.; Xie, H.; Shi, H.; Gao, J.; Juhlin, C.C.; Björnhagen, V.; Höög, A.; Lee, L.; Larsson, C.; Lui, W.O. Merkel cell polyomavirus oncoproteins induce microRNAs that suppress multiple autophagy genes. *Int. J. Cancer* **2019**. [CrossRef]
29. Daily, K.; Coxon, A.; Williams, J.S.; Lee, C.C.R.; Coit, D.G.; Busam, K.J.; Brownell, I. Assessment of cancer cell line representativeness using microarrays for Merkel cell carcinoma. *J. Investig. Derm.* **2015**, *135*, 1138–1146. [CrossRef]
30. Frisch, S.M.; Schaller, M.; Cieply, B. Mechanisms that link the oncogenic epithelial–mesenchymal transition to suppression of anoikis. *J. Cell Sci.* **2013**, *126*, 21–29. [CrossRef]
31. Huang, X.Y.; Yuan, T.Z.; Liang, M.H.; Du, M.J.; Xia, S.; Dittmar, R.; Wang, D.; See, W.; Costello, B.A.; Quevedo, F.; et al. Exosomal miR-1290 and miR-375 as prognostic markers in castration-resistant prostate cancer. *Eur. Urol.* **2015**, *67*, 33–41. [CrossRef] [PubMed]
32. Su, Y.Y.; Sun, L.; Guo, Z.R.; Li, J.C.; Bai, T.T.; Cai, X.X.; Li, W.H.; Zhu, Y.F. Upregulated expression of serum exosomal miR-375 and miR-1307 enhance the diagnostic power of CA125 for ovarian cancer. *J. Ovarian Res.* **2019**, *12*, 6. [CrossRef] [PubMed]
33. Van Gele, M.; Boyle, G.M.; Cook, A.L.; Vandesompele, J.; Boonefaes, T.; Rottiers, P.; Van Roy, N.; De Paepe, A.; Parsons, P.G.; Leonard, J.H.; et al. Gene-expression profiling reveals distinct expression patterns for Classic versus Variant Merkel cell phenotypes and new classifier genes to distinguish Merkel cell from small-cell lung carcinoma. *Oncogene* **2004**, *23*, 2732–2742. [CrossRef] [PubMed]
34. Krutzfeldt, J.; Rajewsky, N.; Braich, R.; Rajeev, K.G.; Tuschl, T.; Manoharan, M.; Stoffel, M. Silencing of microRNAs in vivo with 'antagomirs'. *Nature* **2005**, *438*, 685–689. [CrossRef]
35. Krutzfeldt, J.; Kuwajima, S.; Braich, R.; Rajeev, K.G.; Pena, J.; Tuschl, T.; Manoharan, M.; Stoffel, M. Specificity, duplex degradation and subcellular localization of antagomirs. *Nucleic Acids Res.* **2007**, *35*, 2885–2892. [CrossRef]
36. Xie, J.; Ameres, S.L.; Friedline, R.; Hung, J.-H.; Zhang, Y.; Xie, Q.; Zhong, L.; Su, Q.; He, R.; Li, M. Long-term, efficient inhibition of microRNA function in mice using rAAV vectors. *Nat. Methods* **2012**, *9*, 403–409. [CrossRef]
37. Gentner, B.; Schira, G.; Giustacchini, A.; Amendola, M.; Brown, B.D.; Ponzoni, M.; Naldini, L. Stable knockdown of microRNA in vivo by lentiviral vectors. *Nat. Methods* **2009**, *6*, 63–66. [CrossRef]
38. Aquino-Jarquin, G. Emerging role of CRISPR/Cas9 technology for microRNAs editing in cancer research. *Cancer Res.* **2017**, *77*, 6812–6817. [CrossRef]
39. Chen, Z.; Liu, H.; Yang, H.; Gao, Y.; Zhang, G.; Hu, J. The long noncoding RNA, TINCR, functions as a competing endogenous RNA to regulate PDK1 expression by sponging miR-375 in gastric cancer. *Oncotargets* **2017**, *10*, 3353–3362. [CrossRef]
40. Liu, G.; Huang, K.M.; Jie, Z.W.; Wu, Y.Z.; Chen, J.X.; Chen, Z.Z.; Fang, X.Q.; Shen, S.Y. CircFAT1 sponges miR-375 to promote the expression of Yes-associated protein 1 in osteosarcoma cells. *Mol. Cancer* **2018**, *17*, 170. [CrossRef]
41. Wang, Y. Upregulated lncRNA-HNGA1, a target of miR-375, contributes to aerobic glycolysis of head and neck squamous cell carcinoma through increasing levels of the glucose transporter protein SCL2A1. *Eur. J. Cancer* **2016**, *61*, S14–S15. [CrossRef]
42. Liufu, Z.; Zhao, Y.; Guo, L.; Miao, G.; Xiao, J.; Lyu, Y.; Chen, Y.; Shi, S.; Tang, T.; Wu, C.I. Redundant and incoherent regulations of multiple phenotypes suggest microRNAs' role in stability control. *Genome Res.* **2017**, *27*, 1665–1673. [CrossRef] [PubMed]
43. Schrama, D.; Sarosi, E.M.; Adam, C.; Ritter, C.; Kaemmerer, U.; Klopocki, E.; Konig, E.M.; Utikal, J.; Becker, J.C.; Houben, R. Characterization of six Merkel cell polyomavirus-positive Merkel cell carcinoma cell lines: Integration pattern suggest that large T antigen truncating events occur before or during integration. *Int. J. Cancer* **2019**, *145*, 1020–1032. [CrossRef] [PubMed]

44. Li, J.H.; Liu, S.; Zhou, H.; Qu, L.H.; Yang, J.H. starBase v2.0: Decoding miRNA-ceRNA, miRNA-ncRNA and protein-RNA interaction networks from large-scale CLIP-Seq data. *Nucleic Acids Res.* **2014**, *42*, D92–D97. [CrossRef] [PubMed]
45. Agarwal, V.; Bell, G.W.; Nam, J.W.; Bartel, D.P. Predicting effective microRNA target sites in mammalian mRNAs. *Elife* **2015**, *4*, e05005. [CrossRef]
46. Zhou, Y.Y.; Zhou, B.; Pache, L.; Chang, M.; Khodabakhshi, A.H.; Tanaseichuk, O.; Benner, C.; Chanda, S.K. Metascape provides a biologist-oriented resource for the analysis of systems-level datasets. *Nat. Commun.* **2019**, *10*, 1523. [CrossRef]
47. Subramanian, A.; Tamayo, P.; Mootha, V.K.; Mukherjee, S.; Ebert, B.L.; Gillette, M.A.; Paulovich, A.; Pomeroy, S.L.; Golub, T.R.; Lander, E.S.; et al. Gene set enrichment analysis: A knowledge-based approach for interpreting genome-wide expression profiles. *Proc. Natl. Acad. Sci. USA* **2005**, *102*, 15545–15550. [CrossRef]

cancers

MDPI

Review

Merkel Cell Polyomavirus and Merkel Cell Carcinoma

Valeria Pietropaolo [1], Carla Prezioso [1,2] and Ugo Moens [3,*]

1 Department of Public Health and Infectious Diseases, "Sapienza" University, 00185 Rome, Italy; valeria.pietropaolo@uniroma1.it (V.P.); carla.prezioso@uniroma1.it (C.P.)
2 IRCSS San Raffaele Pisana, Microbiology of Chronic Neuro-Degenerative Pathologies, 00166 Rome, Italy
3 Molecular Inflammation Research Group, Department of Medical Biology, Faculty of Health Sciences, University of Tromsø—The Arctic University of Norway, 9037 Tromsø, Norway
* Correspondence: ugo.moens@uit.no

Received: 9 June 2020; Accepted: 28 June 2020; Published: 3 July 2020

Abstract: Viruses are the cause of approximately 15% of all human cancers. Both RNA and DNA human tumor viruses have been identified, with Merkel cell polyomavirus being the most recent one to be linked to cancer. This virus is associated with about 80% of Merkel cell carcinomas, a rare, but aggressive cutaneous malignancy. Despite its name, the cells of origin of this tumor may not be Merkel cells. This review provides an update on the structure and life cycle, cell tropism and epidemiology of the virus and its oncogenic properties. Putative strategies to prevent viral infection or treat virus-positive Merkel cell carcinoma patients are discussed.

Keywords: biomarkers; cell tropism; signaling pathways; therapy; transgenic mice; tumorigenesis

---

1. Introduction

1.1. Genome MCPyV

Merkel cell polyomavirus (MCPyV) is a naked double-stranded DNA virus belonging to the *Polyomaviridae* family [1]. Its circular genome of ~5400 base-pairs (bp) encompassed three functional domains (Figure 1). The early region includes the "Tumor" (T) antigen gene locus [2], from which, alternatively-spliced RNA transcripts are produced. This region encodes for distinctive gene products: the large T (LT), small (sT), 57kT antigens and a product from an alternate frame of the LT open reading frame (ALTO) [3]. The LT, sT and 57 kT antigens, due to alternative splicing, share a 78 amino acid sequence at their N-terminal region [4].

Similar to other human polyomaviruses (HPyVs), the MCPyV LT antigen contains a number of motifs and domains that play key roles in viral genome replication and transcription, as well as tumorigenesis (Figure 1). The N-terminal half encompasses the DnaJ domain, which consists of the CR1 motif (13–17 amino acids) followed by the HPDKGG, the sequence is responsible for Hsc70 binding [5,6]. The WXXWW sequence found in LT of other PyVs and that binds the mitotic checkpoint serine-threonine protein kinase Bub1 is absent in MCPyV LT [7]. At this position, MCPyV LT has a sequence known as MCPyV T antigen unique region (MUR), containing a binding motif for the vacuolar sorting protein Vam6p [8]. Adjacent to this, the conserved LXCXE retinoblastoma (RB) binding motif is present.

Finally, a nuclear localization signal (NLS) with sequence RKRK is situated in the N-terminal region of LT [9]. The C-terminal region of LT contains an origin binding domain (OBD) and the helicase/ATPase domain [8]. Both the OBD and the helicase/ATPase domain are required for replication of the viral genome. The C-terminal region of LT of other HPyVs binds to p53, a tumor suppressor that regulates the gene expression in response to events such as DNA damage, leading to apoptosis, cell cycle arrest or senescence, and inhibition of angiogenesis, and is usually deregulated in cancer [10]. This p53 binding site is contained in the OBD and helicase/ATPase domain. The possible p53 binding domain in MCPyV LT and its interaction with p53 is discussed in Section 4.2.

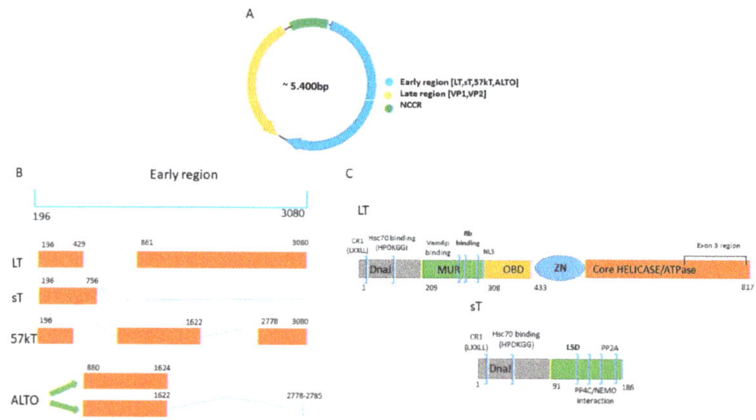

Figure 1. Structure of the MCPyV genome and the early region transcripts and the early proteins large T antigen (LT) and small T antigen (sT) with their functional domains. (**A**) Schematic presentation of the ~5400 bp circular dsDNA genome that includes a non-coding region (NCCR), an early region encoding T antigens that coordinate viral replication, and a late region containing the genes for the viral capsid proteins VP1 and VP2. (**B**) Multiple transcripts are generated from the early region by alternative splicing, including LT, sT, 57 kT antigen (57 kT) and alternative frame of the large T open reading frame (ALTO). (**C**) LT contains the DnaJ domain with a conserved HPDKGG motif, the MCPyV unique region (MUR) with the retinoblastoma protein (RB) binding motif, the nuclear localization signal (NLS), the DNA or origin binding domain (OBD), the zinc-finger domain (ZN) and the helicase/ATPase domain. sT antigen encompasses the DnaJ domain, the LT stabilizing domain (LSD), and interaction domains for the protein phosphatases PP2A and PP4.

MCPyV-positive MCCs (hereafter referred to as VP-MCC) express a C-terminal truncated LT (tLT) due to nonsense mutations or frameshift mutations generating premature stop codons. Tumor-derived tLTs retain the DnaJ region and the RB binding domain, and sometimes the NLS, but lack the OBD and helicase/ATPase domain [5,11] (Figure 1). The C-terminal region contains several elements fundamental for viral replication, hence tLT fails to support viral replication [12]. As for other HPyVs, and in general for other tumor viruses, there is strong selective pressure within tumors to eliminate viral replication capacity [13].

MCPyV LT is rich in potential phosphoacceptor sites (94 serine, 42 threonine, and 23 tyrosine residues). Li et al., found that phosphorylation of LT at S816 by ATM kinase induced apoptosis and thus contribute to anti-tumorigenic properties of the C-terminal domain [14]. Diaz and colleagues identified three additional phosphorylation sites: T271, T297 and T299. Mutation of T271 into alanine did not have an effect on viral replication. LT T297A stimulated replication, whereas LT T299A was unable to do so. The authors demonstrated that phosphorylation of T297 may negatively regulate viral replication by reducing the binding affinity of LT to the viral origin of replication (ORI), while T299 phosphorylation affects both binding to and unwinding of the DNA [15]. Taken together, truncation of the C-terminal region of LT and phosphorylation of specific residues in LT may abrogate viral replication. S220 is another phosphoacceptor site and the effect of its phosphorylation is discussed in Section 4.1. The phosphorylation status of LT in MCC has not been examined.

As a result of alternative splicing of a common precursor transcript, LT and sT share the 80 N-terminal amino acids [8]. The sT antigen contains the DnaJ domain but lacks the RB motif [16] (Figure 1). At its unique C-terminal region, sT encompasses two zinc-binding domains (CXCXXC motif), which provide structural and functional stabilities and two domains rich in cysteine and proline residues responsible for the sT interaction with protein phosphatase 2A (PP2A) (see further) [17]. A unique MCPyV sT domain, not present in sT of other HPyVs, is the LT stabilization domain (LSD) at residues 91–95. This region, as will be discussed later, is involved in inhibition of proteasomal degradation of LT (Figure 1) [18].

The late region encodes the major capsid protein VP1 and the minor capsid protein VP2 (Figure 1). MCPyV does not seem to express VP3 despite an in-frame internal start codon in the *VP2* gene [19]. When expressed in mammalian cells, VP1 (or VP1 and VP2) self-assemble into 45–55 nm diameter virus-like particles (VLPs) that are used in serological assays [20].

Interspersed between the early and late region is the non-coding control region (NCCR), which contains the ORI characterized by a core of 71-bp sufficient to initiate DNA replication (Figure 1). This core region consists of an AT rich tract and eight 5′-GAGGC-3′ LT binding motifs [12]. The NCRR also contains regulatory elements and bidirectional transcriptional promoters required for early and late viral gene expression [21]. The NCCRs of HPyVs such as BKPyV and JCPyV show often rearrangements that affect viral DNA replication, promoter activity, virus production and could help to increase the pathogenic properties of these viruses [22–24]. MCPyV NCCR polymorphism is found, but no specific NCCR architecture seems to be associated with VP-MCC as MCPyV variants with identical NCCR have been isolated from both MCC and non-MCC material [25]. However, MCPyV NCCR variation affects early and late promoter activities in a VN-MCC cell line and in human dermal fibroblast and wild-type LT inhibited both early and late promoter activities in both cell lines, whereas tLT had the opposite effect [25]. A recent study demonstrated the onset of insertions and deletions in the NCCR among an HIV-1-positive population [26]. Whether NCCR variation has an influence on viral replication and pathogenic properties of the virus remains to be investigated.

The molecular characterization of viral genomes has been useful to describe viral lineages associated with specific human populations, as demonstrated for other PyVs [27–29]. Phylogenetic analysis, performed on LT and sT antigens and on VP1, showed that MCPyV sequences circulate in Europe/North America, Africa, Asia, South America and Oceania groups, suggesting the occurrence of a viral divergence followed human migrations around the globe [30]. There is a significant evidence for an ancient and relatively stable association of PyVs with their hosts, suggesting that co-divergence is the main factor during the evolution [31]. However, deviations from co-divergence indicate that additional evolutionary processes are at play. Phylogenetic analysis, about the evolutionary history of MCPyV, showed that the MCPyV LT is most similar to gorilla polyomavirus 1 (GgorgPyV1) and chimpanzee polyomaviruses 2 and 3 (PtrovPyV 2 and 3) [1], raising the possibility that MCPyV stems from a nonhuman primate (including chimpanzees and gorillas) and even ape-specific group of PyVs [31]. Non-human primates still represent an important proportion of the bush meat consumed in West and Central Africa (ca. 12%). Hunting and butchering of bush meat provide the major routes of pathogen and a cross-species transmission events (e.g., human immunodeficiency viruses and severe acute respiratory syndrome coronavirus 2). This could also explain how MCPyV may have been transmitted from apes to humans [32].

1.2. Seroprevalence

MCPyV prevalence study suggests that this virus is chronically shed from human skin representing part of the skin microbiota [33]. The initial exposure to MCPyV, based on the VP1 serology assay, supposedly occurs in early childhood. As reported in a study from Cameroon, significant titers against MCPyV were detected in newborns, although these titers decreased to undetectable levels by 16 months of age [34]. The maternal derived antibodies could represent the reason of the seropositivity in newborns. Moreover, these antibodies, effective in preventing primary infection, could explain why the seroprevalence is lower in children and higher in adults [34]. By 18 months of age, when the maternal antibodies were no longer detectable, children were susceptible to de novo infection and were able to mount an own antibody response. Beginning at 18 months of age, an increasing fraction of children became positive until approximately 80% tested positive at the age of 5 [34]. In a separate cohort from the same study, the correlation of seropositivity was observed between siblings of similar ages, suggesting that siblings likely were exposed to MCPyV at the same time and by each other [35]. These data suggest that transmission may occur via direct contact with the skin or saliva [34,35]. Several studies support the increasing risk with age for exposure and persistent infection by MCPyV [36–39]. A study conducted in Italy, with participants aged from 1 to 100 years old, showed how the seroprevalence for MCPyV rapidly increased with age: from 41.7% in children age from 1 to 4 years old, to 87.6% among young adult (15–19 years old), remaining frequent in adulthood (79–96.2%) [40]. MCPyV seroprevalence studies performed in China (61% overall) and the Czech Republic (63%) yielded similar results with an increasing trend with age [41,42]. Antibodies versus MCPyV LT and sT are detected in about 1% of healthy individuals and they can be present in patients with MCC [43]. Often MCC patients have higher titers of VP1 antibodies than normal healthy individuals [20].

1.3. Cell Tropism: Skin; Replication in Dermal Fibroblasts

Because MCPyV was originally detected in MCC, a tumor believed to originate from Merkel cells (MCs), which are specialized skin cells, and is chronically shed from skin from healthy individuals, it was believed that the virus is dermatotrophic. It is now questioned that MCs are the target of MCPyV infection or productive replication because there are too few MC in the human skin to account for the millions of copies of MCPyV DNA detected on healthy skin [33]. Liu et al., speculated that the natural MCPyV host cells were one of the more abundant cell types in the human skin. They showed that human dermal fibroblasts support productive viral replication [44], and because MCs are situated in the basal layer of the epidermis near dermal fibroblasts, the authors hypothesized that MCPyV actively replicating in the dermal fibroblasts could accidently enter MCs and cause MCC [44]. Likely, MCs could represent a replication environment that supports viral integration and transformation [44]. It has also been demonstrated that MCPyV is capable of expressing LT and VP1 in fibroblast cell lines originating from lung tissue [44]. Hence, an active viral replication of MCPyV might be connected to all fibroblast tissues [44]. MCPyV DNA has been detected in cutaneous swabs [45] and it is possible that infected dermal fibroblasts might die and virions could be carried to the skin surface by the flow of differentiating keratinocytes [46]. This suggests that viral particles can be more widespread from the site of replication and release. This hypothesis is supported by the observation that MCPyV is frequently detected in eyebrow hair bulbs [47]. MCPyV can infect dermal fibroblasts near hair follicles and it is possible that mature virions could be cleared to the surface of human skin through hair follicles and/or associated sebaceous and sweat glands [47].

2. MCPyV and MCC

MCC is a rare, neuroendocrine, cutaneous malignancy that was first described in 1972 by Toker as "trabecular carcinoma of the skin" [48]. The name was later changed to MCC, since the tumor cells were similar to Merkel cells, present in particular around hair follicles and in the basal layer of the epidermis. Although MCC is a rare skin cancer, it is highly aggressive displaying a mortality rate of ~45% [49]. Consequently, MCC has a case-fatality rate higher than observed with melanoma [49]. Almost one third of the patients, at primary diagnosis, present loco regional metastases or lymph node metastases [49]. During the last 10 years, MCC incidence has increased significantly and is expected to increase further, since, the occurrence of this type of cancer, is related with aging (immunosenescence) and exposure to the sun [50]. An important alternative explanation for this finding is that before the large use of CK20 immunostaining, the pathology diagnosis was difficult and may at these ancient times require electronic microscopy, which was frequently not performed. Thus, true MCC were frequently misclassified [51,52]. The correlation between MCC and UV radiation is well documented [53]. Pigmentation of the skin seems to protect against MCC, as black, Asian and Hispanic individuals have considerably lower risk of MCC than white populations. Moreover, the occurrence of MCC is frequent in elderly patients on chronically sun-exposed skin, in individuals treated with UVA photo-chemotherapy and in patients with a history of other skin cancers associated with sun exposure. Melanoma is also linked with a three-fold greater risk of MCC [54]. A molecular UV signature, characterized by DNA mutations that are typically caused by UV damage, such as C to T transitions, has been demonstrated only in a subset of cases of VN-MCCs [55,56]. The association with UV exposure in VP-MCC could be related to other factors, such as UV-induced immunosuppression. In fact, immunodeficiency forms a risk factor in the development of MCC. MCC is more frequent in patients with leukemia [57] or HIV infection [58] and in those who are immunosuppressed, as a result of organ transplantation or other causes [59]. The mortality is higher in immunosuppressed individuals than in immunocompetent patients [60]. These findings emphasize the crucial role of an efficient immune surveillance in the control of tumor growth and progression.

While ultraviolet radiation induced DNA damage is implicated in VN-tumors, the major causative factor of the MCC is considered MCPyV [61]. MCPyV was first identified in 2008, through whole-transcriptome sequencing [62], integrated into the genome of eight out of ten tested MCC cells. The Southern blot patterns of the primary tumor and a metastatic lymph node, isolated from the same patient, demonstrated an identical viral DNA integration at several different chromosomal sites. This important finding indicated that the viral integration was clonal and it was an early, if not initiating event, in VP-MCC oncogenesis process [62]. In addition, a C-terminal tLT form, lacking the OBD and helicase activity of LT required for viral DNA replication, was also observed [62]. Numerous studies have now confirmed that 80% of the examined tumors contain clonally integrated copies of the virus and express tLT [62–64]. MCPyV integration into the host genome occurs by accidental genome fragmentation during viral replication, in random site, without involvement of cellular tumor suppressor genes or oncogenes [56]. Viral integration involved mutations that result in the truncation of LT and a study by Schrama and co-workers suggests that truncating mutations occur before or during integration [65]. In vitro cell studies have demonstrated that expression of full-length LT in VP-MCC causes a specific DNA damage response, which is probably induced by in situ replication of the integrated viral DNA, which in turn is triggered by the binding of LT to the MCPyV ORI. Truncation of LT abolishes viral replication and seems to be necessary for MCC oncogenesis [5,66]. Tumor-derived tLT preserves the N-terminal J domain and LXCXE motif, whereas the DNA binding, helicase and cell growth-inhibitory domains are lost [66]. The tLT could potentiate a stable integration of the MCPyV into the host genome [66]. All VP-MCC tested contain ≥1 viral genome copies/cell [65,67–70], whereas in non-MCC tumors that contain MCPyV, the viral load was at least 2–3 logs lower (reviewed in [61]).

3. Cells of Origin of MCC

It was originally proposed that MCC derived from MCs because of similar immunophenotypes [71]. Both cell types express cytokeratin 20 [72], synaptophysin [73], neural cell adhesion molecule/CD56 [74], and numerous endocrine markers [75]. However, it is more and more unlikely that MC are the cells of origin because several characteristics of MCC argue against MC as the progenitor cell of MCC. Epithelial, fibroblastic, lymphoid, and neural crest origin of MCC has been put forward (Arguments in favor or contra these cell types as origin of MCC are summarized in Table 1.

VP-MCC may also originate from different cell types than VN-tumors. Dermal fibroblasts were suggested since they are permissive for MCPyV infection [44], but also keratinocytes could be the cell of origin of VP-MCC because keratinocyte-specific expression of MCPyV oncoproteins resulted in oncogenic effects [76]. Other studies suggest that VN-MCC derive from epidermal keratinocytes, whereas VP-MCC derive from dermal fibroblasts [77,78].

A recent report supports the assumption that VP-MCC may derive from the epithelial lineage [79]. The authors sequenced a combined tumor of trichoblastoma (neoplasm of epithelial follicular germinative cells) and VP-MCC. Non-integrated viral DNA encoding full-length LT could be amplified from the trichoblastoma, while integrated virus (~20 copies/cell) was detected in the MCC. Remarkably, two different tLT may be expressed in this MCC tumor.

Table 1. Supportive and contradictory arguments for the different hypotheses on cell of origin of MCC. The suspected cell of origin of VN-MCC and VP-MCC is indicated.

Cell of Origin	Supporting Cell of Origin	Arguing Against Cell of Origin
Merkel cell (VN-MCC)	1. Neuroendocrine granules [80] 2. CK20 expression [72] 3. Piezo 2 expression [81] 4. Other neuroendocrine markers such as CD56, chromogranin A, synaptophysin, insulinoma-associated protein 1 [82]	1. Epidermal location [83] 2. Postmitotic cells [84] 3. Diffusely arranged skeleton [82] 4. c-KIT, PAX-5, SCF, BCL2, CD24 are commonly expressed in MCC, but absent in MC [85] 5. Neural cell adhesion molecule L1 (CD171) and neurofilament in MCC but not MC [86] 6. Vasoactive intestinal peptide and metenkephalin expressed in MC but not MCC [87] 7. Diffuse CK20 staining in MC; dot-like staining in MCC [88] 8. CK20-positive MC not infected by MCPyV [82] 9. Mouse models of VN- and VP-MCC using MC-specific Cre drivers do not develop MCC [89]
(Epi)dermal stem cell (VN-MCC)	1. Neuronal cell markers [82] 2. CK14 expression [90] 3. CK19 expression [91,92] 4. SOX-2 expression [93] 5. Mitotic potential [85] 6. Other epidermal markers such as EMA, CK56, and EpCAM [87] 7. VN-MCC harbor UV mutational signature characteristic of epidermal-derived cancers [77]	1. Expression of B-cell markers [85] 2. SOX-2 more widespread expression [82,85] 3. Absence of MCPyV DNA in these cells [85] 4. CK19 also found in MC [94]
Pro/pre-B cell (VP-MCC)	1. B-cell specific lineage factors such as PAX-5, c-KIT, TdT *, SCF, RAG1 [85] 2. Expression Ig in VP-MCC [95] 3. IgH and Igκ rearrangements in VP-MCC [96] 4. MCC regression with idelalisib treatment [97]	1. Neurendocrine granules [80] 2. Location of MCPyV transduction [85]
Skin-derived precursors (VN-MCC)	1. Dermal location [82] 2. Broad differentiation	1. Absence of MCPyV DNA in these cells [85]

Table 1. *Cont.*

Cell of Origin	Supporting Cell of Origin	Arguing Against Cell of Origin
Dermal fibroblasts (VP-MCC)	1. Permissive for MCPyV [44] 2. Mutational burden similar to VP-MCC [77] 3. Mutational signature similar to VP-MCC [77] 4. Only cell type that can be transformed by sT in vitro [98–100]	1. Gene expression profile [85] 2. Neuroendocrine differentiation [85] 3. Expression of B-cell markers [85] 4. Lack of MCPyV DNA in HDF adjacent to VP-MCC [85]
Keratinocytes (VN- and VP-MCC)	1. Keratinocytes and MCs are derived from the same epidermal progenitor cell [101] 2. Transgenic mice expressing LT or sT in keratinocytes can result in oncogenic effects [76] 3. Mutational burden of VN-MCC is in line only with two other cancers, both keratinocyte-derived skin cancers [77] 4. VN-MCC have mutations in *NOTCH1*, *HRAS* and *FAT1*, which are frequent in squamous cell carcinoma [77]	

* TdT expression in 65% of all examined MCC and expression is significantly correlated with the presence of MCPyV.; PAX-5:90% of all examined MCC.

Whole genome sequencing identified six somatic mutations common for both tumors. The trichoblastoma had expression of KRT17 and SOX9, and activation of GLI1 as observed by nuclear localization, markers that are shared with MC progenitors [102–104]. Therefore, the authors suggest that the trichoblatoma cell in which MCPyV integration occurred and led to the development of MCC could be an epithelial progenitor cell of the hair follicle or an already differentiated MC [79].

4. The Oncogenic Mechanisms of MCPyV T Antigens

4.1. MCC Cell Growth Depends on LT But Not sT

Since the early proteins of other HPyVs possess oncogenic potentials in cell cultures and in animal models [105], the role of LT and sT in tumor growth was examined. Knock down of sT and LT (i.e., truncated LT and 57kT which cannot be distinguished in most VP-MCC cell lines) reduced MCC cell proliferation in culture, but also in xenograft mice [98,106,107]. Specific knockdown of only LT was sufficient to generate growth inhibition. Rescue experiments, i.e., expression of T antigens in cells where their endogenous expression was knocked down showed that wild-type sT plus LT could rescue cell growth. The growth promoting property of LT involves binding to RB because mutations in the DnaJ domain, the RB domain, or S220A abrogates LT's ability to promote cell growth [106,108,109].

Ectopic expression of the tLT variant MKL-1 in MCC13 promoted cell cycle progression [109]. However, RNA interference studies showed that sT is dispensable for growth and survival of VP-MCC cell lines [98]. Interestingly, knockdown of the T antigens in the VP-MCC LoKe cell line did not results in any growth inhibition. The authors speculate that additional aberrations enable cell growth even in the absence of T antigens and therefore, in some VN-MCC cases a viral hit-and-run mechanism was possible where MCPyV initiates tumor formation, then disappears, but additional mutations drive tumor progression and maintenance [110]. Studies in mouse and human fibroblasts demonstrated that expression of a tumor-derived tLT has stronger growth promoting activities than wild-type LT and 57kT [111]. Expression of full-length did not induce anchorage-independent growth, whereas tLT proteins induced aggregates in soft agar that did not grow into full colonies, suggesting that tLT has increased cell proliferative capacity compared with the wild-type LT. Expression of the C-terminal 100 amino acids residues inhibited the cell growth of fibroblasts and of the VP-MCC MKL-1 cell line [111]. The mechanism by which this region inhibits cellular growth is not known but is likely to be independent of p53 since neither full-length LT nor 57 kT are able to bind p53. The C-terminal domain may interact with a yet unidentified cellular protein involved in growth regulation. Putative candidates are the cell cycle checkpoint kinase ATM, casein kinase 2β and phosphatidylinositol-5-phosphate 4-kinase type 2β, which are all involved in proliferation and were found to interact with MCPyV LT, but the biological importance of these interactions were not examined, nor was the region of LT required for interaction identified (reviewed in [21]). CRISPR/Cas9 targeting of LT/57kT impaired MS-1 and WaGa cell proliferation, decreased G1/S cell cycle progression and increased apoptosis. Additional targeting of sT did not enhance the effect in LT/57kT mutated cells [112].

4.2. Oncogenic Properties of LT

Cell culture studies revealed that neither full-length nor tumor-derived tLT was able to trigger cellular transformation [99], but LT is required for growth of VP-MCC cells [108,113]. The C-terminal domain of LT causes DNA damage and stimulates host DNA damage response, leading to p53 activation and inhibition of cellular proliferation. Phosphorylation of the C-terminus by ATM kinase induces apoptosis and inhibits proliferation [14,66]. Thus, the C-terminus of LT contains anti-tumorigenic properties and may explain why this region is deleted in VP-MCC. To further elucidate the role of MCPyV LT in MCC tumorigenesis, cellular proteins that interact with LT were identified using different methods [21,114]. However, the biological relevance of these interactions and possible implications for MCPyV-induced cancer have not always been studied.

4.2.1. LT and p53

LT expressed in VP-MCC is truncated in its C-terminal part, which encompasses the p53-binding domain in LT of other HPyVs. As expected, tLT did not interact with p53, but surprisingly neither did full-length LT [111]. In another study, Borchert and co-workers showed that an antibody against p53 could immuno-precipitate full-length, but not tLT [115]. However, LT did not bind p53 directly and LT, but not tLT inhibited p53-mediated transcription. They suggested that full-length LT interacts with a bridging protein that serves as a co-activator in p53-driven transcription. Alternatively, another protein may change the conformation of LT allowing it to bind p53 as has been shown for human papillomavirus E6 protein. E6 forms a complex with E6AP and p53, but neither E6 nor E6AP are separately able to recruit p53. However, E6AP renders the conformation of E6 competent for interaction with p53 [116]. Park et al., reported that expression of tLT in IMR90 lung fibroblasts significantly stimulated transcript levels of p53-responsive genes and increased both total protein and Ser-15 phosphorylation levels of p53 [117]. They showed that the interaction between LT and RB1 lead to increased levels of ARF and activation of p53. ARF is an inhibitor of the E3 ubiquitin protein ligase MDM2, which degrades p53 [118]. Hence, LT can through RB-ARF-MDM2 axis stabilize p53.

4.2.2. LT and Retinoblastoma (RB) Family

Both full length and tLT interact with RB1, although with different strength [111,115,117]. This suggests that LT may usurp RB1, thereby relieving repression of E2F-mediated transcription and induce cell cycle progression into S phase. MCPyV LT did not interact with the p107 and p130 retinoblastoma family members, nor did it interfere with p107-induced and p130-induced cell cycle arrest and repression of E2F responsive genes [111,113,115]. The weaker in vitro oncogenic potentials of MCPyV LT compared to LT of other PyVs may be attributed to its weaker impact on the tumor suppressors p53 and RB.

4.2.3. LT and HSC70

LT interacts with HSC70 via the DnaJ domain and stimulates viral replication [12]. Like other PyVs, it is presumed that MCPyV LT disrupts Rb-E2F family complexes through the action of its DnaJ domain and ATPase activity of Hsc70 [119,120]. The biological significance of the DnaJ domain in sT is unknown as mutations in DnaJ of sT did not interfere with its effect on viral replication or in vitro transformation activity [12,99].

4.2.4. LT and VPS39 Subunit of HOPS Complex/Vam6p

Human Vam6p, a cytoplasmic protein involved in lysosomal processing and clustering, interacts with MCPyV full-length LT as well as MCC-derived tLT [18]. LT and tLT that retains its nuclear localization signal translocate hVam6p to the nucleus and sequester it from involvement in lysosomal trafficking. The physiological consequences of LT:Vam6p interaction are not known, but it might play a role in MCPyV replication rather than tumorigenesis, because VP-MCC have been described that express tLT without a nuclear localization signal [62,121,122].

4.2.5. LT and ATOH1

Sox2 (sex-determining region Y-box 2) and Atoh1 (atonal homolog 1) are critical transcription factors for MC development in mice [123]. Harold and colleagues found that knockdown of all T antigen isoforms in VP-MCC cell lines co-cultured with human keratinocytes promotes a neuronal phenotype in the MCC cells and resulted in reduced expression of ATOH1 and SOX2 [124]. The tLT 339 variant stimulated ATOH1 and SOX2 expression levels, but neither a LT399 retinoblastoma binding deficient mutant nor sT increased expression of ATOH1 and SOX2. Activation of the SOX2-ATOH2 pathway by LT in a retinoblastoma-dependent manner is important for both the manifestation of a Merkel cell phenotype and tumorigenesis. Transcriptional activation by ATOH1 requires E-boxes

(5-CANNTG-3′) and E47 binding site [125], both of which are present in the miR-375 promoter. Indeed, ATOH1 stimulated expression of miR-375 and ectopic expression of tLTs stimulated the activity of a minimal promoter containing three E-box and induced *ATOH1* mRNA and miR-375 in fibroblast MRC-5 cells [126]. Moreover, high transcript levels of *LT* and *ATOH* were detected in the VP-MCC WaGa cells. sT, however, was unable to enhance *ATOH1* mRNA and miRNA-375 levels. The neuroendocrine features of MCC may therefore be linked to MCPyV-induced expression of ATOH1. Whether LT-induced expression of miR-375 is exclusively mediated by ATOH1 or also by an ATOH1-independent mechanism remains to be elucidated. As both *ATOH1* and miR-375 promoters were hypomethylated, LT may stimulate demethylation of these promoters. Finally, strong expression of ATOH1 and miR-375 was also observed in classical VN-MCC cells, indicating a virus-independent mechanism in their expression [126].

4.2.6. LT and Ubiquitin-Specific Protease 7 (Usp7)

All MCPyV T-antigens interact with Usp7, a cellular deubiquitination enzyme [127]. The binding with LT, tLT and 57kT is direct, whereas sT probably interacts indirectly. Binding of Usp7 required the tumor necrosis factor receptor-associated domain of Usp7 and did not alter the ubiquitination levels of the T antigens, but stimulated the binding affinity of LT to the ORI, thereby restricting viral DNA replication. Usp7-mediated restriction of MCPyV replication could promote viral persistence [127]. Whether Usp7:T antigens interaction contributes to MCC tumorigenesis remains elusive. However, interference with other functions of Usp7 such as DNA damage response, epigenetic regulation, and immune response may also play a role in the development of virus-induced MCC [128].

4.2.7. LT and Other Interacting Proteins

Other interaction partners of MCPyV LT are summarized in Table 2. The interaction in VP-MCC has not been validated and the biological consequences of these interactions have not been investigated.

Table 2. MCPyV LT and sT interaction partners and their role in the life cycle of MCPyV. See text for details.

T Antigen	Protein	Functional Class	Biological Role	Reference
sT	abhydrolyse domain containing 12 (ABHD12)	metabolism	unknown	[129]
sT	ankyrin repeat domain 13A (ANKRD13Aa)	protein stability	unknown	[129]
sT	ATPase sarcoplasmic/endoplasmic reticulum Ca^{2+} transporting 2	metabolism	unknown	[129]
LT	ATP binding cassette subfamily A Member 13 (ABCA13)	signaling	unknown	[129]
sT	ATP binding cassette subfamily D member 3 (ABCD3)	signaling	unknown	[129]
LT	ATP binding cassette subfamily D member 13 (ABCD13)	signaling	unknown	[129]
sT	aryl hydrocarbon receptor interacting protein (AIP)	transcription	unknown	[129]
LT	adaptor related protein complex 2 subunit A and M (AP2A1 and M1)	intracellular transport	unknown	[129]
sT	ADAM metallopeptidase domain 9 (ADAM9)	cytoskeleton/extracellular matrix	unknown	[129]
LT	ataxia telangiectasia mutated (ATM kinase)	DNA replication and repair	LT phosphorylation	[14]
LT, sT	BCL2 associated anthanogene 2, 3 and 5 (BAG2, 3 and 5)	protein stability/apoptosis	unknown	[129]
LT	bromodomain protein 4 (Brd4)	cell cycle/DNA replication	viral genome replication	[130,131]
sT	cadherin 1 (CDH1)	cytoskeleton/extracellular matrix	unknown	[129]
LT	casein kinase 2 beta (CK2β)	Signaling	unknown	[129]
sT	cathepsin B (CTSB)	protein stability/modification	unknown	[129]

Table 2. *Cont.*

T Antigen	Protein	Functional Class	Biological Role	Reference
LT	caveolae associated protein 2 (CAVIN2)	intracellular transport	unknown	[129]
sT	CCHC-type Zinc finger nucleic acid binding protein (CNBP)	transcription	unknown	[129]
sT	cell surface glycoprotein 44 (CD44)	cell-cell interaction, cell adhesion, migration	unknown	[129]
sT	cell division cycle 20 (CDC20)	cell cycle	sT-mediated phosphorylation of 4E-BP1	[129,132,133]
sT	coatomer protein complex subunit γ2	intracellular transport	unknown	[129]
sT	2′, 3′-cyclic nucleotide 3′ phosphodiesterase (CNP)	nucleotide metabolism	unknown	[129]
LT	DEAD-box helicase (DDX24)	post-transcription/translation	unknown	[129]
sT	heat shock protein 40 members A1 and B4 (DnaJA1 and B4)	chaperone	unknown	[129]
LT, sT	heat shock protein 40 member C7 (DnaJC7)	chaperone	unknown	[129]
LT	transcription factors E2F3 and 4 (E2F3 and 4)	transcription	unknown	[129]
sT	EGF containing fibulin extracellular matrix protein 2 (EFEM2)	cytoskeleton/extracellular matrix	unknown	[129]
sT	eukaryotic translation initiation factor 4E binding protein 1 (eIF-4EBP1)	translation	disregulated cap-dependent translation which promotes tumorigenesis	[99,133]
LT, sT	emerin (EMD)	cytoskeleton	unknown	[129]
LT	family with sequence similarity 71 member E2 (FAM71E2)	unknown	unknown	[129]
sT	F-box and WD repeat domain containing 7 (Fbxw7)	protein stability	tumorigenic properties of the virus (stabilization of LT and cellular proteins)	[134,135]
LT	general transcription factor IIIC subunit 1 (GTF3C1)	transcription	unknown	[129]
LT	high density lipoprotein binding protein (HDLBP)	metabolism	unknown	[129]
LT, sT	heat shock protein 70 (HSPA1 and A4)	chaperone	cell cycle progression	[129]
sT	insulin like growth factor 2 receptor (IGF2R)	signaling	unknown	[129]
LT, sT	inhibitor of nuclear factor kappa-B kinase-interacting protein (IκBIP)	signaling	unknown	[129]
LT	karyopherin subunit α2, 3 and 4 (KPNA2, 3 and 4)	intracellular transport	unknown	[129]
sT	lysyl oxidase (LOX)	metabolism	unknown	[129]
LT	microtubulin-associated protein 4 (MAP4)	cytoskeleton	unknown	[129]
sT	membrane bound O-acetyltransferase domain containing 7	metabolism/plasma membrane lipid organization	unknown	[129]
LT	mediator complex subunit 14 (MED14)	transcription	unknown	[129]
sT	matrix metalloproteinase 14	extracellular matrix	unknown	[129]
sT	myelin protein zero like 1 (MPZL1)	signaling	unknown	[129]
sT	mitochondrial carrier 2 (MTCH2)	metabolism	unknown	[129]
sT	myoferlin (MYOF)	membrane morphology	unknown	[129]
sT	NF-kappa-B essential modulator (NEMO=IKBKG)	signaling	inhibition NFκB signaling; immune evasion	[136]
sT	Notch 2 receptor (NOTCH2)	signaling	unknown	[129]
sT	nuclear receptor binding SET domain protein1 (NSD1)	transcription	unknown	[129]
LT	prolyl 4-hydroxylase subunit alpha 3 (P4HA3)	metabolism	unknown	[129]
sT	prolyl 4-hydroxylase subunit β (P4HB)	metabolism	unknown	[129]

Table 2. *Cont.*

T Antigen	Protein	Functional Class	Biological Role	Reference
sT	platelet-derived growth factor receptor subunit β (PDGFRβ)	signaling	unknown	[129]
LT	PGAM family member 5, mitochondrial Ser/Thr protein phosphatase (PGAM5)	signaling	unknown	[129]
sT	progesterone receptor membrane component 2 (PGRMC2)	signaling	unknown	[129]
LT	phosphatidylinositol-5-phosphate 4-kinase type 2 beta (PIP4K2β)	signaling	unknown	[129]
LT	protein phosphatase 2 scaffold subunit α (PP2AR1α)	signaling	unknown	[129]
sT	protein phosphatase 2 catalytic subunit α and β (PPP2CA and CB)	Signaling	mutation in PP2A binding site had no effect on the known activities of sT	[17,136,137]
sT	PRA1 domain family member 2 (PRAF2)	intracellular transport	unknown	[129]
sT	protein phosphatase 2 regulatory subunit Aα and Aβ (PP2R1A and B)	signaling	unknown	[138]
sT	protein phosphatase regulatory subunit 1 (PP4R1)	signaling	microtubule destabilization and cell motility (metastasis?); inhibition NFκB signaling (immune evasion?)	[17,136,139, 140]
sT	protein phosphatase Mg^{2+}/Mn^{2+} dependent 1A, 1B and 1G (PPM1A, B and G)	signaling	unknown	[129]
sT	proteasome 26S ATPase 2,3 and 4 (PSMC2, 3 and 4)	protein stability	unknown	[129]
LT	caveolae associated protein 1 (PTRF)	transcription	unknown	[129]
sT	pituitary tumor-transforming gene 1 protein-interacting protein (PTTPG1P)	intracellular transport	unknown	[129]
sT	Rab18 (RAB18)	signaling	unknown	[129]
LT	Retinoblastoma protein 1 (RB1)	cell cycle	cell cycle progression	[111,115,117]
sT	ribonuclease/angiogenin inhibitor 1 (RNH1)	transcription/translation	unknown	[129]
sT	ribosomal protein L21	translation	unknown	[129]
sT	ribosomal protein S27 like	translation	unknown	[129]
LT	recitulon 4 (RTN4)	intracellular transport	unknown	[129]
LT	sphingosine-1-phosphate lyase 1 (SGPL1)	metabolism	unknown	[129]
sT	secreted protein acidic and cysteine rich (SPARC)	extracellular matrix	unknown	[129]
sT	sulfide quinone oxidoreductase (SQRDL)	metabolism	unknown	[129]
LT	signal recognition particle 14 (SRP14)	intracellular transport	unknown	[129]
LT, sT	signal recognition particle receptor subunit b (SRPRB)	intracellular transport	unknown	[129]
sT	ser/thr kinase 38 (STK38)	signaling	unknown	[129]
LT, sT	STIP1 homology and U-box containing protein 1 (STUB1)	protein stability	unknown	[129]
sT	surfeit 4 (SURF4)	intracellular transport	unknown	[129]
LT	Ubiquitin-specific protease (USP7)	protein stability	inhibition viral DNA replication	[127]
LT	transcription elongation factor B subunit 1 (TCEB1)	transcription	unknown	[129]
LT	transcription factor DP1 (TFDP1)	transcription	unknown	[129]
sT	translocase of inner mitochondrial membrane 8A (TIMM8A)	intracellular transport	unknown	[129]
sT	transmembrane protein 165 (TMEM165)	protein glycosylation	unknown	[129]
sT	thioredoxin related transmembrane protein 3 (TMX3)	protein folding	unknown	[129]

Table 2. *Cont.*

T Antigen	Protein	Functional Class	Biological Role	Reference
sT	toll interacting protein (TOLLIP)	signaling	unknown	[129]
LT	tripartite motif containing 38 (TRIM38)	protein stability	unknown	[129]
LT	testis-specific Y-encoded-like protein 1 (TSPYL1)	transcription	unknown	[129]
sT	tubulin α1 (TUBA1B)	protein folding and gap junctions	unknown	[17]
sT	tubulin β2α (TUBB2A)	mitosis and intracellular transport	unknown	[17]
LT, sT	upregulated during skeletal muscle growth 5 (USMG5)	nucleotide synthesis	unknown	[129]
LT	VPS39 subunit Of HOPS complex (Vam6p)	intracellular transport	role in DNA replication (?)	[18,129]
LT	VAMP associated proteins A and B (VAPA and VAPB)	intracellular transport	unknown	[129]
sT	vitamin K epoxide reductase complex subunit 1 (VKORC1)	metabolism	unknown	[129]
LT	vacuolar protein sorting-associated protein 11 homolog (VSP11)	intracellular transport	unknown	[129]

LT and sT induce microRNAs that target mRNAs for proteins involved in autophagy.

Autophagy plays an important role in cancer and in immune evasion [141–143]. Silencing LT or LT+sT in VP-MCC cell lines reduced the expression of miR-30a-3p, miR-30a-5p and miRNA-375, while ectopic expression of tLT or sT in VN-MCC cells increased the levels of these miRNAs. Induced expression of miR-30a-3p, miR-30a-5p and miRNA-375 required the DnaJ domain [144]. Target mRNA of these miRNAs encode the autophagy proteins ATG7, SQSTM1/p62 and BECN1. The authors showed that sT and tLT, but not wild-type LT suppressed autophagy processes in MCC cells and protein levels of ATG7 and SQSTM1/p62 were lower in VP-MCC compared with VN-MCC. Hence, T antigens-mediated suppression of autophagy might protect cancer cells from cell death and contribute to tumorigenesis [144].

4.3. The Role of 57kT and ALTO in VP-MCC

Whether 57 kT and ALTO are implication in MCPyV-induced tumorigenesis remains to be established. The 57kT protein retains the RB binding domain and the CR1 and DnaJ binding motifs. Immortalized human fibroblasts Bj-hTERT expressing 57kT grew slower than control cells and when LT cDNA was stably expressed in mouse and human fibroblasts, the 57kT form was preferentially expressed. Expression of 57 kT has never been detected in VP-MCC [68,69], but due to truncation in the LT gene, LT and 57kT cannot be distinguished in most MCC using the antibodies currently available. The role of 57kT in MCC remains unsolved. Deletion of ALTO did not abrogate viral replication and is dispensable for MCPyV-driven tumor cell proliferation [3,108], but the function of this protein remains elusive.

4.4. The Oncogenic Properties of sT

MCPyV sT is sufficient to fully transform Rat-1 and NIH3T3 mouse fibroblasts [98–100]. Knockdown of sT expression in VP-MCC cell lines causes cells to stop proliferating, but did not result in cell death. Co-expression of full-length or tLT did not enhance sT-induced colony formation compared with expression of sT alone [99].

4.4.1. sT and Transgenic Mice

Considering the non-transforming potentials of LT in cell culture and that sT can induce transformation, sT, but not LT transgenic mice models have been generated. Verhaegen et al., generated a transgenic mouse model in which sT expression was regulated by the epidermis-specific keratin-5 promoter [134]. Analysis of embryos revealed that sT promotes neoplastic transformation in epithelia in a PP2A-independent, but LSD-dependent manner. Adult animals developed lesions strongly resembling squamous cell carcinoma in situ. However, expression of sT alone does not appear to be sufficient to drive epidermal cells in MCC in a mouse model. The same group generated K5-tLT, K5-sT+tLT, K5-st+Atoh1, K5-tLT+Atoh1, K5-sT+tLT+Atoh1, and K5-Atoh1 transgenic mice [145]. The tLT embryo had no apparent phenotype, co-expression of sT+Atoh1 resulted in MCC-like tumors, and co-expression of tLT did not noticeably altered the phenotype of sT or sT+Atoh1 mice. These studies indicate that Atoh1-induced differentiation of epidermal cells into neuroendocrine lineage together with sT as the viral oncogenic driver can result in MCC development. Transgenic mice co-expressing sT and tLT under control of the keratinocyte-specific K14 promoter developed hyperplasia, hyperkeratosis and acanthosis, and some mice develop papillomas, but not MCC [76]. Shuda and c-workers developed a sT-Δp53-Atoh1 transgenic mice which allowed sT expression in MC cells [89]. Although these mice have increased embryonic MC precursor proliferation, they did not develop MCC.

Taken together, in vitro and animal studies and the detection of sT in the absence of LT in some VP-MCC indicate that sT may be more involved in the oncogenic process, whereas LT is required to maintain the tumor cell growth [98,99]. However, studies in the genuine cells of origin of VP-MCC are required to determine the requirements of sT and LT in cell growth and oncogenesis.

4.4.2. sT and Eukaryotic Translation Initiation Factor 4E Binding Protein (4E-BP1)

Transcription initiation factor 4E-BP1, a downstream target of the Akt-mTOR pathway, binds in its unphosphorylated or hypo-phosphorylated form eukaryotic initiation factor 4E (eIF4E), thereby preventing assembly of eIF4F onto capped mRNA and inhibiting translation [146]. sT interacts with 4E-BP1 and expression of sT, but not LT promoted 4E-BP1 phosphorylation [99,133]. sT-induced phosphorylation of 4E-BP1 is accomplished by sT interacting with Cdc20 and possibly Cdc20 homolog 1 (Cdh1), which activates the CDK1/cyclin B1 complex and CDK1 and phosphorylate 4E-BP1 [132,133]. 4E-BP1 hyperphosphorylation was required for sT-induced transformation of rodent cells [99,133]. The importance of sT-mediated 4E-BP1 phosphorylation in MCPyV-induced MCC is not completely understood, but sT-induced hyperphosphorylation of 4E-BP1 can dysregulate cap-dependent translation, an event that has been shown to promote tumorigenesis [147].

4.4.3. sT and E3 Ubiquitin Ligases

Binding of sT to E3 ubiquitin ligase complex SCF^{Fbw7} led to inactivation of the enzymatic activity and stabilization of LT, which is a substrate of SCF^{Fbw7} [135]. Binding occurs through LSD and loss of net positive charge in the LSD abrogated sT:SCF^{Fbw7} interaction [100]. sT-induced stabilization of LT stimulates viral replication and transformation of rodent fibroblasts cell cultures by sT is SCF^{Fbw7}-dependent [100,148], and increased protein levels of SCF^{Fbw7} substrates Mcl-1, c-Jun, mTOR and cyclin E in sT transgenic mice [134]. sT also interacts with the E3 ubiquitin ligases Cdc20-anaphase promoting complex [17] and β-TrCP [149] and this stimulated genome instability [135]. Inactivation of E3 ubiquitin ligases by sT may be therefore be an important contributor in MCPyV-induced transformation and tumorigenesis. However, Dye and colleagues failed to detect interaction between sT and SCF^{Fbw7} and sT and β-TrCP and no increased c-Myc levels were observed when sT was overexpressed. They also demonstrated that sT-mediated stabilization of LT did not require SCF^{Fbw7} [150]. The reason for the discrepancies between the different students is presently unknown. sT can form a complex with the E3 ubiquitin ligase STIP1 homology and U-box containing protein 1 (STUB1) [129]. This E3 ubiquitin ligase plays also a role in innate and adaptive immunity [151], but the biological implications of sT:STUB1 interaction in MCPyV replication and MCC remain to be determined.

4.4.4. sT and N-myc Downstream Regulated Gene-1 (NDRG1)

Stable expressing the entire MCPyV early region in human immortalized keratinocytes resulted in >1.5-fold up- or down-regulated of 325 genes [152]. Of these, 73 had decreased expression and the majority encodes proteins involved in cell senescence, DNA repair, signal transduction, and cell cycle regulation, including HIST1H1C. Upregulation of HIST1H1C was also confirmed in VP- and VN-MCC cell lines, MCC tumors, and in sT expressing human fibroblasts expressing compared with normal fibroblasts [153–155]. Of the upregulated genes, many encode proteins implicated in cell cycle regulation and signaling pathways, including CDK4, cyclins D2 and D3, CDC25, FOXQ1, DUSP10, and CTSH. One gene that was specifically down-regulated by MCPyV, but and not by other HPyVs and SV40 was the *N-myc downstream regulated gene-1* (*NDRG1*). NDRG1 is a known tumor suppressor and metastasis suppressor [156]. Knock-down of sT+LT in MKL-1, MKL-2, MS-1 and CVG-1 increased NDRG1 levels in all four cell lines, and decreased cyclin D1 and CDK2 levels in MKL-2, MS-1, and CVG-1 cells. Overexpression of NDRG1 in MKL-2 reduced cyclin D1 and CDK2 levels, but not in MKL-1 cells. The different status of transformation of may explain the difference between MKL-1 and the other VP-MCC cell lines. Depletion of sT alone or sT+LT resulted in comparable increase in *NDRG1* mRNA levels, suggesting that sT is sufficient. Overexpression of NDRG1 in keratinocytes stably expressing MCPyV early region or in MKL-1 and MKL-2 cells inhibits cellular proliferation and migration. Taken, together these observations indicate that MCPyV-mediated repression of NDRG1 participates in MCC tumorigenesis and that sT may be the main contributor. The expression levels of NDRG1 have not yet been examined in VN- and VP-MCC. In a study in 91 MCC tumors (30 VN and 61 VP), cyclin D1 expression was only detected in two tumors, both of which were MCPyV negative [157].

4.4.5. sT and p53

LT indirectly activates p53 (see above) and sT can stabilize LT, yet co-expression of LT and sT reduced p53 activation [148]. MCPyV sT can inhibit p53 activity indirectly by binding to and activating the transcription factor MYCL and the histone acetylase complex EP400 [117]. The MYCL:EP400 complex controls transcription of *MDM2* and *CSNK1A1* genes. The latter encodes casein kinase 1α which activates MDM4, an inhibitor of p53 [158]. The activation of p53 by LT may exert anti-tumorigenic effect, while sT-mediated inhibition of p53 favors pro-tumorigenesis. The relative concentrations of LT and sT, but also the strength of impact of LT and sT on p53 will determine the outcome. VP-MCC cells have been shown to express high levels of MDM4 [117]. Accordingly, p53 levels were found to be lower in VP-MCC cell lines compared to VN-MCC cell lines [117,159,160]. Examination of MCC revealed that mutations in *TP53* gene are almost exclusively detected in VN-MCC, but only 7% of VP-MCC expressed detectable p53 levels and an inverse correlation between p53 expression and viral DNA copy number was observed [157,161]. One study reported that p53 levels were variable between patients, with no obvious differences between VN- and VP-tumors [162]. The expression levels of MDM4 in VN- and VP-MCC biopsies have not yet been examined. Another consequence of the interaction of sT with MYCL:EP400 complex that may be involved in tumorigenesis was recently published. This complex stimulates the expression of components of the lysine-specific histone demethylase 1 (LSD1) complex that acts as a transcriptional repressor [117]. Treatment of VP-MCC cell lines with LSD1 inhibitors completely blocked colony formation in soft agar, and LSD1 inhibitors reduced the growth of MCC in vitro and in xenograft models using VP-MCC cells. Hence, sT-mediated activation of the LSD1 complex seems to play a pivotal role in VP-MCC, and LSD1 inhibitor could be used to treat VP-MCC patients.

4.4.6. sT and Protein Phosphatases

Because aberrant or loss of enzymatic activity of protein phosphatases (PPs) can lead to transformation and their role in cancers, PPs are considered tumor suppressors and are targeted by several tumor viruses [163–167]. MCPyV sT interacts with PP1A, 1B and 1G [17,129]. The biological consequences of sT:PP1 interaction have not been determined, but RB is a PP1 substrate. Inhibition of PP1 by sT may therefore results in hyperphosphorylation of RB, release of repression of E2F target genes, and drive to enter the S-phase [168].

PP2A exists as a heterotrimer composed of a structural subunit A, a regulatory subunit B, and a catalytic C subunit [169]. MCPyV sT binds the structural subunit Aβ and Aα, and the catalytic subunits Cα and Cβ. This binding reduced the catalytic activity of the enzyme [17,136–138]. sT's binding to PP2A excluded the regulatory subunit B56α, but not other B subunits [17]. The biological implications of the sT:PP2A interaction are not known because mutations in sT that prevented PP2A binding had no effect on sT's transforming activity [99], nor did it impede sT-induced skin hyperplasia in transgenic mice [134].

MCPyV sT was reported to interact with PP4 [17,136,139,140], and this interaction promotes microtubule destabilization and stimulates cell motility and filopodium formation [139,140]. The sT:PP4 association also interferes with the NFκB pathway. The transcription factor NFκB is retained in an inactive state in the cytoplasm through interaction with inhibitor of κB (IκB). Activation of the NFκB pathway occurs after phosphorylation of IκB by IκB kinase (IKK) and subsequent degradation of IκB. IKK is a trimeric complex that consists of IKKα, IKKβ, and IKKγ or NFκB essential modulator (NEMO). Release of NFκB allows nuclear translocation where it affects transcription of NFκB-responsive genes [170]. NFκB target genes encode proteins involved in inflammation, immune responses, including antiviral response [171,172]. Griffith and colleagues demonstrated that sT associates with a PP4R1-PP4C complex, which stimulates the interaction between NEMO and the protein phosphatase PP4C-PP4R1 complex. Consequently, NEMO-mediated recruitment of PP4C to the IKK complex reduces IKK phosphorylation, with subsequent inhibition of IκB and failure to release, activate (phosphorylate), and translocate NFκB to the nucleus [136]. Thus, MCPyV may affect inflammatory and immune responses by interfering with the NFκB pathway. However, the importance of the sT:NFκB interaction in tumorigenesis is questioned because a significantly higher expression of pSer-536 RelA/p65 subunit of NFκB was observed in VP- (n = 24) compared to VN-MCC (n = 17). The phosphorylated p65 form was exclusively detected in the nucleus [173].

4.4.7. sT and Sheddases

MCPyV sT stimulates expression of the sheddases ADAM10 and 17, proteins involved in cell signaling, inflammation, and tumor formation and progression [174]. The exact mechanism by which sT enhances ADAM 10 and ADAM17 expression is not known, but sT increases expression of the transcription factors ACAD8, PPARG, and ITGB3BP that activate the ADAM10 promoter [155]. ADAM 10 and 17 protein levels are higher VP-MCC tumors compared to VN-MCC, suggesting that sT-induced sheddase expression may contribute to MCC progression [174].

4.4.8. sT and Metabolism

Ectopic expression of sT in normal human fibroblasts IMR90 resulted in significantly perturbed metabolism with elevated aerobic glycosylation and upregulation of transcription of metabolite transport genes [155]. Proteins whose transcripts were significantly upregulated included monocarboxylate lactate transporter 1 (MCT1), glucose transporter GLUT1, and GLUT3. Inhibition of MCT1 activity suppressed the growth of VP-MCC cell lines and impaired MCPyV-dependent transformation of IMR90 cells. The authors showed that MYCL cooperates with the tumor derived MCPyV early region (expressing sT and tLT) to induce expression of MCT1 and knockdown of the p65 subunit of NFκB reduced sT, as well as sT+MYCL stimulated MCT1 expression. Taken together, these data suggest that sT-mediated changes in the metabolic state are implicated in virus-induced MCC tumorigenesis. MCT1 expression levels in VN- and VP-MCCs have not been examined, but inhibitors of MCT1 could be considered to treat VP-MCC.

4.4.9. sT and Other Interaction Partners

Other cellular proteins reported to interact with MCPyV sT are shown in Table 2. The interaction in genuine host cells for MCPyV and in VP-MCC has not been confirmed, nor has the physiological relevance of these interactions been explored.

4.5. Effect of MCPyV on Signaling Pathways in MCC

4.5.1. The Phosphatidyl-3-Kinase/AKT/Mammalian Target of the Rapamycin (PI3K/AKT/mTOR) Pathway

The PI3K/AKT/mTOR pathway, which plays pivotal roles in cell growth, motility, survival, metabolism, and angiogenesis is often the target of viral infections [175,176]. Strong staining with phosphoT308 AKT antibodies was observed in most of the MCC samples examined, but there was no significant correlation between phosphoAKT and MCPyV status [177,178]. Another study reported AKT phosphorylation in 4 VN-MCC cell lines, but not in VP-MCC cell lines [179]. However, three of the tested VN-MCC cell lines (MCC13, MCC26 and UIOS) are non-classical MCC cell lines High expression levels of PI3Kα and PI3Kδ were observed in respectively 20% and 52% of archival MCC specimens ($n = 50$) [180]. The viral status in the MCC samples was not described, but PI3Kα transcript levels were detected in 2 VN and 2 VP-MCC cell lines, while one of the VP-MCC cell lines (MKL-1) had no detectable PI3Kδ mRNA levels. This suggests that the expression levels of PI3K do not depend on the presence of MCPyV, which is underscored by the finding that silencing of LT and sT in four MCPyV positive MCC cell lines had no effect on AKT phosphorylation [177]. Taken together, the results indicate that activation of AKT in MCC is not caused by MCPyV. A well-known substrate of the PI3K/AKT/mTORC1 pathway is 4E-BP1 and its interaction with sT was discussed earlier.

4.5.2. Protein Kinase C Pathway

Protein kinase C (PKC) is family of serine/threonine kinases that comprises PKCα, βI, βII, γ, δ, ε, η, θ, ζ and ι [181]. Because PKCε plays critical roles in cancer [182], its activation (i.e., phosphorylation of Ser729) was examined in 8 VP-MCC and three VN-MCC specimens [183]. Seven of the VP-MCCs were positive for phospho-PKCε, whereas only one of the three VN-MCC samples expressed phospho-PKCε. These results suggest a correlation between PKCε activation and MCPyV positivity in MCC. However, relative few samples were examined and the involvement of MCPyV in PKCε activation remains to be proven.

4.5.3. Notch Pathway

There are four human Notch receptors (NOTCH1-4) and their ligands include Jagged 1 and 2, and Delta-like proteins [184]. Relative expression levels of NOTCH1, NOTCH2, NOTCH3, and Jagged 1 were compared in 19 VN- and 19 VP-MCC tumors [185]. NOTCH3 expression was higher in VP-MCC compared to VN-MCC, while the opposite was found for Jagged 1. Patients with higher NOTCH3 expression had better overall survival, whereas expression of NOTCH1 and NOTCH2 was not associated with MCPyV status or prognosis. Whether MCPyV proteins are implicated in the upregulation of NOTCH3 and downregulation of Jagged 1 remains to be investigated. MCPyV sT can bind NOTCH2, but the functional implication of this interaction is not known [129]. sT may also activate the NOTCH pathway through stimulating the expression of ADAM10 [174].

4.5.4. Hedgehog Signaling Pathway

Patched 1 (PTCH1) is the receptor for the hedgehog ligand of which 3 are found in humans: sonic (SHH), Indian (IHH), and desert (DHH) hedgehog [186]. Expression of SHH and IHH was monitored in 29 VP-MCCs and 21 VN MCCs. A significant higher expression of SHH and IHH was observed in the VP-MCCs than in VN-MCCs [187].

4.5.5. Apoptotic Pathway

Expression of pro-survival proteins Bcl-2, $BclX_L$, Bcl-w, Mcl-1 and A1 has been investigated in both VN- and VP-MCC. High expression of these anti-apoptotic proteins was measured in most MCC and no correlation was found with the viral status of the tumor [159,188–192]. Despite high Bcl-2 levels in most tumors, a phase II clinical trial with Bcl-2 antisense RNA G3139 showed very little efficacy in 12 MCC patients [193].

5. Immune Evasion of VP-MCC

More than 90% of the MCC patients are immunocompetent and VP-MCC tumors are highly antigenic, yet they evade immunological destruction [57,63,194]. MCPyV can escape detection by the immune system by different mechanisms. Down-regulating major histocompatibility complex class 1 (HLA class 1) was observed in 84% of MCC tumors, and HLA class 1 expression was significantly lower in VP-MCC than in VN-MCC [195]. MCPyV-specific T cells and MCC-infiltration lymphocytes express elevated levels of multiple markers of exhaustion such as programmed death 1 (PD-1) and T cell immunoglobulin and mucin domain-containing protein 3 (TIM-3) [196]. Moreover, the level of vascular E-selectin is reduced in >50% of the examined MCC ($n = 56$; viral status not determined) and this negatively affects the ability of lymphocytes to migrate into the tumor microenvironment [197]. Programmed death ligand -1 (PD-L1) may be aberrantly expressed by tumor cells, creating a shield against immune attack [198]. Immunohistochemical staining of 8 VN-MCC and 34 VP-MCC showed that none of the VN tumors expressed PD-L1, while 50% of the VP-MCCs were positive for PD-L1 [199]. Another study on 14 MCC (6 VN and 8 VP) reported that 1 VN-MCC had few (1%) PD-L1 positive tumors cells, whereas 7 of the VP-MCC were PD-L1 positive with 2–7.5% of the cells expressing PD-L1 [200]. It is not known whether MCPyV can affect the expression of PD-L1, but upregulation of PD-L1 has been observed in persistent infection with the oncoviruses hepatitis B and C [201].

MCPyV can also avoid the innate immune system because its early region downregulated the expression of TRL9 in the B lymphocyte RPMI-8226 cell line by targeting the transcription factor C/EBPβ [202]. sT could also reduce TLR9 expression, but the mechanism is not known, but it may operate by stabilizing LT [148]. A study on 128 MCC patients revealed that decreased expression of TLR 9 correlated strongly with MCPyV positivity of the tumor, while expression of TLR2, 4, 5, and 7 did not correlate with the viral status of the tumor [203].

The interference with the NFκB pathway by MCPyV sT was discussed earlier. However, another putative mechanism by which MCPyV can interfere with this pathway is through the interaction of LT with bromodomain protein 4 (Brd4). Brd4 acts as a transcriptional and epigenetic regulator [204], and can interfere with the NFκB pathway by interacting with IκB [205]. Brd4 stimulates MCPyV DNA replication by interacting with MCPyV LT and recruitment of replication factor C [130]. Arora et al., showed that also tLT binds Brd4 and that co-expression of Brdr4 in combination with either LT, sT, or tLT did not stimulated MCPyV promoter activity in U2-OS cells [131]. However, Brd4 in combination with LT+sT, but not with tLT+sT, enhanced promoter activity. Studies by our group showed that full-length LT inhibiting the activity of early as well late promoter from 8 different MCPyV variants in MCC13 and immortalized human dermal fibroblasts, whereas truncated variants stimulated their cognate promoter in both cell lines. The effect of sT on MCPyV promoter activity was not examined [25]. The study by Arora and colleagues was done in U2-OS cells and it was not specified if early or late promoter activity was monitored and from which virus strain the promoter was derived. Moreover, they used tLT referred to as tLT21 and tLT168, while we used tumor-derived tLT and tested their effect on the corresponding promoter. Whether the MCPyV LT:Brd4 interaction interferes with NFκB signaling pathway and contributes the virus-induced tumorigenesis remains to be investigated.

Cytokines trigger inflammatory and immune responses upon viral infection [206,207], and play a pivotal role in tumorigenesis [208,209]. A study in BJ human foreskin fibroblasts showed that tLT or tLT+sT induced IL-1β, IL-6, IL-8, and CXCL1 levels, but their expression levels have not been monitored in VN- and VP-MCC cell lines or tumor tissue [210]. Prokineticins are chemokine-like proteins that possess angiogenic and immunoregulatory activities [211]. VP-MCCs had higher prokineticin-2 mRNA levels than the virus-negative tumors [212]. Our group found that chemokine (C-C motif) ligand 17/thymus and activation-regulated (CCL17/TARC) is upregulated in VP-MCC cell lines compared to VN-MCC cell lines. Full-length and tLT, but not sT, enhanced the CCL17/TARC promoter activity and increased protein levels [213]. The exact mechanisms by which MCPyV may affect cytokine expression and their possible role in MCC remain to be determined. Another study reported that sT downregulates IL2, IL-8, CCL20 and CXCL9 expression in the VN-MCC cell MCC-13 [136], but expression levels in VN- and VP-MCC tumors have not been compared. Stimulator of interferon genes (STING) is a signaling molecule that controls type I interferon and other proinflammatory cytokines production [214]. STING protein was undetectable in VP MKL-1, MKL-2 and MS-1 cells, but not in non-classical VN MCC13, MCC26 and UISO cells. Five MCC tumors (virus state not mentioned) also stained negative with STING antibodies [215]. STING silencing may help MCC tumor cells to escape immune eradication. More VN- and VP-MCC should be scrutinized to establish whether STING is specifically silenced in the VP-MCC and the potential role of T antigens in silencing STING should be explored. Postsurgical adjuvant radiation is common in the treatment of MCC patients and although adjuvant radiotherapy can improve locoregional control with reduced recurrence rate of the tumor, it may not affect overall survival [216–219]. Whether VN- and VP-MCC patients display different sensitivity to radiation is not known, but a previous study had shown that absence of STING impaired radiation-mediated tumor regression [220]. Because it was recently reported that STING expression is downregulated in VP-MCC cells [215], VP-MCC may be less sensitive to radiotherapy than VN-MCC.

6. Specific Biomarker for VP-MCC

Apart from detection of viral DNA, RNA, and protein, diagnostic markers that specifically discriminate VP-MCC from VN-MCC are lacking. Likewise, biomarkers to predict disease progression and response to therapy of VP-MCC are lacking. However, the presence of antibodies against LT and sT may be used a diagnostic and prognostic marker. While most individuals have antibodies against MCPyV VP1 (see Section 1.2), only ~1% of healthy patients had low titer antibodies against viral T antigens, whereas 41% of MCC patients had such antibodies [221]. The viral status of all MCC patients was not known, but for those patients it was known, serology for the LT/sT much more closely reflected the virus status of the tumor. In addition, the titers of T antigens antibodies decreased rapidly in patients whose cancer did not recur, whereas they rose with disease progression. So antibodies against LT/sT can predict if the patient has a VP tumor, but these antibodies can also be used to monitor the development of the disease and whether the patient respond to treatment or not.

Some putative markers will be discussed in this section, although most of them do not seem to be very specific and more VN- and VP-MCC patients need to be studied to validate their usefulness.

Several studies have shown that p63 may be an adverse prognostic factor as high levels have been linked to a worse prognosis [222–225], but the viral status in the MCC tumors was not always described. In one follow up study, the presence of MCPyV was examined, but no correlation between p63 expression and viral presence was found [226]. The chromatin architectural factor DEK was found to be expressed in 15/15 MCC tumors examined, but the viral presence or the clinical stage of the tumors was not identified [227]. This protein is also overexpressed in other cancers (reviewed in [228]), so that it is not a specific MCC biomarker. K homology domain-containing protein overexpressed in cancer (KOC=IMP3) is overexpressed in 90% of the MCC samples (n = 20) and expression correlated with metastasis, but the relationship with MCPyV was not investigated [229]. KOC is a prognostic marker in pancreatic cancers and melanomas [230,231] and might be a prognostic marker for MCC. Other proteins examined in MCC include vitamin D receptor, the inhibitory ligand of the Notch receptor Delta-like protein 3 (DDL3), HIF-1α and its target genes GLUT-1, MCT4, CAIX, and vascular endothelial growth factor receptor 3 (VEGFR-3), and P-cadherin [232–238]. However, the viral status of the tumor was not known (vitamin D receptor), no difference between VN- and VP-MCC was found (GLUT-1, MCT4, CAIX), or there was a tendency to higher expression in VP-MCC, but the difference was not significant (DDL3, HIF1α, P-cadherin). VEGFR-3 was found in all MCCs, but significantly higher in VP-MCC [238]. The value of VEGFR3 as a biomarker is controversial because other studies failed to detect VEGFR-3 in MCC [232,233]. The inconsistency, lack of virus status and limited number of samples of these studies have failed to identify a bona fide biomarker for VP-MCC.

MicroRNA as VP-MCC Biomarkers

MicroRNA (miR) are small RNA molecules that inhibit gene expression at a post-transcriptional level by preventing translation or inducing degrading of their target mRNA. Because of their stability, presence in all body fluids, and sometimes disease-specific expression, they can be useful prognostic and diagnostic markers in cancer. Several groups have examined miR expression in MCC (reviewed in [239]), but miR-375 in particular has been more extensively studied. This miR is enriched in VP-MCC compared to VN-MCC [240], while another study found that the miR-375 levels were enhanced in MCC cell tumors but not associated with the viral status [241,242]. The MCPyV genome itself encodes a single miR (mcv-miR-M1) that is complementary to a sequence in the *LT* gene adjacent to the RB binding motif. However, mcv-miR-M1 was detected in ~50% of 38 tested VP-MCC and the expression levels were low [243]. It seems unlikely that mcv-miR-M1 contributes to MCVPyV-induced tumorigenesis and its use as biomarker for VP-MCC is doubtful.

7. VP-MCC Specific Therapy

Current MCC treatment include surgery, radiation therapy, chemotherapy and immunotherapy. The standard option is surgery followed by radiation [63,64,244]. However, immunotherapy-based strategies is rapidly becoming a preferred therapy in several cancers, including MCC. Three types of immune checkpoint inhibitors are currently being applied on MCC patients: pembrolizumab and nivolumab, antibodies against programmed cell death-1 (PD-1), avelumab, an antibody against the ligand of PD-1, PD-L1, and the CTLA-4 antibody ipilimumab. All these antibodies are approved by US Food and Drug Administration for treatment of different cancers [245]. Clinical trials with pembrolizumab in MCC patients showed 56% objective response rate (ORR) and the ORR in VP-MCC patients (n = 16) and VN-MCC patients (n = 9) was 62% and 44% respectively and 59% (n = 32) and 53% (n = 18), respectively [246,247]. The ORR to nivolumab was also regardless of the viral state of the tumor [248]. The ORR in VP-MCC patients (n = 23) and VN-MCC patients was 26% and 46%, respectively of the VN-MCC patients (n = 13), while in another study including 46 VP-MCC and 31 VN-MCC patients, the ORR was 28% and 36%, respectively [249]. Treatment of five MCC patients with ipilimumab indicated a positive activity of ipilimumab, but the viral state of the patients was not reported [250]. In conclusion, the response of MCC patients to checkpoint inhibitors seems to be independent of the viral state of the tumor, urging the development of VP-MCC specific therapy.

7.1. Vaccines

Serological studies in different countries demonstrated that most of the adult population have antibodies against MCPyV VP1 and infection occurs early in childhood (see Section 1.2). Although prophylactic vaccination with purified VP1 or virus-like particles at early age can prevent viral infection and later on development of VP-MCC, no such vaccines exist for the moment. The reason for this is probably that although the virus establishes a life-long infection in ~80% of the people virus, only a minority will develop VP-MCC and unfortunately, this makes the development of a MCPyV vaccine not very lucrative for pharmaceutical and biotechnology companies. Another option is preventive vaccination with T antigens. Approximately 1% of non-MCC patients have low titer of antibodies against these oncoprotein probably because of low viral activity or latent infection and the fact that LT is a nuclear protein, reducing its processing and presentation by HLA class 1 [251]. However, seroprevalence and titers increase significantly in MCC patients [221]. A possible pitfall with LT vaccination is that HLA class 1 is downregulated in 84% of MCC tumors, and HLA class 1 expression was significantly lower in VP-MCC than in VN-MCC [195], suggesting reduced antigen presentation. Studies with LT/sT proteins as prophylactic vaccines are lacking, but DNA vaccines coding sT or the N-terminal domain of LT have been tried out in mice. DNA vaccination generated a specific T cell immune response in mice and potent protective and therapeutic antitumor effects in a preclinical murine MCC tumor model [252–254]. Therapeutic vaccination to improve the immune system of MCC patient offers another alternative. For an excellent recent review, the reader is referred to [251]. Therapeutic vaccination with MCPyV T antigens can support improperly primed T cells and stimulate naive CD T cells. However, administration of whole sT protein or tumor variants of LT may promote tumorigenesis. This can be circumvented by using peptides of these proteins or mutant forms (e.g., a non-RB binding LT) that lost their oncogenic properties.

7.2. CRISPR/Cas9-Based Methodology

Depletion of T antigens expression by the CRISPR/Cas9 technology was shown to impair proliferation and induce apoptosis of VP-MCC cell lines [112]. Development of a CRISPR/Cas9-based therapeutic tool against T antigens in VP-MCC is a possibility. Challenges of the method to overcome are delivery of the gRNA, efficiency, and accuracy of the mutation.

7.3. RNA Interference Based Treatment

shRNA-mediated knockdown of T antigens expression has been successfully used in cell culture. Intratumorally injected shRNA against viral oncoproteins either as nanoparticles or vector-based are being studied [255,256] and clinical trials are being performed (e.g., NCT01505153). (Sub)cutaneously located primary MCC tumors should be easily accessible for injections with shRNA.

7.4. Anti-Viral Drugs

No specific inhibitors of PyV LT or sT exist, although small molecules that inhibit the ATPase activity of SV40, BKPyV and JCPyV have been described [257,258]. Since MCC tumors express tLT devoided of the ATPase domain, these drugs are not applicable for VP-MCC. However, some molecules not specific for MCPyV show anti-viral activity in cell culture and xenograft models. The malaria drug artesunate reduces growth and survival of VP-MCC cells in vitro and VP-MCC tumors in a xenograft mouse model, but had no or little effect on primary fibroblasts, melanoma cell lines, and non-classical VN-MCC cell lines [259]. Artesunate down-regulated expression of LT, but only in one VP-MCC cell line was the sensitivity towards artesunate reduced upon knockdown of LT expression. Artesunate has entered clinical trials with solid tumors cancer patients other than MCC patients and was well tolerated and modest clinical activity was observed [260]. DDL3 could be a therapeutic target in patients with VP-MCC because DLL3 expression is higher than in VP-MCC [237]. Treatment of a 67-year-old patient with metastatic MCC who received three doses of DLL3-targeting antibody-drug conjugate rovalpituzumab tesirine (Rova-T) had partial response with 57% decreased of the target lesions [237]. It was not specified whether this patient had VP-MCC. Nuclear expression of survivin in MCC is associated with poor prognosis [261]. MCPyV LT upregulates survivin levels and expression of survivin is necessary for VP-MCC cells to survive [159]. The survivin inhibitor YM155 reduced growth of some VP-MCC xenograft tumors and was nontoxic in mice, suggesting YM155 is an attractive drug to treat VP-MCC patients [159,262]. Several clinical trials using survivin vaccination are registered (e.g., NCT00108875, NCT02851056, NCT00573495, NCT03879694), but none include MCC patients. MCPyV LT's interaction with HSC70 is important for inactivation of RB1 [12]. MAL3-101, a specific inhibitor of DnaJ-stimulated HSC70 ATPase activity [263], induces apoptosis in some MCC cell lines and inhibits tumor growth of xenografted VP-MCC WaGa cells without toxic side effects [264]. However, MAL3-101 triggered apoptosis of MCC cells irrespective of the presence of MCPyV, but cells with high HSC70 expression levels were more sensitive. MAL3-101 appears to be a candidate to treat MCC, independent of the viral state, but to our best knowledge, no clinical trials are ongoing. VP-MCC have higher expression of ADAM10 and 17 compared to VN-MCC [174]. TIM-3 is shed by both ADAM 10 and 17 and blocking TIM-3 by antibodies reduced PD-1 expression and increased cytokine production [265]. Therefore, TIM-3 seems to dampen the immune system [266]. Moreover, ADAM 10 cleaves HLA class 1 [267]. Thus, ADAM 10 and 17 inhibitors may stimulate the immune system and could be used for the treatment of VP-MCC. VP-MCC cell lines and MCC tumors do not express STING (see Section 4; [215]). Treating MKL-1 and MS-1 cells that stably express STING responsive to the STING agonist DMXAA not only restored the induction of interferons and proinflammatory cytokines and chemokines, but stimulated PD-L1 expression, T cell migration and activation, and triggered cell death in vitro [215]. The authors suggested that introducing STING by e.g., an adenovirus-based vector in MCC together with DMXAA could be used to treat VP-MCC patients.

8. Conclusions and Future Perspectives

The first HPyVs were discovered in 1971 and despite their ability to transform cells and induce tumors in animals, their role in human cancer remains unclear [268–271]. It was not until 2008 when the lab of Chang and Moore isolated MCPyV that the first HPyV that can cause cancer was identified [62]. Together with raccoon polyomavirus, they are the only two PyVs known to cause cancer in their natural host [272]. Despite our increase in understanding MCPyV's role in MCC, many important questions remain unsolved. The uncertainty about the genuine cell(s) of origin of VP-MCC has hampered studies to scrutinize the exact roles of the T antigens in tumorigenesis. Transgenic mice studies have failed to ubiquitously demonstrate that sT can cause MCC. Research questions related to the biology of this virus (route of infection, transmission, spreading, cell tropism, replication) need to be solved. Efficient cell cultures for MCPyV are lacking, although human dermal fibroblasts can sustain viral replication [44]. Another enigma is why only about 0.5–1 individuals/100,000/year develop MCC with 80% of them being VP (for recent reviews see [49,64]), although most people are infected with MCPyV (see Section 1.2) and seem to chronically shed virus from the skin [33]. MCPyV induced MCC might just be an unfortunate, non-intendent event. An animal model to study virus-induced MCC is lacking (sT transgenic mice do not develop MCC and xenograft studies are usually performed in immune deficient mice). Cases of MCC has been described in other mammals, including cat, dog and steer [273–278]. It is not known whether a polyomavirus might be involved in these MCC, but bovine and canine polyomaviruses have been described [279,280], while LiPyV, originally isolated from human skin [281], was detected in feces of cats [282]. However, the sT and LT of bovine PyV, dog PyV and LiPyV share <50% homology with the corresponding proteins of MCPyV. T antigens of gorilla and chimpanzee PyVs, which are phylogenetically more closely related [1], are 80% identical to the MCPyV T antigens and DnaJ, RB, MUR, LSD, and PP4 domains are conserved, but MCC has not been described in the apes. If MCC in any of these animals has a polyomavirus etiology, they could be used as model systems to improve our knowledge on virus-induced MCC and to test out novel therapeutic strategies. Other gaps of knowledge are related to the clinics. VP-MCC specific biomarkers that can be used in diagnosis, prognosis, and response to treatment are currently lacking so that determining the viral state of the tumor depends on detecting the presence of viral DNA, RNA or T antigens in biopsies. Specific therapy for VP-MCC does not yet exist and will require identification of potential therapeutic targets in VP-MCC. Proteomics of VN- and VP-MCC may allow identification of tumor-specific proteins that can be targets for therapy or useful biomarkers. The discovery of MCPyV as a causative agent of MCC has stirred up MCC research and the next decennia will certainly further increase our knowledge and lead to the development of improved treatment for this aggressive cancer.

Author Contributions: Conceptualization, V.P., C.P. and U.M.; writing—original draft preparation, V.P., C.P. and U.M.; writing—review and editing, V.P., C.P. and U.M. All authors have read and agreed to the published version of the manuscript.

Funding: The APC was funded by UiT, The Arctic University of Norway.

Acknowledgments: Carla Prezioso was supported by Italian Ministry of Health (starting Grant: SG-2018-12366194).

Conflicts of Interest: The authors declare no conflict of interest.

Abbreviations

ALTO	Alternate frame of the LT open reading frame
HLA-1	Major histocompatibility complex class 1
HPyV	Human polyomavirus
LSD	Large T antigen stabilization domain
LT	Large T antigen
MC	Merkel cell
MCC	Merkel cell carcinoma
MCPyV	Merkel cell polyomavirus
miR	MicroRNA
MUR	MCPyV T antigen unique region
NCCR	Non-coding control region
NLS	Nuclear localization signal
OBD	Origin binding domain
ORI	Origin of replication
ORR	Objective response rate
PD-1	Programmed death 1
PD-L1	Programmed death ligand 1
PP	Protein phosphatase
PyV	Polyomavirus
RB	Retinoblastoma protein
sT	Small t antigen
TLR	Toll-like receptor
tLT	Truncated LT
VN-MCC	Virus-negative Merkel cell carcinoma
VP-MCC	Virus-positive Merkel cell carcinoma

References

1. Calvignac-Spencer, S.; Feltkamp, M.C.; Daugherty, M.D.; Moens, U.; Ramqvist, T.; Johne, R.; Ehlers, B. A taxonomy update for the family Polyomaviridae. *Arch. Virol.* **2016**, *161*, 1739–1750. [CrossRef] [PubMed]
2. Gjoerup, O.; Chang, Y. Update on human polyomaviruses and cancer. *Adv. Cancer Res.* **2010**, *106*, 1–51. [CrossRef] [PubMed]
3. Carter, J.J.; Daugherty, M.D.; Qi, X.; Bheda-Malge, A.; Wipf, G.C.; Robinson, K.; Roman, A.; Malik, H.S.; Galloway, D.A. Identification of an overprinting gene in Merkel cell polyomavirus provides evolutionary insight into the birth of viral genes. *Proc. Natl. Acad. Sci. USA* **2013**, *110*, 12744–12749. [CrossRef]
4. Coursaget, P.; Samimi, M.; Nicol, J.T.; Gardair, C.; Touzé, A. Human Merkel cell polyomavirus: Virological background and clinical implications. *APMIS* **2013**, *121*, 755–769. [CrossRef] [PubMed]
5. Shuda, M.; Feng, H.; Kwun, H.J.; Rosen, S.T.; Gjoerup, O.; Moore, P.S.; Chang, Y. T antigen mutations are a human tumor-specific signature for Merkel cell polyomavirus. *Proc. Natl. Acad. Sci. USA* **2008**, *105*, 16272–16277. [CrossRef] [PubMed]
6. An, P.; Saenz Robles, M.T.; Pipas, J.M. Large T antigens of polyomaviruses: Amazing molecular machines. *Annu. Rev. Microbiol.* **2012**, *66*, 213–236. [CrossRef]
7. Cotsiki, M.; Lock, R.L.; Cheng, Y.; Williams, G.L.; Zhao, J.; Perera, D.; Freire, R.; Entwistle, A.; Golemis, E.A.; Roberts, T.M.; et al. Simian virus 40 large T antigen targets the spindle assembly checkpoint protein Bub1. *Proc. Natl. Acad. Sci. USA* **2004**, *101*, 947–952. [CrossRef]
8. Wendzicki, J.A.; Moore, P.S.; Chang, Y. Large T and small T antigens of Merkel cell polyomavirus. *Curr. Opin. Virol.* **2015**, *11*, 38–43. [CrossRef]
9. Nakamura, T.; Sato, Y.; Watanabe, D.; Ito, H.; Shimonohara, N.; Tsuji, T.; Nakajima, N.; Suzuki, Y.; Matsuo, K.; Nakagawa, H.; et al. Nuclear localization of Merkel cell polyomavirus large T antigen in Merkel cell carcinoma. *Virology* **2010**, *398*, 273–279. [CrossRef]
10. Liu, J.; Zhang, C.; Hu, W.; Feng, Z. Tumor suppressor p53 and its mutants in cancer metabolism. *Cancer Lett.* **2015**, *356*, 197–203. [CrossRef]

11. Van Ghelue, M.; Khan, M.T.; Ehlers, B.; Moens, U. Genome analysis of the new human polyomaviruses. *Rev. Med. Virol.* **2012**, *22*, 354–377. [CrossRef] [PubMed]
12. Kwun, H.J.; Guastafierro, A.; Shuda, M.; Meinke, G.; Bohm, A.; Moore, P.S.; Chang, Y. The minimum replication origin of merkel cell polyomavirus has a unique large T-antigen loading architecture and requires small T-antigen expression for optimal replication. *J. Virol.* **2009**, *83*, 12118–12128. [CrossRef] [PubMed]
13. Chang, Y.; Moore, P.S. Merkel cell carcinoma: A virus-induced human cancer. *Annu. Rev. Pathol.* **2012**, *7*, 123–144. [CrossRef]
14. Li, J.; Diaz, J.; Wang, X.; Tsang, S.H.; You, J. Phosphorylation of Merkel cell polyomavirus large tumor antigen at serine 816 by ATM kinase induces apoptosis in host cells. *J. Biol. Chem.* **2015**, *290*, 1874–1884. [CrossRef]
15. Diaz, J.; Wang, X.; Tsang, S.H.; Jiao, J.; You, J. Phosphorylation of large T antigen regulates merkel cell polyomavirus replication. *Cancers* **2014**, *6*, 1464–1486. [CrossRef] [PubMed]
16. Baez, C.F.; Brandão Varella, R.; Villani, S.; Delbue, S. Human Polyomaviruses: The battle of large and small tumor antigens. *Virology* **2017**, *8*. [CrossRef]
17. Kwun, H.J.; Shuda, M.; Camacho, C.J.; Gamper, A.M.; Thant, M.; Chang, Y.; Moore, P.S. Restricted protein phosphatase 2A targeting by Merkel cell polyomavirus small T antigen. *J. Virol.* **2015**, *89*, 4191–4200. [CrossRef]
18. Liu, X.; Hein, J.; Richardson, S.C.; Basse, P.H.; Toptan, T.; Moore, P.S.; Gjoerup, O.V.; Chang, Y. Merkel cell polyomavirus large T antigen disrupts lysosome clustering by translocating human Vam6p from the cytoplasm to the nucleus. *J. Biol. Chem.* **2011**, *286*, 17079–17090. [CrossRef]
19. Schowalter, R.M.; Buck, C.B. The Merkel cell polyomavirus minor capsid protein. *PLoS Pathog.* **2013**, *9*, e1003558. [CrossRef]
20. Pastrana, D.V.; Tolstov, Y.L.; Becker, J.C.; Moore, P.S.; Chang, Y.; Buck, C.B. Quantitation of human seroresponsiveness to Merkel cell polyomavirus. *PLoS Pathog.* **2009**, *5*, e1000578. [CrossRef]
21. Moens, U.; Krumbholz, A.; Ehlers, B.; Zell, R.; Johne, R.; Calvignac-Spencer, S.; Lauber, C. Biology, evolution, and medical importance of polyomaviruses: An update. *Infect. Genet. Evol.* **2017**, *54*, 18–38. [CrossRef] [PubMed]
22. Gosert, R.; Rinaldo, C.H.; Funk, G.A.; Egli, A.; Ramos, E.; Drachenberg, C.B.; Hirsch, H.H. Polyomavirus BK with rearranged noncoding control region emerge in vivo in renal transplant patients and increase viral replication and cytopathology. *J. Exp. Med.* **2008**, *205*, 841–852. [CrossRef] [PubMed]
23. Gosert, R.; Kardas, P.; Major, E.O.; Hirsch, H.H. Rearranged JC virus noncoding control regions found in progressive multifocal leukoencephalopathy patient samples increase virus early gene expression and replication rate. *J. Virol.* **2010**, *84*, 10448–10456. [CrossRef]
24. Delbue, S.; Elia, F.; Carloni, C.; Tavazzi, E.; Marchioni, E.; Carluccio, S.; Signorini, L.; Novati, S.; Maserati, R.; Ferrante, P. JC virus load in cerebrospinal fluid and transcriptional control region rearrangements may predict the clinical course of progressive multifocal leukoencephalopathy. *J. Cell Physiol.* **2012**, *227*, 3511–3517. [CrossRef] [PubMed]
25. Abdulsalam, I.; Rasheed, K.; Sveinbjørnsson, B.; Ehlers, B.; Moens, U. Promoter activity of Merkel cell Polyomavirus variants in human dermal fibroblasts and a Merkel cell carcinoma cell line. *Virol. J.* **2020**, *17*, 54. [CrossRef]
26. Prezioso, C.; Obregon, F.; Ambroselli, D.; Petrolo, S.; Checconi, P.; Rodio, D.M.; Coppola, L.; Nardi, A.; Vito, C.; Sarmati, L.; et al. Merkel cell Polyomavirus (MCPyV) in the context of immunosuppression: Genetic analysis of noncoding control region (NCCR) variability among a HIV-1-positive population. *Viruses* **2020**, *12*, E507. [CrossRef]
27. Kitchen, A.; Miyamoto, M.M.; Mulligan, C.J. Utility of DNA viruses for studying human host history: Case study of JC virus. *Mol. Phylogenet. Evol.* **2008**, *46*, 673–682. [CrossRef]
28. Krumbholz, A.; Bininda-Emonds, O.R.; Wutzler, P.; Zell, R. Evolution of four BK virus subtypes. *Infect. Genet. Evol.* **2008**, *8*, 632–643. [CrossRef]
29. Mes, T.H.; van Doornum, G.J.; Schutten, M. Population genetic tests suggest that the epidemiologies of JCV and BKV are strikingly different. *Infect. Genet. Evol.* **2010**, *10*, 397–403. [CrossRef]
30. Torres, C.; Barrios, M.E.; Cammarata, R.V.; Victoria, M.; Fernandez-Cassi, X.; Bofill-Mas, S.; Colina, R.; Blanco Fernández, M.D.; Mbayed, V.A. Phylodynamics of Merkel-cell polyomavirus and human polyomavirus 6: A long-term history with humans. *Mol. Phylogenet. Evol.* **2018**, *126*, 210–220. [CrossRef]

31. Madinda, N.F.; Ehlers, B.; Wertheim, J.O.; Akoua-Koffi, C.; Bergl, R.A.; Boesch, C.; Akonkwa, D.B.; Eckardt, W.; Fruth, B.; Gillespie, T.R.; et al. Assessing host-virus codivergence for close relatives of Merkel cell polyomavirus infecting african great apes. *J. Virol.* **2016**, *90*, 8531–8541. [CrossRef] [PubMed]
32. Leendertz, F.H.; Scuda, N.; Cameron, K.N.; Kidega, T.; Zuberbühler, K.; Leendertz, S.A.; Couacy-Hymann, E.; Boesch, C.; Calvignac, S.; Ehlers, B. African great apes are naturally infected with polyomaviruses closely related to Merkel cell polyomavirus. *J. Virol.* **2011**, *85*, 916–924. [CrossRef] [PubMed]
33. Schowalter, R.M.; Pastrana, D.V.; Pumphrey, K.A.; Moyer, A.L.; Buck, C.B. Merkel cell polyomavirus and two previously unknown polyomaviruses are chronically shed from human skin. *Cell Host Microbe* **2010**, *7*, 509–515. [CrossRef] [PubMed]
34. Martel-Jantin, C.; Pedergnana, V.; Nicol, J.T.; Leblond, V.; Trégouët, D.A.; Tortevoye, P.; Plancoulaine, S.; Coursaget, P.; Touzé, A.; Abel, L.; et al. Merkel cell polyomavirus infection occurs during early childhood and is transmitted between siblings. *J. Clin. Virol.* **2013**, *58*, 288–291. [CrossRef] [PubMed]
35. Van der Meijden, E.; Bialasiewicz, S.; Rockett, R.J.; Tozer, S.J.; Sloots, T.P.; Feltkamp, M.C. Different serologic behavior of MCPyV, TSPyV, HPyV6, HPyV7 and HPyV9 polyomaviruses found on the skin. *PLoS ONE* **2013**, *8*, e81078. [CrossRef]
36. Gaudette, L.A.; Illing, E.M.; Hill, G.B. Canadian Cancer Statistics 1991. *Health Rep.* **1991**, *3*, 107–135.
37. Tolstov, Y.L.; Pastrana, D.V.; Feng, H.; Becker, J.C.; Jenkins, F.J.; Moschos, S.; Chang, Y.; Buck, C.B.; Moore, P.S. Human Merkel cell polyomavirus infection II. MCV is a common human infection that can be detected by conformational capsid epitope immunoassays. *Int. J. Cancer* **2009**, *125*, 1250–1256. [CrossRef]
38. Carter, J.J.; Paulson, K.G.; Wipf, G.C.; Miranda, D.; Madeleine, M.M.; Johnson, L.G.; Lemos, B.D.; Lee, S.; Warcola, A.H.; Iyer, J.G.; et al. Association of Merkel cell polyomavirus-specific antibodies with Merkel cell carcinoma. *J. Natl. Cancer Inst.* **2009**, *101*, 1510–1522. [CrossRef]
39. Pastrana, D.V.; Brennan, D.C.; Cuburu, N.; Storch, G.A.; Viscidi, R.P.; Randhawa, P.S.; Buck, C.B. Neutralization serotyping of BK polyomavirus infection in kidney transplant recipients. *PLoS Pathog.* **2012**, *8*, e1002650. [CrossRef]
40. Nicol, J.T.; Robinot, R.; Carpentier, A.; Carandina, G.; Mazzoni, E.; Tognon, M.; Touze, A.; Coursaget, P. Age-specific seroprevalences of merkel cell polyomavirus, human polyomaviruses 6, 7, and 9, and trichodysplasia spinulosa-associated polyomavirus. *Clin. Vaccine Immunol.* **2013**, *20*, 363–368. [CrossRef]
41. Sroller, V.; Hamšíková, E.; Ludvíková, V.; Vochozková, P.; Kojzarová, M.; Fraiberk, M.; Saláková, M.; Morávková, A.; Forstová, J.; Němečková, S. Seroprevalence rates of BKV, JCV, and MCPyV polyomaviruses in the general Czech Republic population. *J. Med. Virol.* **2014**, *86*, 1560–1568. [CrossRef] [PubMed]
42. Zhang, C.; Liu, F.; He, Z.; Deng, Q.; Pan, Y.; Liu, Y.; Zhang, C.; Ning, T.; Guo, C.; Liang, Y.; et al. Seroprevalence of Merkel cell polyomavirus in the general rural population of Anyang, China. *PLoS ONE* **2014**, *9*, e106430. [CrossRef]
43. Paulson, K.G.; Lewis, C.W.; Redman, M.W.; Simonson, W.T.; Lisberg, A.; Ritter, D.; Morishima, C.; Hutchinson, K.; Mudgistratova, L.; Blom, A.; et al. Viral oncoprotein antibodies as a marker for recurrence of Merkel cell carcinoma: A prospective validation study. *Cancer* **2017**, *123*, 1464–1474. [CrossRef] [PubMed]
44. Liu, W.; Yang, R.; Payne, A.S.; Schowalter, R.M.; Spurgeon, M.E.; Lambert, P.F.; Xu, X.; Buck, C.B.; You, J. Identifying the target cells and mechanisms of Merkel Cell Polyomavirus infection. *Cell Host Microbe* **2016**, *19*, 775–787. [CrossRef] [PubMed]
45. Hashida, Y.; Nakajima, K.; Nakajima, H.; Shiga, T.; Tanaka, M.; Murakami, M.; Matsuzaki, S.; Naganuma, S.; Kuroda, N.; Seki, Y.; et al. High load of Merkel cell polyomavirus DNA detected in the normal skin of Japanese patients with Merkel cell carcinoma. *J. Clin. Virol.* **2016**, *82*, 101–107. [CrossRef]
46. Liu, W.; Krump, N.A.; Buck, C.B.; You, J. Merkel cell Polyomavirus infection and detection. *J. Vis. Exp.* **2019**, *144*. [CrossRef]
47. Peretti, A.; Borgogna, C.; Rossi, D.; De Paoli, L.; Bawadekar, M.; Zavattaro, E.; Boldorini, R.; De Andrea, M.; Gaidano, G.; Gariglio, M. Analysis of human β-papillomavirus and Merkel cell polyomavirus infection in skin lesions and eyebrow hair bulbs from a cohort of patients with chronic lymphocytic leukaemia. *Br. J. Dermatol.* **2014**, *171*, 1525–1528. [CrossRef]
48. Toker, C. Trabecular carcinoma of the skin. *Arch. Dermatol.* **1972**, *105*, 107–110. [CrossRef]
49. Becker, J.C.; Stang, A.; DeCaprio, J.A.; Cerroni, L.; Lebbe, C.; Veness, M.; Nghiem, P. Merkel cell carcinoma. *Nat. Rev. Dis Primers* **2017**, *3*, 17077. [CrossRef]

50. Paulson, K.G.; Park, S.Y.; Vandeven, N.A.; Lachance, K.; Thomas, H.; Chapuis, A.G.; Harms, K.L.; Thompson, J.A.; Bhatia, S.; Stang, A.; et al. Merkel cell carcinoma: Current US incidence and projected increases based on changing demographics. *J. Am. Acad. Dermatol.* **2018**, *78*, 457–463. [CrossRef]
51. Harms, P.W.; Collie, A.M.; Hovelson, D.H.; Cani, A.K.; Verhaegen, M.E.; Patel, R.M.; Fullen, D.R.; Omata, K.; Dlugosz, A.A.; Tomlins, S.A.; et al. Next generation sequencing of Cytokeratin 20-negative Merkel cell carcinoma reveals ultraviolet-signature mutations and recurrent TP53 and RB1 inactivation. *Mod. Pathol.* **2016**, *29*, 240–248. [CrossRef] [PubMed]
52. Pulitzer, M.P.; Amin, B.D.; Busam, K.J. Merkel cell carcinoma: Review. *Adv. Anat. Pathol.* **2009**, *16*, 135–144. [CrossRef] [PubMed]
53. Agelli, M.; Clegg, L.X. Epidemiology of primary Merkel cell carcinoma in the United States. *J. Am. Acad. Dermatol.* **2003**, *49*, 832–841. [CrossRef]
54. Howard, R.A.; Dores, G.M.; Curtis, R.E.; Anderson, W.F.; Travis, L.B. Merkel cell carcinoma and multiple primary cancers. *Cancer Epidemiol. Biomarkers Prev.* **2006**, *15*, 1545–1549. [CrossRef] [PubMed]
55. Goh, G.; Walradt, T.; Markarov, V.; Blom, A.; Riaz, N.; Doumani, R.; Stafstrom, K.; Moshiri, A.; Yelistratova, L.; Levinsohn, J.; et al. Mutational landscape of MCPyV-positive and MCPyV-negative Merkel cell carcinomas with implications for immunotherapy. *Oncotarget* **2016**, *7*, 3403–3415. [CrossRef] [PubMed]
56. Starrett, G.J.; Marcelus, C.; Cantalupo, P.G.; Katz, J.P.; Cheng, J.; Akagi, K.; Thakuria, M.; Rabinowits, G.; Wang, L.C.; Symer, D.E.; et al. Merkel cell Polyomavirus exhibits dominant control of the tumor genome and transcriptome in virus-associated Merkel cell carcinoma. *mBio* **2017**, *8*, e02079. [CrossRef]
57. Heath, M.; Jaimes, N.; Lemos, B.; Mostaghimi, A.; Wang, L.C.; Peñas, P.F.; Nghiem, P. Clinical characteristics of Merkel cell carcinoma at diagnosis in 195 patients: The AEIOU features. *J. Am. Acad. Dermatol.* **2008**, *58*, 375–381. [CrossRef]
58. Engels, E.A.; Frisch, M.; Goedert, J.J.; Biggar, R.J.; Miller, R.W. Merkel cell carcinoma and HIV infection. *Lancet* **2002**, *359*, 4974–4998. [CrossRef]
59. Clarke, C.A.; Robbins, H.A.; Tatalovich, Z.; Lynch, C.F.; Pawlish, K.S.; Finch, J.L.; Hernandez, B.Y.; Fraumeni, J.F.J.; Madeleine, M.M.; Engels, E.A. Risk of merkel cell carcinoma after solid organ transplantation. *J. Natl. Cancer Inst.* **2015**, *107*, dju382. [CrossRef]
60. Paulson, K.G.; Iyer, J.G.; Blom, A.; Warton, E.M.; Sokil, M.; Yelistratova, L.; Schuman, L.; Nagase, K.; Bhatia, S.; Asgari, M.M.; et al. Systemic immune suppression predicts diminished Merkel cell carcinoma-specific survival independent of stage. *J. Investig. Dermatol.* **2013**, *133*, 642–646. [CrossRef]
61. Csoboz, B.; Rasheed, K.; Sveinbjørnsson, B.; Moens, U. Merkel cell polyomavirus and non-Merkel cell carcinomas: Guilty or circumstantial evidence? *APMIS* **2020**, *128*, 104–120. [CrossRef] [PubMed]
62. Feng, H.; Shuda, M.; Chang, Y.; Moore, P.S. Clonal integration of a polyomavirus in human Merkel cell carcinoma. *Science* **2008**, *319*, 1096–1100. [CrossRef] [PubMed]
63. Harms, P.W. Update on Merkel cell carcinoma. *Clin. Lab. Med.* **2017**, *3*, 17077. [CrossRef]
64. Harms, P.W.; Harms, K.L.; Moore, P.S.; DeCaprio, J.A.; Nghiem, P.; Wong, M.K.K. Brownell I: The biology and treatment of Merkel cell carcinoma: Current understanding and research priorities. *Nat. Rev. Clin. Oncol.* **2018**, *15*, 763–776. [CrossRef]
65. Schrama, D.; Sarosi, E.M.; Adam, C.; Ritter, C.; Kaemmerer, U.; Klopocki, E.; König, E.M.; Utikal, J.; Becker, J.C.; Houben, R. Characterization of six Merkel cell polyomavirus-positive Merkel cell carcinoma cell lines: Integration pattern suggest that large T antigen truncating events occur before or during integration. *Int. J. Cancer* **2019**, *145*, 1020–1032. [CrossRef] [PubMed]
66. Li, J.; Wang, X.; Diaz, J.; Tsang, S.H.; Buck, C.B.; You, J. Merkel cell polyomavirus large T antigen disrupts host genomic integrity and inhibits cellular proliferation. *J. Virol.* **2013**, *87*, 9173–9188. [CrossRef] [PubMed]
67. Fischer, N.; Brandner, J.; Fuchs, F.; Moll, I.; Grundhoff, A. Detection of Merkel cell polyomavirus (MCPyV) in Merkel cell carcinoma cell lines: Cell morphology and growth phenotype do not reflect presence of the virus. *Int. J. Cancer* **2010**, *126*, 2133–2142. [CrossRef]
68. Shuda, M.; Arora, R.; Kwun, H.J.; Feng, H.; Sarid, R.; Fernández-Figueras, M.T.; Tolstov, Y.; Gjoerup, O.; Mansukhani, M.M.; Swerdlow, S.H.; et al. Human Merkel cell polyomavirus infection I. MCV T antigen expression in Merkel cell carcinoma, lymphoid tissues and lymphoid tumors. *Int. J. Cancer* **2009**, *125*, 1243–1249. [CrossRef] [PubMed]

69. Rodig, S.J.; Cheng, J.; Wardzala, J.; DoRosario, A.; Scanlon, J.J.; Laga, A.C.; Martinez-Fernandez, A.; Barletta, J.A.; Bellizzi, A.M.; Sadasivam, S.; et al. Improved detection suggests all Merkel cell carcinomas harbor Merkel polyomavirus. *J. Clin. Investig.* **2012**, *122*, 4645–4653. [CrossRef]
70. Velásquez, C.; Amako, Y.; Harold, A.; Toptan, T.; Chang, Y.; Shuda, M. Characterization of a Merkel Cell Polyomavirus-positive Merkel Cell carcinoma cell line CVG-1. *Front. Microbiol.* **2018**, *9*, 713. [CrossRef]
71. Gould, V.E.; Moll, R.; Moll, I.; Lee, I.; Franke, W.W. Neuroendocrine (Merkel) cells of the skin: Hyperplasias, dysplasias, and neoplasms. *Lab. Investig.* **1985**, *52*, 334–353.
72. Moll, I.; Kuhn, C.; Moll, R. Cytokeratin 20 is a general marker of cutaneous Merkel cells while certain neuronal proteins are absent. *J. Investig. Dermatol.* **1995**, *104*, 910–915. [CrossRef] [PubMed]
73. Ortonne, J.P.; Petchot-Bacque, J.P.; Verrando, P.; Pisani, A.; Pautrat, G.; Bernerd, F. Normal Merkel cells express a synaptophysin-like immunoreactivity. *Dermatologica* **1988**, *177*, 1–10. [CrossRef] [PubMed]
74. Kurokawa, M.; Nabeshima, K.; Akiyama, Y.; Maeda, S.; Nishida, T.; Nakayama, F.; Amano, M.; Ogata, K.; Setoyama, M. CD56: A useful marker for diagnosing Merkel cell carcinoma. *J. Dermatol. Sci.* **2003**, *31*, 219–224. [CrossRef]
75. Foschini, M.P.; Eusebi, V. Divergent differentiation in endocrine and nonendocrine tumors of the skin. *Semin. Diagn. Pathol.* **2000**, *17*, 162–168.
76. Spurgeon, M.E.; Lambert, P.F. Merkel cell polyomavirus: A newly discovered human virus with oncogenic potential. *Virology* **2013**, *435*, 118–130. [CrossRef] [PubMed]
77. Sunshine, J.C.; Jahchan, N.S.; Sage, J.; Choi, J. Are there multiple cells of origin of Merkel cell carcinoma? *Oncogene* **2018**, *37*, 1409–1416. [CrossRef] [PubMed]
78. Yang, A.; Cordoba, C.; Cheung, K.; Konya, J. Merkel cell carcinoma in situ: New insights into the cells of origin. *Australasian J. Dermatol.* **2019**, *60*, e311–e313. [CrossRef] [PubMed]
79. Kervarrec, T.; Aljundi, M.; Appenzeller, S.; Samimi, M.; Maubec, E.; Cribier, B.; Deschamps, L.; Sarma, B.; Sarosi, E.M.; Berthon, P.; et al. Polyomavirus-positive Merkel cell carcinoma derived from a Trichoblastoma suggests an epithelial origin of this Merkel cell carcinoma. *J. Investig. Dermatol.* **2020**, *140*, 976–985. [CrossRef]
80. Munde, P.B.; Khandekar, S.P.; Dive, A.M.; Sharma, A. Pathophysiology of merkel cell. *J. Oral Maxillofac. Pathol.* **2013**, *17*, 408–412. [CrossRef]
81. Harms, P.W.; Patel, R.M.; Verhaegen, M.E.; Giordano, T.J.; Nash, K.T.; Johnson, C.N.; Daignault, S.; Thomas, D.G.; Gudjonsson, J.E.; Elder, J.T.; et al. Distinct gene expression profiles of viral- and nonviral-associated merkel cell carcinoma revealed by transcriptome analysis. *J. Investig. Dermatol.* **2013**, *133*, 936–945. [CrossRef] [PubMed]
82. Tilling, T.; Moll, I. Which are the cells of origin in merkel cell carcinoma? *J. Skin Cancer* **2012**, *2012*, 680410. [CrossRef] [PubMed]
83. Moll, I.; Roessler, M.; Brandner, J.M.; Eispert, A.C.; Houdek, P.; Moll, R. Human Merkel cells–aspects of cell biology, distribution and functions. *Eur. J. Cell Biol.* **2005**, *84*, 259–271. [CrossRef] [PubMed]
84. Moll, I.; Zieger, W.; Schmelz, M. Proliferative Merkel cells were not detected in human skin. *Arch. Dermatol. Res.* **1996**, *288*, 1841–1887. [CrossRef] [PubMed]
85. Sauer, C.M.; Haugg, A.M.; Chteinberg, E.; Rennspiess, D.; Winnepenninckx, V.; Speel, E.J.; Becker, J.C.; Kurz, A.K.; Zur Hausen, A. Reviewing the current evidence supporting early B-cells as the cellular origin of Merkel cell carcinoma. *Crit. Rev. Oncol. Hematol.* **2017**, *116*, 99–105. [CrossRef]
86. Deichmann, M.; Kurzen, H.; Egner, U.; Altevogt, P.; Hartschuh, W. Adhesion molecules CD171 (L1CAM) and CD24 are expressed by primary neuroendocrine carcinomas of the skin (Merkel cell carcinomas). *J. Cutan. Pathol.* **2003**, *30*, 363–368. [CrossRef]
87. Kervarrec, T.; Samimi, M.; Guyétant, S.; Sarma, B.; Chéret, J.; Blanchard, E.; Berthon, P.; Schrama, D.; Houben, R.; Touzé, A. Histogenesis of Merkel cell carcinoma: A Comprehensive review. *Front. Oncol.* **2019**, *9*, 451. [CrossRef]
88. Scott, M.P.; Helm, K.F. Cytokeratin 20: A marker for diagnosing Merkel cell carcinoma. *Am. J. Dermatopathol.* **1999**, *21*, 16–20. [CrossRef]
89. Shuda, M.; Guastafierro, A.; Geng, X.; Shuda, Y.; Ostrowski, S.M.; Lukianov, S.; Jenkins, F.J.; Honda, K.; Maricich, S.M.; Moore, P.S.; et al. Merkel Cell Polyomavirus Small T Antigen induces cancer and Embryonic Merkel Cell proliferation in a transgenic mouse model. *PLoS ONE* **2015**, *10*, e0142329. [CrossRef]
90. Hwang, J.H.; Alanen, K.; Dabbs, K.D.; Danyluk, J.; Silverman, S. Merkel cell carcinoma with squamous and sarcomatous differentiation. *J. Cutan. Pathol.* **2008**, *35*, 955–959. [CrossRef]

91. Abbas, O.; Bhawan, J. Expression of stem cell markers nestin and cytokeratin 15 and 19 in cutaneous malignancies. *J. Eur. Acad. Dermatol. Venereol.* **2011**, *25*, 311–316. [CrossRef] [PubMed]
92. Lemasson, G.; Coquart, N.; Lebonvallet, N.; Boulais, N.; Galibert, M.D.; Marcorelles, P.; Misery, L. Presence of putative stem cells in Merkel cell carcinomas. *J. Eur. Acad. Dermatol. Venereol.* **2012**, *26*, 789–795. [CrossRef]
93. Laga, A.C.; Lai, C.Y.; Zhan, Q.; Huang, S.J.; Velazquez, E.F.; Yang, Q.; Hsu, M.Y.; Murphy, G.F. Expression of the embryonic stem cell transcription factor SOX2 in human skin: Relevance to melanocyte and merkel cell biology. *Am. J. Pathol.* **2010**, *176*, 903–913. [CrossRef] [PubMed]
94. Fradette, J.; Godbout, M.J.; Michel, M.; Germain, L. Localization of Merkel cells at hairless and hairy human skin sites using keratin 18. *Biochem. Cell Biol.* **1995**, *73*, 635–639. [CrossRef] [PubMed]
95. Murakami, I.; Takata, K.; Matsushita, M.; Nonaka, D.; Iwasaki, T.; Kuwamoto, S.; Kato, M.; Mohri, T.; Nagata, K.; Kitamura, Y.; et al. Immunoglobulin expressions are only associated with MCPyV-positive Merkel cell carcinomas but not with MCPyV-negative ones: Comparison of prognosis. *Am. J. Surg. Pathol.* **2014**, *38*, 1627–1635. [CrossRef] [PubMed]
96. Zur Hausen, A.; Rennspiess, D.; Winnepenninckx, V.; Speel, E.J.; Kurz, A.K. Early B-cell differentiation in Merkel cell carcinomas: Clues to cellular ancestry. *Cancer Res.* **2013**, *73*, 4982–4987. [CrossRef]
97. Shiver, M.B.; Mahmoud, F.; Gao, L. Response to Idelalisib in a patient with stage IV Merkel-cell carcinoma. *N. Engl. J. Med.* **2015**, *373*, 1580–1582. [CrossRef]
98. Angermeyer, S.; Hesbacher, S.; Becker, J.C.; Schrama, D.; Houben, R. Merkel cell polyomavirus-positive Merkel cell carcinoma cells do not require expression of the viral small T antigen. *J. Investig. Dermatol.* **2013**, *133*, 2059–2064. [CrossRef]
99. Shuda, M.; Kwun, H.J.; Feng, H.; Chang, Y.; Moore, P.S. Human Merkel cell polyomavirus small T antigen is an oncoprotein targeting the 4E-BP1 translation regulator. *J. Clin. Investig.* **2011**, *121*, 3623–3634. [CrossRef]
100. Nwogu, N.; Ortiz, L.E.; Kwun, H.J. Surface charge of Merkel cell polyomavirus small T antigen determines cell transformation through allosteric FBW7 WD40 domain targeting. *Oncogenesis* **2020**, *9*, 53. [CrossRef]
101. Woo, S.H.; Stumpfova, M.; Jensen, U.B.; Lumpkin, E.A.; Owens, D.M. Identification of epidermal progenitors for the Merkel cell lineage. *Development* **2010**, *137*, 3965–3971. [CrossRef] [PubMed]
102. Larouche, D.; Tong, X.; Fradette, J.; Coulombe, P.A.; Germain, L. Vibrissa hair bulge houses two populations of skin epithelial stem cells distinct by their keratin profile. *FASEB J.* **2008**, *22*, 1404–1415. [CrossRef]
103. Brownell, I.; Guevara, E.; Bai, C.B.; Loomis, C.A.; Joyner, A.L. Nerve-derived sonic hedgehog defines a niche for hair follicle stem cells capable of becoming epidermal stem cells. *Cell Stem Cell* **2011**, *8*, 552–565. [CrossRef] [PubMed]
104. Nguyen, M.B.; Cohen, I.; Kumar, V.; Xu, Z.; Bar, C.; Dauber-Decker, K.L.; Tsai, P.C.; Marangoni, P.; Klein, O.D.; Hsu, Y.C.; et al. FGF signalling controls the specification of hair placode-derived SOX9 positive progenitors to Merkel cells. *Nat. Commun.* **2018**, *9*, 2333. [CrossRef] [PubMed]
105. DeCaprio, J.A.; Garcea, R.L. A cornucopia of human polyomaviruses. *Nat. Rev. Microbiol.* **2013**, *11*, 264–276. [CrossRef] [PubMed]
106. Houben, R.; Shuda, M.; Weinkam, R.; Schrama, D.; Feng, H.; Chang, Y.; Moore, P.S.; Becker, J.C. Merkel cell polyomavirus-infected Merkel cell carcinoma cells require expression of viral T antigens. *J. Virol.* **2010**, *84*, 7064–7072. [CrossRef]
107. Houben, R.; Adam, C.; Baeurle, A.; Hesbacher, S.; Grimm, J.; Angermeyer, S.; Henzel, K.; Hauser, S.; Elling, R.; Bröcker, E.B.; et al. An intact retinoblastoma protein-binding site in Merkel cell polyomavirus large T antigen is required for promoting growth of Merkel cell carcinoma cells. *Int. J. Cancer* **2012**, *130*, 847–856. [CrossRef]
108. Houben, R.; Angermeyer, S.; Haferkamp, S.; Aue, A.; Goebeler, M.; Schrama, D.; Hesbacher, S. Characterization of functional domains in the Merkel cell polyoma virus Large T antigen. *Int. J. Cancer* **2015**, *136*, E290–E300. [CrossRef]
109. Schrama, D.; Hesbacher, S.; Angermeyer, S.; Schlosser, A.; Haferkamp, S.; Aue, A.; Adam, C.; Weber, A.; Schmidt, M.; Houben, R. Serine 220 phosphorylation of the Merkel cell polyomavirus large T antigen crucially supports growth of Merkel cell carcinoma cells. *Int. J. Cancer* **2016**, *138*, 1153–1162. [CrossRef]
110. Houben, R.; Grimm, J.; Willmes, C.; Weinkam, R.; Becker, J.C.; Schrama, D. Merkel cell carcinoma and Merkel cell polyomavirus: Evidence for hit-and-run oncogenesis. *J. Investig. Dermatol.* **2012**, *132*, 254–256. [CrossRef]
111. Cheng, J.; Rozenblatt-Rosen, O.; Paulson, K.G.; Nghiem, P.; DeCaprio, J.A. Merkel cell polyomavirus large T antigen has growth-promoting and inhibitory activities. *J. Virol.* **2013**, *87*, 6118–6126. [CrossRef]

112. Temblador, A.; Topalis, D.; Andrei, G.; Snoeck, R. CRISPR/Cas9 Editing of the Polyomavirus Tumor antigens inhibits merkel cell carcinoma growth in vitro. *Cancers* **2019**, *11*, 1260. [CrossRef]
113. Hesbacher, S.; Pfitzer, L.; Wiedorfer, K.; Angermeyer, S.; Borst, A.; Haferkamp, S.; Scholz, C.J.; Wobser, M.; Schrama, D.; Houben, R. RB1 is the crucial target of the Merkel cell polyomavirus Large T antigen in Merkel cell carcinoma cells. *Oncotarget* **2016**, *7*, 32956–32968. [CrossRef] [PubMed]
114. Moens, U.; Macdonald, A. Effect of the large and small T-Antigens of human Polyomaviruses on signaling pathways. *Int. J. Mol. Sci.* **2019**, *20*, 3914. [CrossRef] [PubMed]
115. Borchert, S.; Czech-Sioli, M.; Neumann, F.; Schmidt, C.; Wimmer, P.; Dobner, T.; Grundhoff, A.; Fischer, N. High-affinity Rb binding, p53 inhibition, subcellular localization, and transformation by wild-type or tumor-derived shortened Merkel cell polyomavirus large T antigens. *J. Virol.* **2014**, *88*, 3144–3160. [CrossRef]
116. Martinez-Zapien, D.; Ruiz, F.X.; Poirson, J.; Mitschler, A.; Ramirez, J.; Forster, A.; Cousido-Siah, A.; Masson, M.; Vande Pol, S.; Podjarny, A.; et al. Structure of the E6/E6AP/p53 complex required for HPV-mediated degradation of p53. *Nature* **2016**, *529*, 541–545. [CrossRef]
117. Park, D.E.; Cheng, J.; Berrios, C.; Montero, J.; Cortés-Cros, M.; Ferretti, S.; Arora, R.; Tillgren, M.L.; Gokhale, P.C.; DeCaprio, J.A. Dual inhibition of MDM2 and MDM4 in virus-positive Merkel cell carcinoma enhances the p53 response. *Proc. Natl. Acad. Sci. USA* **2019**, *116*, 1027–1032. [CrossRef]
118. Sherr, C.J. Divorcing ARF and p53: An unsettled case. *Nat. Rev. Cancer* **2006**, *6*, 663–673. [CrossRef] [PubMed]
119. Sullivan, C.S.; Cantalupo, P.; Pipas, J.M. The molecular chaperone activity of simian virus 40 large T antigen is required to disrupt Rb-E2F family complexes by an ATP-dependent mechanism. *Mol. Cell Biol.* **2000**, *20*, 6233–6243. [CrossRef]
120. Moens, U.; Van Ghelue, M.; Johannessen, M. Oncogenic potentials of the human polyomavirus regulatory proteins. *Cell Mol. Life Sci.* **2007**, *64*, 1656–1678. [CrossRef]
121. Martel-Jantin, C.; Filippone, C.; Cassar, O.; Peter, M.; Tomasic, G.; Vielh, P.; Brière, J.; Petrella, T.; Aubriot-Lorton, M.H.; Mortier, L.; et al. Genetic variability and integration of Merkel cell polyomavirus in Merkel cell carcinoma. *Virology* **2012**, *426*, 134–142. [CrossRef]
122. Matsushita, M.; Iwasaki, T.; Kuwamoto, S.; Kato, M.; Nagata, K.; Murakami, I.; Kitamura, Y.; Hayashi, K. Merkel cell polyomavirus (MCPyV) strains in Japanese merkel cell carcinomas (MCC) are distinct from Caucasian type MCPyVs: Genetic variability and phylogeny of MCPyV genomes obtained from Japanese MCPyV-infected MCCs. *Virus Genes* **2014**, *48*, 233–242. [CrossRef] [PubMed]
123. Perdigoto, C.N.; Bardot, E.S.; Valdes, V.J.; Santoriello, F.J.; Ezhkova, E. Embryonic maturation of epidermal Merkel cells is controlled by a redundant transcription factor network. *Development* **2014**, *141*, 4690–4696. [CrossRef] [PubMed]
124. Harold, A.; Amako, Y.; Hachisuka, J.; Bai, Y.; Li, M.Y.; Kubat, L.; Gravemeyer, J.; Franks, J.; Gibbs, J.R.; Park, H.J.; et al. Conversion of Sox2-dependent Merkel cell carcinoma to a differentiated neuron-like phenotype by T antigen inhibition. *Proc. Natl. Acad. Sci. USA* **2019**, *116*, 20104–20114. [CrossRef] [PubMed]
125. Aguado-Llera, D.; Goormaghtigh, E.; de Geest, N.; Quan, X.J.; Prieto, A.; Hassan, B.A.; Gómez, J.; Neira, J.L. The basic helix-loop-helix region of human neurogenin 1 is a monomeric natively unfolded protein which forms a "fuzzy" complex upon DNA binding. *Biochemistry* **2010**, *49*, 1577–1589. [CrossRef]
126. Fan, K.; Gravemeyer, J.; Ritter, C.; Rasheed, K.; Gambichler, T.; Moens, U.; Shuda, M.; Schrama, D.; Becker, J.C. MCPyV large T antigen-induced Atonal Homolog 1 Is a lineage-dependency Oncogene in Merkel cell carcinoma. *J. Investig. Dermatol.* **2020**, *140*, 56–65. [CrossRef]
127. Czech-Sioli, M.; Siebels, S.; Radau, S.; Zahedi, R.P.; Schmidt, C.; Dobner, T.; Grundhoff, A.; Fischer, N. The Ubiquitin-specific Protease Usp7, a novel Merkel cell Polyomavirus large T-antigen interaction partner, modulates viral DNA replication. *J. Virol.* **2020**, *94*, e01638-19. [CrossRef]
128. Bhattacharya, S.; Chakraborty, D.; Basu, M.; Ghosh, M.K. Emerging insights into HAUSP (USP7) in physiology, cancer and other diseases. *Signal. Transduct. Target. Ther.* **2018**, *3*, 17. [CrossRef]
129. Rozenblatt-Rosen, O.; Deo, R.C.; Padi, M.; Adelmant, G.; Calderwood, M.A.; Rolland, T.; Grace, M.; Dricot, A.; Askenazi, M.; Tavares, M.; et al. Interpreting cancer genomes using systematic host network perturbations by tumour virus proteins. *Nature* **2012**, *487*, 491–495. [CrossRef]
130. Wang, X.; Li, J.; Schowalter, R.M.; Jiao, J.; Buck, C.B.; You, J. Bromodomain protein Brd4 plays a key role in Merkel cell polyomavirus DNA replication. *PLoS Pathog.* **2012**, *8*, e1003021. [CrossRef]
131. Arora, R.; Vats, A.; Chimankar, V. MCV Truncated Large T antigen interacts with BRD4 in tumors. *Matters* **2019**, *2019*. [CrossRef] [PubMed]

132. Shuda, M.; Velásquez, C.; Cheng, E.; Cordek, D.G.; Kwun, H.J.; Chang, Y.; Moore, P.S. CDK1 substitutes for mTOR kinase to activate mitotic cap-dependent protein translation. *Proc. Natl. Acad. Sci. USA* **2015**, *112*, 5875–5882. [CrossRef]
133. Velásquez, C.; Cheng, E.; Shuda, M.; Lee-Oesterreich, P.J.; Pogge von Strandmann, L.; Gritsenko, M.A.; Jacobs, J.M.; Moore, P.S.; Chang, Y. Mitotic protein kinase CDK1 phosphorylation of mRNA translation regulator 4E-BP1 Ser83 may contribute to cell transformation. *Proc. Natl. Acad. Sci. USA* **2016**, *113*, 8466–8471. [CrossRef]
134. Verhaegen, M.E.; Mangelberger, D.; Harms, P.W.; Vozheiko, T.D.; Weick, J.W.; Wilbert, D.M.; Saunders, T.L.; Ermilov, A.N.; Bichakjian, C.K.; Johnson, T.M.; et al. Merkel cell polyomavirus small T antigen is oncogenic in transgenic mice. *J. Investig. Dermatol.* **2015**, *135*, 1415–1424. [CrossRef] [PubMed]
135. Kwun, H.J.; Wendzicki, J.A.; Shuda, Y.; Moore, P.S.; Chang, Y. Merkel cell polyomavirus small T antigen induces genome instability by E3 ubiquitin ligase targeting. *Oncogene* **2017**, *36*, 6784–6792. [CrossRef] [PubMed]
136. Griffiths, D.A.; Abdul-Sada, H.; Knight, L.M.; Jackson, B.R.; Richards, K.; Prescott, E.L.; Peach, A.H.; Blair, G.E.; Macdonald, A.; Whitehouse, A. Merkel cell polyomavirus small T antigen targets the NEMO adaptor protein to disrupt inflammatory signaling. *J. Virol.* **2013**, *87*, 13853–13867. [CrossRef] [PubMed]
137. Abdul-Sada, H.; Müller, M.; Mehta, R.; Toth, R.; Arthur, J.S.C.; Whitehouse, A.; Macdonald, A. The PP4R1 sub-unit of protein phosphatase PP4 is essential for inhibition of NF-κB by merkel polyomavirus small tumour antigen. *Oncotarget* **2017**, *8*, 25418–25432. [CrossRef]
138. Cheng, J.; Park, D.E.; Berrios, C.; White, E.A.; Arora, R.; Yoon, R.; Branigan, T.; Xiao, T.; Westerling, T.; Federation, A.; et al. Merkel cell polyomavirus recruits MYCL to the EP400 complex to promote oncogenesis. *PLoS Pathog.* **2017**, *13*, e1006668. [CrossRef]
139. Knight, L.M.; Stakaityte, G.; Wood, J.J.; Abdul-Sada, H.; Griffiths, D.A.; Howell, G.J.; Wheat, R.; Blair, G.E.; Steven, N.M.; Macdonald, A.; et al. Merkel cell polyomavirus small T antigen mediates microtubule destabilization to promote cell motility and migration. *J. Virol.* **2015**, *89*, 35–47. [CrossRef]
140. Stakaitytė, G.; Nwogu, N.; Dobson, S.J.; Knight, L.M.; Wasson, C.W.; Salguero, F.J.; Blackbourn, D.J.; Blair, G.E.; Mankouri, J.; Macdonald, A.; et al. Merkel cell Polyomavirus small T antigen drives cell motility via Rho-GTPase-induced Filopodium formation. *J. Virol.* **2018**, *92*, e00940-17. [CrossRef]
141. White, E. The role for autophagy in cancer. *J. Clin. Investig.* **2015**, *125*, 42–46. [CrossRef] [PubMed]
142. Janji, B.; Viry, E.; Moussay, E.; Paggetti, J.; Arakelian, T.; Mgrditchian, T.; Messai, Y.; Noman, M.Z.; van Moer, K.; Hasmim, M.; et al. The multifaceted role of autophagy in tumor evasion from immune surveillance. *Oncotarget* **2016**, *7*, 17591–17607. [CrossRef] [PubMed]
143. Santana-Codina, N.; Mancias, J.D.; Kimmelman, A.C. The role of autophagy in cancer. *Annu. Rev. Cancer Biol.* **2017**, *1*, 19–39. [CrossRef] [PubMed]
144. Kumar, S.; Xie, H.; Shi, H.; Gao, J.; Juhlin, C.C.; Björnhagen, V.; Höög, A.; Lee, L.; Larsson, C.; Lui, W.O. Merkel cell polyomavirus oncoproteins induce microRNAs that suppress multiple autophagy genes. *Int. J. Cancer* **2020**, *146*, 1652–1666. [CrossRef]
145. Verhaegen, M.E.; Mangelberger, D.; Harms, P.W.; Eberl, M.; Wilbert, D.M.; Meireles, J.; Bichakjian, C.K.; Saunders, T.L.; Wong, S.Y.; Dlugosz, A.A. Merkel cell Polyomavirus small T antigen initiates Merkel cell carcinoma-like tumor development in mice. *Cancer Res.* **2017**, *77*, 3151–3157. [CrossRef] [PubMed]
146. Qin, X.; Jiang, B.; Zhang, Y. 4E-BP1, a multifactor regulated multifunctional protein. *Cell Cycle* **2016**, *15*, 781–786. [CrossRef]
147. Bjornsti, M.A.; Houghton, P.J. Lost in translation: Dysregulation of cap-dependent translation and cancer. *Cancer Cell* **2004**, *5*, 519–523. [CrossRef] [PubMed]
148. Kwun, H.J.; Shuda, M.; Feng, H.; Camacho, C.J.; Moore, P.S.; Chang, Y. Merkel cell polyomavirus small T antigen controls viral replication and oncoprotein expression by targeting the cellular ubiquitin ligase SCFFbw7. *Cell Host Microbe* **2013**, *14*, 125–135. [CrossRef] [PubMed]
149. Kwun, H.J.; Chang, Y.; Moore, P.S. Protein-mediated viral latency is a novel mechanism for Merkel cell polyomavirus persistence. *Proc. Natl. Acad. Sci. USA* **2017**, *114*, E4040–E4047. [CrossRef]
150. Dye, K.N.; Welcker, M.; Clurman, B.E.; Roman, A.; Galloway, D.A. Merkel cell polyomavirus Tumor antigens expressed in Merkel cell carcinoma function independently of the ubiquitin ligases Fbw7 and β-TrCP. *PLoS Pathog.* **2019**, *15*, e1007543. [CrossRef]

151. Zhan, S.; Wang, T.; Ge, W. Multiple functions of the E3 ubiquitin ligase CHIP in immunity. *Int. Rev. Immunol.* **2017**, *36*, 300–312. [CrossRef] [PubMed]
152. Gupta, P.; Shahzad, N.; Harold, A.; Shuda, M.; Venuti, A.; Romero-Medina, M.C.; Pacini, L.; Brault, L.; Robitaille, A.; Taverniti, V.; et al. Merkel Cell Polyomavirus Downregulates N-myc Downstream-Regulated Gene 1, Leading to Cellular Proliferation and Migration. *J. Virol.* **2020**, *94*, e00899-19. [CrossRef] [PubMed]
153. Masterson, L.; Thibodeau, B.J.; Fortier, L.E.; Geddes, T.J.; Pruetz, B.L.; Malhotra, R.; Keidan, R.; Wilson, G.D. Gene expression differences predict treatment outcome of merkel cell carcinoma patients. *J. Skin Cancer* **2014**, *2014*, 596459. [CrossRef] [PubMed]
154. Daily, K.; Coxon, A.; Williams, J.S.; Lee, C.R.; Coit, D.G.; Busam, K.J.; Brownell, I. Assessment of cancer cell line representativeness using microarrays for Merkel cell carcinoma. *J. Investig. Dermatol.* **2015**, *135*, 1138–1146. [CrossRef]
155. Berrios, C.; Padi, M.; Keibler, M.A.; Park, D.E.; Molla, V.; Cheng, J.; Lee, S.M.; Stephanopoulos, G.; Quackenbush, J.; DeCaprio, J.A. Merkel cell Polyomavirus small T antigen promotes pro-Glycolytic metabolic perturbations required for transformation. *PLoS Pathog.* **2016**, *12*, e1006020. [CrossRef] [PubMed]
156. Melotte, V.; Qu, X.; Ongenaert, M.; van Criekinge, W.; de Bruïne, A.P.; Baldwin, H.S.; van Engeland, M. The N-myc downstream regulated gene (NDRG) family: Diverse functions, multiple applications. *FASEB J.* **2010**, *24*, 4153–4166. [CrossRef]
157. Sihto, H.; Kukko, H.; Koljonen, V.; Sankila, R.; Böhling, T.; Joensuu, H. Merkel cell polyomavirus infection, large T antigen, retinoblastoma protein and outcome in Merkel cell carcinoma. *Clin. Cancer Res.* **2011**, *17*, 4806–4813. [CrossRef]
158. Pei, D.; Zhang, Y.; Zheng, J. Regulation of p53: A collaboration between Mdm2 and Mdmx. *Oncotarget* **2012**, *3*, 228–235. [CrossRef]
159. Arora, R.; Shuda, M.; Guastafierro, A.; Feng, H.; Toptan, T.; Tolstov, T.; Normolle, D.; Vollmer, L.L.; Vogt, A.; Dömling, A.; et al. Survivin is a therapeutic target in Merkel cell carcinoma. *Sci. Transl. Med.* **2012**, *4*, 133ra156. [CrossRef]
160. Houben, R.; Dreher, C.; Angermeyer, S.; Borst, A.; Utikal, J.; Haferkamp, S.; Peitsch, W.K.; Schrama, D.; Hesbacher, S. Mechanisms of p53 restriction in Merkel cell carcinoma cells are independent of the Merkel cell polyoma virus T antigens. *J. Investig. Dermatol.* **2013**, *133*, 2453–2460. [CrossRef]
161. Waltari, M.; Sihto, H.; Kukko, H.; Koljonen, V.; Sankila, R.; Böhling, T.; Joensuu, H. Association of Merkel cell polyomavirus infection with tumor p53, KIT, stem cell factor, PDGFR-alpha and survival in Merkel cell carcinoma. *Int. J. Cancer* **2011**, *129*, 619–628. [CrossRef] [PubMed]
162. Schrama, D.; Peitsch, W.K.; Zapatka, M.; Kneitz, H.; Houben, R.; Eib, S.; Haferkamp, S.; Moore, P.S.; Shuda, M.; Thompson, J.F.; et al. Merkel cell polyomavirus status is not associated with clinical course of Merkel cell carcinoma. *J. Investig. Dermatol.* **2011**, *131*, 1631–1638. [CrossRef]
163. Guergnon, J.; Godet, A.N.; Galioot, A.; Falanga, P.B.; Colle, J.H.; Cayla, X.; Garcia, A. PP2A targeting by viral proteins: A widespread biological strategy from DNA/RNA tumor viruses to HIV-1. *Biochim. Biophys. Acta* **2011**, *1812*, 1498–1507. [CrossRef]
164. Ruvolo, P.P. The broken "Off" switch in cancer signaling: PP2A as a regulator of tumorigenesis, drug resistance, and immune surveillance. *BBA Clin.* **2016**, *6*, 87–99. [CrossRef]
165. Luo, W.; Xu, C.; Ayello, J.; Dela Cruz, F.; Rosenblum, J.M.; Lessnick, S.L.; Cairo, M.S. Protein phosphatase 1 regulatory subunit 1A in ewing sarcoma tumorigenesis and metastasis. *Oncogene* **2018**, *37*, 798–809. [CrossRef]
166. Fowle, H.; Zhao, Z.; Graña, X. PP2A holoenzymes, substrate specificity driving cellular functions and deregulation in cancer. *Adv. Cancer Res.* **2019**, *144*, 55–93. [CrossRef]
167. Park, J.; Lee, D.H. Functional roles of protein phosphatase 4 in multiple aspects of cellular physiology: A friend and a foe. *BMB Rep.* **2020**, *53*, 181–190. [CrossRef]
168. Kolupaeva, V.; Janssens, V. PP1 and PP2A phosphatases–cooperating partners in modulating retinoblastoma protein activation. *FEBS J.* **2013**, *280*, 627–643. [CrossRef]
169. Janssens, V.; Longin, S.; Goris, J. PP2A holoenzyme assembly: In cauda venenum (the sting is in the tail). *Trends Biochem. Sci.* **2008**, *33*, 113–121. [CrossRef]
170. Gilmore, T.D. Introduction to NF-kappaB: Players, pathways, perspectives. *Oncogene* **2006**, *25*, 6680–6684. [CrossRef]
171. Baeuerle, P.A.; Henkel, T. Function and activation of NF-kappa B in the immune system. *Annu. Rev. Immunol.* **1994**, *12*, 141–179. [CrossRef]

172. Pahl, H.L. Activators and target genes of Rel/NF-kappaB transcription factors. *Oncogene* **1999**, *18*, 6853–6866. [CrossRef] [PubMed]
173. Matsushita, M.; Iwasaki, T.; Nonaka, D.; Kuwamoto, S.; Nagata, K.; Kato, M.; Kitamura, Y.; Hayashi, K. Higher expression of activation-induced Cytidine Deaminase is significantly associated with Merkel cell Polyomavirus-negative Merkel cell carcinomas. *Yonago Acta Med.* **2017**, *60*, 145–153. [CrossRef] [PubMed]
174. Nwogu, N.; Boyne, J.R.; Dobson, S.J.; Poterlowicz, K.; Blair, G.E.; Macdonald, A.; Mankouri, J.; Whitehouse, A. Cellular sheddases are induced by Merkel cell polyomavirus small tumour antigen to mediate cell dissociation and invasiveness. *PLoS Pathog.* **2018**, *14*, e1007276. [CrossRef] [PubMed]
175. Buchkovich, N.J.; Yu, Y.; Zampieri, C.A.; Alwine, J.C. The TORrid affairs of viruses: Effects of mammalian DNA viruses on the PI3K-Akt-mTOR signalling pathway. *Nat. Rev. Microbiol.* **2008**, *6*, 266–275. [CrossRef]
176. Engelman, J.A. Targeting PI3K signalling in cancer: Opportunities, challenges and limitations. *Nat. Rev. Cancer* **2009**, *9*, 550–562. [CrossRef]
177. Hafner, C.; Houben, R.; Baeurle, A.; Ritter, C.; Schrama, D.; Landthaler, M.; Becker, J.C. Activation of the PI3K/AKT pathway in Merkel cell carcinoma. *PLoS ONE* **2012**, *7*, e31255. [CrossRef]
178. Iwasaki, T.; Matsushita, M.; Nonaka, D.; Kuwamoto, S.; Kato, M.; Murakami, I.; Nagata, K.; Nakajima, H.; Sano, S.; Hayashi, K. Comparison of Akt/mTOR/4E-BP1 pathway signal activation and mutations of PIK3CA in Merkel cell polyomavirus-positive and Merkel cell polyomavirus-negative carcinomas. *Hum. Pathol.* **2015**, *46*, 210–216. [CrossRef]
179. Nardi, V.; Song, Y.; Santamaria-Barria, J.A.; Cosper, A.K.; Lam, Q.; Faber, A.C.; Boland, G.M.; Yeap, B.Y.; Bergethon, K.; Scialabba, V.L.; et al. Activation of PI3K signaling in Merkel cell carcinoma. *Clin. Cancer Res.* **2012**, *18*, 1227–1236. [CrossRef]
180. Fang, B.; Kannan, A.; Zhao, S.; Nguyen, Q.H.; Ejadi, S.; Yamamoto, M.; Camilo Barreto, J.; Zhao, H.; Gao, L. Inhibition of PI3K by copanlisib exerts potent antitumor effects on Merkel cell carcinoma cell lines and mouse xenografts. *Sci. Rep.* **2020**, *10*, 8867. [CrossRef]
181. Newton, A.C. Protein kinase C: Perfectly balanced. *Crit. Rev. Biochem. Mol. Biol.* **2018**, *53*, 208–230. [CrossRef]
182. Jain, K.; Basu, A. The Multifunctional Protein Kinase C-ε in Cancer Development and Progression. *Cancers* **2014**, *6*, 860–878. [CrossRef]
183. Costa, A.; Mackelfresh, J.; Gilbert, L.; Bonner, M.Y.; Arbiser, J.L. Activation of Protein Kinase C ε in Merkel Cell Polyomavirus-Induced Merkel Cell Carcinoma. *JAMA Dermatol.* **2017**, *153*, 931–932. [CrossRef]
184. Kovall, R.A.; Gebelein, B.; Sprinzak, D.; Kopan, R. The Canonical Notch Signaling Pathway: Structural and Biochemical Insights into Shape, Sugar, and Force. *Dev. Cell* **2017**, *41*, 228–241. [CrossRef]
185. Wardhani, L.O.; Matsushita, M.; Kuwamoto, S.; Nonaka, D.; Nagata, K.; Kato, M.; Kitamura, Y.; Hayashi, K. Expression of Notch 3 and Jagged 1 Is Associated With Merkel Cell Polyomavirus Status and Prognosis in Merkel Cell Carcinoma. *Anticancer Res.* **2019**, *39*, 319–329. [CrossRef]
186. Hooper, J.E.; Scott, M.P. Communicating with Hedgehogs. *Nat. Rev. Mol. Cell Biol.* **2005**, *6*, 306–317. [CrossRef]
187. Kuromi, T.; Matsushita, M.; Iwasaki, T.; Nonaka, D.; Kuwamoto, S.; Nagata, K.; Kato, M.; Akizuki, G.; Kitamura, Y.; Hayashi, K. Association of expression of the hedgehog signal with Merkel cell polyomavirus infection and prognosis of Merkel cell carcinoma. *Hum. Pathol.* **2017**, *69*, 8–14. [CrossRef]
188. Kennedy, M.M.; Blessing, K.; King, G.; Kerr, K.M. Expression of bcl-2 and p53 in Merkel cell carcinoma. An immunohistochemical study. *Am. J. Dermatopathol* **1996**, *18*, 273–277. [CrossRef]
189. Feinmesser, M.; Halpern, M.; Fenig, E.; Tsabari, C.; Hodak, E.; Sulkes, J.; Brenner, B.; Okon, E. Expression of the apoptosis-related oncogenes bcl-2, bax, and p53 in Merkel cell carcinoma: Can they predict treatment response and clinical outcome? *Hum. Pathol.* **1999**, *30*, 1367–1372. [CrossRef]
190. Brunner, M.; Thurnher, D.; Pammer, J.; Geleff, S.; Heiduschka, G.; Reinisch, C.M.; Petzelbauer, P.; Erovic, B.M. Expression of VEGF-A/C, VEGF-R2, PDGF-alpha/beta, c-kit, EGFR, Her-2/Neu, Mcl-1 and Bmi-1 in Merkel cell carcinoma. *Mod. Pathol.* **2008**, *21*, 876–884. [CrossRef]
191. Sahi, H.; Koljonen, V.; Kavola, H.; Haglund, C.; Tukiainen, E.; Sihto, H.; Böhling, T. Bcl-2 expression indicates better prognosis of Merkel cell carcinoma regardless of the presence of Merkel cell polyomavirus. *Virchows Arch.* **2012**, *461*, 553–559. [CrossRef] [PubMed]
192. Verhaegen, M.E.; Mangelberger, D.; Weick, J.W.; Vozheiko, T.D.; Harms, P.W.; Nash, K.T.; Quintana, E.; Baciu, P.; Johnson, T.M.; Bichakjian, C.K.; et al. Merkel cell carcinoma dependence on bcl-2 family members for survival. *J. Investig. Dermatol.* **2014**, *134*, 2241–2250. [CrossRef] [PubMed]

193. Shah, M.H.; Varker, K.A.; Collamore, M.; Zwiebel, J.A.; Coit, D.; Kelsen, D.; Chung, K.Y. G3139 (Genasense) in patients with advanced merkel cell carcinoma. *Am. J. Clin. Oncol* **2009**, *32*, 174–179. [CrossRef] [PubMed]
194. Church, C.D.; Nghiem, P. How does the Merkel polyomavirus lead to a lethal cancer? Many answers, many questions, and a new mouse model. *J. Investig. Dermatol.* **2015**, *135*, 1221–1224. [CrossRef] [PubMed]
195. Paulson, K.G.; Tegeder, A.; Willmes, C.; Iyer, J.G.; Afanasiev, O.K.; Schrama, D.; Koba, S.; Thibodeau, R.; Nagase, K.; Simonson, W.T.; et al. Downregulation of MHC-I expression is prevalent but reversible in Merkel cell carcinoma. *Cancer Immunol. Res.* **2014**, *2*, 1071–1079. [CrossRef] [PubMed]
196. Afanasiev, O.K.; Yelistratova, L.; Miller, N.; Nagase, K.; Paulson, K.; Iyer, J.G.; Ibrani, D.; Koelle, D.M.; Nghiem, P. Merkel polyomavirus-specific T cells fluctuate with merkel cell carcinoma burden and express therapeutically targetable PD-1 and Tim-3 exhaustion markers. *Clin. Cancer Res.* **2013**, *19*, 5351–5360. [CrossRef]
197. Afanasiev, O.K.; Nagase, K.; Simonson, W.; Vandeven, N.; Blom, A.; Koelle, D.M.; Clark, R.; Nghiem, P. Vascular E-selectin expression correlates with CD8 lymphocyte infiltration and improved outcome in Merkel cell carcinoma. *J. Investig. Dermatol.* **2013**, *133*, 2065–2073. [CrossRef]
198. Dong, H.; Strome, S.E.; Salomao, D.R.; Tamura, H.; Hirano, F.; Flies, D.B.; Roche, P.C.; Lu, J.; Zhu, G.; Tamada, K.; et al. Tumor-associated B7-H1 promotes T-cell apoptosis: A potential mechanism of immune evasion. *Nat. Med.* **2002**, *8*, 793–800. [CrossRef]
199. Lipson, E.J.; Vincent, J.G.; Loyo, M.; Kagohara, L.T.; Luber, B.S.; Wang, H.; Xu, H.; Nayar, S.K.; Wang, T.S.; Sidransky, D.; et al. PD-L1 expression in the Merkel cell carcinoma microenvironment: Association with inflammation, Merkel cell polyomavirus and overall survival. *Cancer Immunol. Res.* **2013**, *1*, 54–63. [CrossRef]
200. Mitteldorf, C.; Berisha, A.; Tronnier, M.; Pfaltz, M.C.; Kempf, W. PD-1 and PD-L1 in neoplastic cells and the tumor microenvironment of Merkel cell carcinoma. *J. Cutan. Pathol.* **2017**, *44*, 740–746. [CrossRef]
201. Schönrich, G.; Raftery, M.J. The PD-1/PD-L1 Axis and virus infections: A delicate balance. *Front. Cell Infect. Microbiol.* **2019**, *9*, 207. [CrossRef] [PubMed]
202. Shahzad, N.; Shuda, M.; Gheit, T.; Kwun, H.J.; Cornet, I.; Saidj, D.; Zannetti, C.; Hasan, U.; Chang, Y.; Moore, P.S.; et al. The T antigen locus of Merkel cell polyomavirus downregulates human Toll-like receptor 9 expression. *J. Virol.* **2013**, *87*, 13009–13019. [CrossRef] [PubMed]
203. Jouhi, L.; Koljonen, V.; Böhling, T.; Haglund, C.; Hagström, J. The expression of Toll-like receptors 2, 4, 5, 7 and 9 in Merkel cell carcinoma. *Anticancer Res.* **2015**, *35*, 1843–1849.
204. Wu, S.Y.; Chiang, C.M. The double bromodomain-containing chromatin adaptor Brd4 and transcriptional regulation. *J. Biol. Chem.* **2007**, *282*, 13141–13145. [CrossRef] [PubMed]
205. Hajmirza, A.; Emadali, A.; Gauthier, A.; Casasnovas, O.; Gressin, R.; Callanan, M.B. BET family protein BRD4: An emerging actor in NFκB signaling in inflammation and cancer. *Biomedicines* **2018**, *6*, 16. [CrossRef] [PubMed]
206. Malmgaard, L. Induction and regulation of IFNs during viral infections. *J. Interferon Cytokine Res.* **2004**, *24*, 439–454. [CrossRef] [PubMed]
207. Chen, I.Y.; Ichinohe, T. Response of host inflammasomes to viral infection. *Trends Microbiol.* **2015**, *23*, 55–63. [CrossRef]
208. Dranoff, G. Cytokines in cancer pathogenesis and cancer therapy. *Nat. Rev. Cancer* **2004**, *4*, 11–22. [CrossRef]
209. Germano, G.; Allavena, P.; Mantovani, A. Cytokines as a key component of cancer-related inflammation. *Cytokine* **2008**, *43*, 374–379. [CrossRef]
210. Richards, K.F.; Guastafierro, A.; Shuda, M.; Toptan, T.; Moore, P.S.; Chang, Y. Merkel cell polyomavirus T antigens promote cell proliferation and inflammatory cytokine gene expression. *J. Gen. Virol.* **2015**, *96*, 3532–3544. [CrossRef]
211. Monnier, J.; Samson, M. Prokineticins in angiogenesis and cancer. *Cancer Lett.* **2010**, *296*, 144–149. [CrossRef]
212. Lauttia, S.; Sihto, H.; Kavola, H.; Koljonen, V.; Böhling, T.; Joensuu, H. Prokineticins and Merkel cell polyomavirus infection in Merkel cell carcinoma. *Br. J. Cancer* **2014**, *110*, 1446–1455. [CrossRef] [PubMed]
213. Rasheed, K.; Abdulsalam, I.; Fismen, S.; Grimstad, O.; Sveinbjornsson, B.; Moens, U. CCL17/TARC and CCR4 expression in Merkel cell carcinoma. *Oncotarget* **2018**, *9*, 31432–31447. [CrossRef]
214. Barber, G.N. STING: Infection, inflammation and cancer. *Nat. Rev. Immunol.* **2015**, *15*, 760–770. [CrossRef]
215. Liu, W.; Kim, G.B.; Krump, N.A.; Zhou, Y.; Riley, J.L.; You, J. Selective reactivation of STING signaling to target Merkel cell carcinoma. *Proc. Natl. Acad. Sci. USA* **2020**, *117*, 13730–13739. [CrossRef] [PubMed]

216. Jouary, T.; Leyral, C.; Dreno, B.; Doussau, A.; Sassolas, B.; Beylot-Barry, M.; Renaud-Vilmer, C.; Guillot, B.; Bernard, P.; Lok, C.; et al. Groupe de Cancérologie Cutanée of the Société Francaise de Dermatologie. Adjuvant prophylactic regional radiotherapy versus observation in stage I Merkel cell carcinoma: A multicentric prospective randomized study. *Ann. Oncol.* **2012**, *23*, 1074–1080. [CrossRef]
217. Hasan, S.; Liu, L.; Triplet, J.; Li, Z.; Mansur, D. The role of postoperative radiation and chemoradiation in merkel cell carcinoma: A systematic review of the literature. *Front. Oncol.* **2013**, *3*, 276. [CrossRef]
218. Rush, Z.; Fields, R.C.; Lee, N.; Brownell, I. Radiation therapy in the management of Merkel cell carcinoma: Current perspectives. *Expert Rev. Dermatol.* **2011**, *6*, 395–404. [CrossRef]
219. Miles, B.A.; Goldenberg, D. Merkel cell carcinoma: Do you know your guidelines? *Head Neck* **2016**, *38*, 647–652. [CrossRef]
220. Deng, L.; Liang, H.; Xu, M.; Yang, X.; Burnette, B.; Arina, A.; Li, X.D.; Mauceri, H.; Beckett, M.; Darga, T.; et al. STING-dependent Cytosolic DNA sensing promotes radiation-induced type I Interferon-Dependent antitumor immunity in immunogenic tumors. *Immunity* **2014**, *41*, 843–852. [CrossRef]
221. Paulson, K.G.; Carter, J.J.; Johnson, L.G.; Cahill, K.W.; Iyer, J.G.; Schrama, D.; Becker, J.C.; Madeleine, M.M.; Nghiem, P.; Galloway, D.A. Antibodies to merkel cell polyomavirus T antigen oncoproteins reflect tumor burden in merkel cell carcinoma patients. *Cancer Res.* **2010**, *70*, 8388–8397. [CrossRef] [PubMed]
222. Asioli, S.; Righi, A.; Volante, M.; Eusebi, V.; Bussolati, G. p63 expression as a new prognostic marker in Merkel cell carcinoma. *Cancer* **2007**, *110*, 640–647. [CrossRef]
223. Hall, B.J.; Pincus, L.B.; Yu, S.S.; Oh, D.H.; Wilson, A.R.; McCalmont, T.H. Immunohistochemical prognostication of Merkel cell carcinoma: p63 expression but not polyomavirus status correlates with outcome. *J. Cutan. Pathol.* **2012**, *39*, 911–917. [CrossRef] [PubMed]
224. Stetsenko, G.Y.; Malekirad, J.; Paulson, K.G.; Iyer, J.G.; Thibodeau, R.M.; Nagase, K.; Schmidt, M.; Storer, B.E.; Argenyi, Z.B.; Nghiem, P. p63 expression in Merkel cell carcinoma predicts poorer survival yet may have limited clinical utility. *Am. J. Clin. Pathol.* **2013**, *140*, 838–844. [CrossRef] [PubMed]
225. Fleming, K.E.; Ly, T.Y.; Pasternak, S.; Godlewski, M.; Doucette, S.; Walsh, N.M. Support for p63 expression as an adverse prognostic marker in Merkel cell carcinoma: Report on a Canadian cohort. *Hum. Pathol.* **2014**, *45*, 952–960. [CrossRef]
226. Asioli, S.; Righi, A.; de Biase, D.; Morandi, L.; Caliendo, V.; Picciotto, F.; Macripò, G.; Maletta, F.; di Cantogno, L.V.; Chiusa, L.; et al. Expression of p63 is the sole independent marker of aggressiveness in localised (stage I-II) Merkel cell carcinomas. *Mod. Pathol.* **2011**, *24*, 1451–1461. [CrossRef]
227. Patel, R.M.; Walters, L.L.; Kappes, F.; Mehra, R.; Fullen, D.R.; Markovitz, D.M.; Ma, L. DEK expression in Merkel cell carcinoma and small cell carcinoma. *J. Cutan. Pathol.* **2012**, *39*, 753–757. [CrossRef]
228. Sandén, C.; Gullberg, U. The DEK oncoprotein and its emerging roles in gene regulation. *Leukemia* **2015**, *29*, 1632–1636. [CrossRef]
229. Pryor, J.G.; Simon, R.A.; Bourne, P.A.; Spaulding, B.O.; Scott, G.A.; Xu, H. Merkel cell carcinoma expresses K homology domain-containing protein overexpressed in cancer similar to other high-grade neuroendocrine carcinomas. *Hum. Pathol.* **2009**, *40*, 238–243. [CrossRef]
230. Pryor, J.G.; Bourne, P.A.; Yang, Q.; Spaulding, B.O.; Scott, G.A.; Xu, H. IMP-3 is a novel progression marker in malignant melanoma. *Mod. Pathol.* **2008**, *21*, 431–437. [CrossRef]
231. Johnson, B.; Khalil, M.; Blansfield, J.; Lin, F.; Zhu, S.; Kirchner, H.L.; Weir, A.B. 3rd. Investigating the prognostic value of KOC (K homology domain containing protein overexpressed in cancer) overexpression after curative intent resection of pancreatic ductal adenocarcinoma. *J. Gastrointest. Oncol.* **2016**, *7*, E113–E117. [CrossRef] [PubMed]
232. Kase, S.; Yoshida, K.; Osaki, M.; Adachi, H.; Ito, H.; Ohno, S. Expression of erythropoietin receptor in human Merkel cell carcinoma of the eyelid. *Anticancer Res.* **2006**, *26*, 4535–4537. [PubMed]
233. Werchau, S.; Toberer, F.; Enk, A.; Dammann, R.; Helmbold, P. Merkel cell carcinoma induces lymphatic microvessel formation. *J. Am. Acad. Dermatol.* **2012**, *67*, 215–225. [CrossRef] [PubMed]
234. Vlahova, L.; Doerflinger, Y.; Houben, R.; Becker, J.C.; Schrama, D.; Weiss, C.; Goebeler, M.; Helmbold, P.; Goerdt, S.; Peitsch, W.K. P-cadherin expression in Merkel cell carcinomas is associated with prolonged recurrence-free survival. *Br. J. Dermatol.* **2012**, *166*, 1043–1052. [CrossRef] [PubMed]
235. Samimi, M.; Touzé, A.; Laude, H.; Le Bidre, E.; Arnold, F.; Carpentier, A.; Gardair, C.; Carlotti, A.; Maubec, E.; Dupin, N.; et al. Vitamin D deficiency is associated with greater tumor size and poorer outcome in Merkel cell carcinoma patients. *J. Eur. Acad. Dermatol. Venereol.* **2014**, *28*, 298–308. [CrossRef]

236. Rand, J.; Balzer, B.L.; Frishberg, D.P.; Essner, R.; Shon, W. Prevalence of Delta-Like Protein 3 expression in Merkel cell carcinoma. *J. Am. Acad. Dermatol.* **2019**. [CrossRef]
237. Xie, H.; Kaye, F.J.; Isse, K.; Sun, Y.; Ramoth, J.; French, D.M.; Flotte, T.J.; Luo, Y.; Saunders, L.R.; Mansfield, A.S. Delta-like Protein 3 Expression and Targeting in Merkel Cell Carcinoma. *The Oncol.* 2020. [CrossRef]
238. Toberer, F.; Haenssle, H.A.; Heinzel-Gutenbrunner, M.; Enk, A.; Hartschuh, W.; Helmbold, P.; Kutzner, H. Metabolic reprogramming and angiogenesis in primary cutaneous Merkel cell carcinoma: Expression of hypoxia inducible factor-1α and its central downstream factors. *J. Eur. Acad. Dermatol. Venereol.* **2020**. [CrossRef]
239. Konstatinell, A.; Coucheron, D.H.; Sveinbjørnsson, B.; Moens, U. MicroRNAs as Potential Biomarkers in Merkel Cell Carcinoma. *Int. J. Mol. Sci.* **2018**, *19*, 1873. [CrossRef]
240. Xie, H.; Lee, L.; Caramuta, S.; Höög, A.; Browaldh, N.; Björnhagen, V.; Larsson, C.; Lui, W.O. MicroRNA expression patterns related to merkel cell polyomavirus infection in human merkel cell carcinoma. *J. Investig. Dermatol.* **2014**, *134*, 507–517. [CrossRef]
241. Renwick, N.; Cekan, P.; Masry, P.A.; McGeary, S.E.; Miller, J.B.; Hafner, M.; Li, Z.; Mihailovic, A.; Morozov, P.; Brown, M.; et al. Multicolor microRNA FISH effectively differentiates tumor types. *J. Clin. Investig.* **2013**, *123*, 2694–2702. [CrossRef] [PubMed]
242. Fan, K.; Ritter, C.; Nghiem, P.; Blom, A.; Verhaegen, M.E.; Dlugosz, A.; Ødum, N.; Woetmann, A.; Tothill, R.W.; Hicks, R.J.; et al. Circulating Cell-Free miR-375 as surrogate marker of tumor burden in merkel cell carcinoma. *Clin. Cancer Res.* **2018**, *24*, 5873–5882. [CrossRef] [PubMed]
243. Lee, S.; Paulson, K.G.; Murchison, E.P.; Afanasiev, O.K.; Alkan, C.; Leonard, J.H.; Byrd, D.R.; Hannon, G.J.; Nghiem, P. Identification and validation of a novel mature microRNA encoded by the Merkel cell polyomavirus in human Merkel cell carcinomas. *J. Clin. Virol.* **2011**, *52*, 272–275. [CrossRef]
244. Becker, J.C.; Stang, A.; Hausen, A.Z.; Fischer, N.; DeCaprio, J.A.; Tothill, R.W.; Lyngaa, R.; Hansen, U.K.; Ritter, C.; Nghiem, P.; et al. Epidemiology, biology and therapy of Merkel cell carcinoma: Conclusions from the EU project IMMOMEC. *Cancer Immunol. Immunother.* **2018**, *67*, 341–351. [CrossRef] [PubMed]
245. Vaddepally, R.K.; Kharel, P.; Pandey, R.; Garje, R.; Chandra, A.B. Review of Indications of FDA-approved immune checkpoint inhibitors per NCCN Guidelines with the level of evidence. *Cancers* **2020**, *12*. [CrossRef] [PubMed]
246. Nghiem, P.T.; Bhatia, S.; Lipson, E.J.; Kudchadkar, R.R.; Miller, N.J.; Annamalai, L.; Berry, S.; Chartash, E.K.; Daud, A.; Fling, S.P.; et al. PD-1 Blockade with Pembrolizumab in advanced Merkel-cell carcinoma. *N. Engl. J. Med.* **2016**, *374*, 2542–2552. [CrossRef]
247. Nghiem, P.; Bhatia, S.; Lipson, E.J.; Sharfman, W.H.; Kudchadkar, R.R.; Brohl, A.S.; Friedlander, P.A.; Daud, A.; Kluger, H.M.; Reddy, S.A. Durable tumor regression and overall survival in patients with advanced Merkel cell carcinoma receiving Pembrolizumab as first-line therapy. *J. Clin. Oncol.* **2019**, *37*, 693–702. [CrossRef]
248. Topalian, S.L.; Bhatia, S.; Amin, A.; Kudchadkar, R.R.; Sharfman, W.H.; Lebbé, C.; Delord, J.P.; Dunn, L.A.; Shinohara, M.M.; Kulikauskas, R.; et al. Neoadjuvant Nivolumab for Patients With Resectable Merkel Cell Carcinoma in the CheckMate 358 Trial. *J. Clin. Oncol.* **2020**. [CrossRef]
249. D'Angelo, S.P.; Bhatia, S.; Brohl, A.S.; Hamid, O.; Mehnert, J.M.; Terheyden, P.; Shih, K.C.; Brownell, I.; Lebbé, C.; Lewis, K.D.; et al. Avelumab in patients with previously treated metastatic Merkel cell carcinoma: Long-term data and biomarker analyses from the single-arm phase 2 JAVELIN Merkel 200 trial. *J. Immunother. Cancer* **2020**, *8*, e000674. [CrossRef]
250. Winkler, J.K.; Dimitrakopoulou-Strauss, A.; Sachpekidis, C.; Enk, A.; Hassel, J.C. Ipilimumab has efficacy in metastatic Merkel cell carcinoma: A case series of five patients. *J. Eur. Acad. Dermatol. Venereol.* **2017**, *31*, e389–e391. [CrossRef]
251. Tabachnick-Cherny, S.; Pulliam, T.; Church, C.; Koelle, D.M.; Nghiem, P. Polyomavirus-driven Merkel cell carcinoma: Prospects for therapeutic vaccine development. *Mol. Carcinogs.* **2020**, *59*, 807–821. [CrossRef] [PubMed]
252. Zeng, Q.; Gomez, B.P.; Viscidi, R.P.; Peng, S.; He, L.; Ma, B.; Wu, T.C.; Hung, C.F. Development of a DNA vaccine targeting Merkel cell polyomavirus. *Vaccine* **2012**, *30*, 1322–1329. [CrossRef] [PubMed]
253. Gomez, B.; He, L.; Tsai, Y.C.; Wu, T.C.; Viscidi, R.P.; Hung, C.F. Creation of a Merkel cell polyomavirus small T antigen-expressing murine tumor model and a DNA vaccine targeting small T antigen. *Cell Biosci.* **2013**, *3*, 29. [CrossRef] [PubMed]

254. Gomez, B.P.; Wang, C.; Viscidi, R.P.; Peng, S.; He, L.; Wu, T.C.; Hung, C.F. Strategy for eliciting antigen-specific CD8+ T cell-mediated immune response against a cryptic CTL epitope of merkel cell polyomavirus large T antigen. *Cell Biosci.* **2012**, *2*, 36. [CrossRef]
255. Togtema, M.; Jackson, R.; Grochowski, J.; Villa, P.L.; Mellerup, M.; Chattopadhyaya, J.; Zehbe, I. Synthetic siRNA targeting human papillomavirus 16 E6: A perspective on in vitro nanotherapeutic approaches. *Nanomedicine* **2018**, *13*, 455–474. [CrossRef]
256. Rosa, J.; Suzuki, I.; Kravicz, M.; Caron, A.; Pupo, A.V.; Praça, F.G.; Bentley, M. Current non-viral siRNA delivery systems as a promising treatment of skin diseases. *Curr. Pharm. Des.* **2018**, *24*, 2644–2663. [CrossRef]
257. Seguin, S.P.; Ireland, A.W.; Gupta, T.; Wright, C.M.; Miyata, Y.; Wipf, P.; Pipas, J.M.; Gestwicki, J.E.; Brodsky, J.L. A screen for modulators of large T antigen's ATPase activity uncovers novel inhibitors of Simian Virus 40 and BK virus replication. *Antiviral Res.* **2012**, *96*, 70–81. [CrossRef]
258. Randhawa, P.; Zeng, G.; Bueno, M.; Salgarkar, A.; Lesniak, A.; Isse, K.; Seyb, K.; Perry, A.; Charles, I.; Hustus, C.; et al. Inhibition of large T antigen ATPase activity as a potential strategy to develop anti-polyomavirus JC drugs. *Antiviral Res.* **2014**, *112*, 113–119. [CrossRef]
259. Sarma, B.; Willmes, C.; Angerer, L.; Adam, C.; Becker, J.C.; Kervarrec, T.; Schrama, D.; Houben, R. Artesunate Affects T Antigen expression and survival of virus-positive Merkel cell carcinoma. *Cancers* **2020**, *12*, 919. [CrossRef]
260. Deeken, J.F.; Wang, H.; Hartley, M.; Cheema, A.K.; Smaglo, B.; Hwang, J.J.; He, A.R.; Weiner, L.M.; Marshall, J.L.; Giaccone, G.; et al. A phase I study of intravenous artesunate in patients with advanced solid tumor malignancies. *Cancer Chemother. Pharmacol.* **2018**, *81*, 587–596. [CrossRef]
261. Kim, J.; McNiff, J.M. Nuclear expression of survivin portends a poor prognosis in Merkel cell carcinoma. *Mod. Pathol.* **2008**, *21*, 764–769. [CrossRef] [PubMed]
262. Dresang, L.R.; Guastafierro, A.; Arora, R.; Normolle, D.; Chang, Y.; Moore, P.S. Response of Merkel cell polyomavirus-positive merkel cell carcinoma xenografts to a survivin inhibitor. *PLoS ONE* **2013**, *8*, e80543. [CrossRef] [PubMed]
263. Fewell, S.W.; Smith, C.M.; Lyon, M.A.; Dumitrescu, T.P.; Wipf, P.; Day, B.W.; Brodsky, J.L. Small molecule modulators of endogenous and co-chaperone-stimulated Hsp70 ATPase activity. *J. Biol. Chem.* **2004**, *279*, 51131–51140. [CrossRef]
264. Adam, C.; Baeurle, A.; Brodsky, J.L.; Wipf, P.; Schrama, D.; Becker, J.C.; Houben, R. The HSP70 modulator MAL3-101 inhibits Merkel cell carcinoma. *PLoS ONE* **2014**, *9*, e92041. [CrossRef]
265. Ocaña-Guzman, R.; Torre-Bouscoulet, L.; Sada-Ovalle, I. TIM-3 Regulates Distinct Functions in Macrophages. *Front. Immunol.* **2016**, *7*, 229. [CrossRef] [PubMed]
266. Monney, L.; Sabatos, C.A.; Gaglia, J.L.; Ryu, A.; Waldner, H.; Chernova, T.; Manning, S.; Greenfield, E.A.; Coyle, A.J.; Sobel, R.A.; et al. Th1-specific cell surface protein Tim-3 regulates macrophage activation and severity of an autoimmune disease. *Nature* **2002**, *415*, 536–541. [CrossRef]
267. Carey, B.W.; Kim, D.Y.; Kovacs, D.M. Presenilin/gamma-secretase and alpha-secretase-like peptidases cleave human MHC Class I proteins. *Biochem. J.* **2007**, *401*, 121–127. [CrossRef]
268. Anzivino, E.; Rodio, D.M.; Mischitelli, M.; Bellizzi, A.; Sciarra, A.; Salciccia, S.; Gentile, V.; Pietropaolo, V. High Frequency of JCV DNA Detection in prostate cancer tissues. *Cancer Genom. Proteom.* **2015**, *12*, 189–200.
269. Tognon, M.; Provenzano, M. New insights on the association between the prostate cancer and the small DNA tumour virus, BK polyomavirus. *J. Transl. Med.* **2015**, *13*, 387. [CrossRef]
270. Delbue, S.; Comar, M.; Ferrante, P. Review on the role of the human Polyomavirus JC in the development of tumors. *Infect. Agent Cancer* **2017**, *12*, 10. [CrossRef]
271. Levican, J.; Acevedo, M.; León, O.; Gaggero, A.; Aguayo, F. Role of BK human polyomavirus in cancer. *Infect. Agent Cancer* **2018**, *13*, 12. [CrossRef] [PubMed]
272. Dela Cruz, F.N., Jr.; Giannitti, F.; Li, L.; Woods, L.W.; Del Valle, L.; Delwart, E.; Pesavento, P.A. Novel polyomavirus associated with brain tumors in free-ranging raccoons, western United States. *Emerg. Infect. Dis.* **2013**, *19*, 77–84. [CrossRef] [PubMed]
273. Konno, A.; Nagata, M.; Nanko, H. Immunohistochemical diagnosis of a Merkel cell tumor in a dog. *Vet. Pathol.* **1998**, *35*, 538–540. [CrossRef] [PubMed]
274. Bagnasco, G.; Properzi, R.; Porto, R.; Nardini, V.; Poli, A.; Abramo, F. Feline cutaneous neuroendocrine carcinoma (Merkel cell tumour): Clinical and pathological findings. *Vet. Dermatol.* **2003**, *14*, 111–115. [CrossRef]
275. Ozaki, K.; Narama, I. Merkel cell carcinoma in a cat. *J. Vet. Med. Sci.* **2009**, *71*, 1093–1096. [CrossRef]

276. Joiner, K.S.; Smith, A.N.; Henderson, R.A.; Brawner, W.R.; Spangler, E.A.; Sartin, E.A. Multicentric cutaneous neuroendocrine (Merkel cell) carcinoma in a dog. *Vet. Pathol.* **2010**, *47*, 1090–1094. [CrossRef]
277. Teh, A.P.P.; Izzati, U.Z.; Diep, N.V.; Hirai, T.; Yamaguchi, R. Merkel Cell Carcinoma in a Steer. *J. Comp. Pathol.* **2018**, *158*, 17–21. [CrossRef]
278. Sumi, A.; Chambers, J.K.; Doi, M.; Kudo, T.; Omachi, T.; Uchida, K. Clinical features and outcomes of Merkel cell carcinoma in 20 cats. *Vet. Comp. Oncol.* **2018**, *16*, 554–561. [CrossRef]
279. Schuurman, R.; Sol, C.; van der Noordaa, J. The complete nucleotide sequence of bovine polyomavirus. *J. Gen. Virol.* **1990**, *71*, 1723–1735. [CrossRef]
280. Delwart, E.; Kapusinszky, B.; Pesavento, P.A.; Estrada, M.; Seguin, M.A.; Leutenegger, C.M. Genome sequence of canine Polyomavirus in respiratory secretions of dogs with pneumonia of unknown etiology. *Genome Announc.* **2017**, *5*, e00615-17. [CrossRef]
281. Gheit, T.; Dutta, S.; Oliver, J.; Robitaille, A.; Hampras, S.; Combes, J.D.; McKay-Chopin, S.; Le Calvez-Kelm, F.; Fenske, N.; Cherpelis, B.; et al. Isolation and characterization of a novel putative human polyomavirus. *Virology* **2017**, *506*, 45–54. [CrossRef] [PubMed]
282. Fahsbender, E.; Altan, E.; Estrada, M.; Seguin, M.A.; Young, P.; Leutenegger, C.M.; Delwart, E. Lyon-IARC Polyomavirus DNA in feces of Diarrheic cats. *Microbiol. Resour. Announc.* **2019**, *8*, e00550-19. [CrossRef] [PubMed]

cancers

MDPI

Review

Management Recommendations for Merkel Cell Carcinoma—A Danish Perspective

Simon Naseri [1,*], Torben Steiniche [1], Morten Ladekarl [2], Marie Louise Bønnelykke-Behrndtz [3], Lisbet R. Hölmich [4], Seppo W. Langer [5], Alessandro Venzo [6], Elizaveta Tabaksblat [7], Siri Klausen [8], Mathilde Skaarup Larsen [8], Niels Junker [9] and Annette H. Chakera [4]

1 Department of Pathology, Aarhus University Hospital, 8200 Aarhus, Denmark; torbstei@rm.dk
2 Department of Oncology, Clinical Cancer Research Center, Aalborg University Hospital, 9000 Aalborg, Denmark; morten.ladekarl@rn.dk
3 Department of Plastic and Reconstructive Surgery, Aarhus University Hospital, 8200 Aarhus, Denmark; louiseboennelykke@gmail.com
4 Department of Plastic Surgery, Herlev & Gentofte Hospital, Department of Clinical Medicine, Copenhagen University, 2730 Herlev, Denmark; lisbet.rosenkrantz.hoelmich@regionh.dk (L.R.H.); annette.hougaard.chakera@regionh.dk (A.H.C.)
5 Department of Oncology, Copenhagen University Hospital, Rigshospitalet, 2100 Copenhagen, Denmark; seppo.langer@regionh.dk
6 Department of Plastic Surgery and Burns Treatment, Rigshospitalet, Copenhagen University Hospital, 2100 Copenhagen, Denmark; alessandro.venzo@regionh.dk
7 Department of Oncology, Aarhus University Hospital, 8200 Aarhus, Denmark; eliza.mit.tab@auh.rm.dk
8 Department of Pathology, Herlev Hospital, 2730 Herlev, Denmark; siri.klausen@regionh.dk (S.K.); Mathilde.skaarup.larsen@regionh.dk (M.S.L.)
9 Department of Oncology, Herlev Hospital, 2730 Herlev, Denmark; niels.Junker@regionh.dk
* Correspondence: s.nasari@rn.dk

Received: 4 February 2020; Accepted: 25 February 2020; Published: 28 February 2020

Abstract: Merkel cell carcinoma (MCC) is a rare malignant neuroendocrine carcinoma of the skin with a poor prognosis and an apparent increase in incidence. Due to its rarity, evidence-based guidelines are limited, and there is a lack of awareness among clinicians. This review constitutes the consensus management recommendations developed by the Danish MCC expert group and is based on a systematic literature search. Patients with localized disease are recommended surgical excision and adjuvant radiotherapy to the primary site; however, this may be omitted in patients with MCC with low risk features. Patients with regional lymph node involvement are recommended complete lymph node removal and adjuvant radiotherapy in case of extracapsular disease. Metastatic disease was traditionally treated with chemotherapy, however, recent clinical trials with immune therapy have been promising. Immune checkpoint inhibitors targeting the programmed cell death protein 1(PD-1)/programmed death-ligand 1(PD-L1) axis should therefore be strongly considered as first-line treatment for fit patients. A 5-year follow-up period is recommended involving clinical exam every 3 months for 2 years and every 6 months for the following 3 years and PET-CT one to two times a year or if clinically indicated. These national recommendations are intended to offer uniform patient treatment and hopefully improve prognosis.

Keywords: Merkel cell carcinoma; diagnosis; treatment; review; guideline

1. Introduction

Merkel cell carcinoma (MCC) is a rare and highly aggressive neuroendocrine malignancy of the skin. Research in MCC has recently gained traction due to successful clinical trials with immune checkpoint inhibitors and the discovery of the Merkel cell polyoma virus (MCPyV) [1,2]. The incidence in Denmark has increased 5.4-fold from 0.06 cases/100,000 in 1986 to 0.31 cases/100,000 in 2002 [3]. The median age of patients at diagnosis is 77 years, with the majority being male (62%) [4]. MCC is primarily caused by UV-radiation (24%) and/or the presence of MCPyV (76%) [1,5,6]. Immunosuppression also plays a role and increases the risk of MCC, as seen in individuals with chronic lymphatic leukemia (30-fold), the human immunodeficiency virus/acquired immunodeficiency syndrome (HIV/AIDS, 13-fold) and organ transplants recipients (10-fold) [7]. Currently, the 5-year overall survival of MCC is approximately 40%, making it more deadly than melanoma [8]. The current review constitutes the national Merkel cell carcinoma management recommendations developed by the Danish MCC expert group.

2. Diagnosis

2.1. Clinical Features

MCC usually presents as a nonspecific, firm, rapidly growing, painless red, purple or pink exophytic cutaneous nodule (Figure 1) [9]. The tumor is most often located in the head and neck area (43%), followed by the extremities (39%) [8]. As the presentation is often nonspecific and clinical diagnosis is only correct in 1% of cases, the AEIOU acronym has been developed to increase diagnostic accuracy by defining clinical features of MCC: Asymptomatic/lack of tenderness, Expanding rapidly, Immune suppression, Older than 50 years, and Ultraviolet-exposed site on a person with fair skin. Approximately 89% of patients have ≥3 of these features. Common benign misdiagnoses include cyst/acneiform lesion, lipoma, fibroma, vascular lesion or insect bite [9].

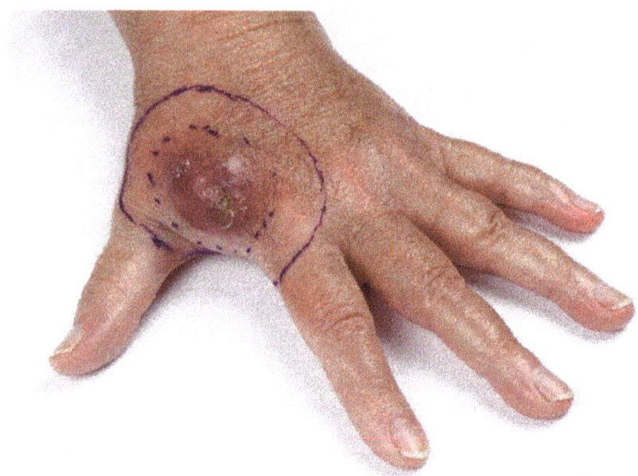

Figure 1. A red, nodular primary MCC on the left hand.

2.2. Pathology

Histology: MCC is often located in the dermis extending into the subcutaneous tissue. The cells are small, uniform and basaloid with granulated chromatin, scant cytoplasm and high mitotic rates [10]. Necrosis, increased vascularization, immune cell infiltration, solar elastosis and collision tumors (squamous cell carcinoma, basal cell carcinoma, actinic keratosis, Bowen disease) are common features [10]. Having the appearance of a small round blue cell tumor, MCC has many differential diagnoses, the most important being basal and squamous cell carcinomas, metastatic small cell carcinomas from other sites, melanoma, non-Hodgkin lymphomas and anaplastic adnexal carcinomas [11].

Immunohistochemistry (Table 1): MCC most commonly expresses epithelial markers, more specifically CK20, with a paranuclear dot-like staining. It shows neuroendocrine features with expression of neuroendocrine markers such as CD56 (88.2% positivity), chromogranin A (84.1% positivity) and synaptophysin (92% positivity) [12].

Table 1. Immunohistochemical markers in MCC and common differential diagnosis (Llombart et al. [11]).

Cancer Type	CK (AE1/AE3)	CK20	CK7	TTF-1	CD56	LCA	S-100	CD-99	Chromogranin A	Synaptophysin
Merkel cell carcinoma	+	+	−	−	+	−	−	+	+	+
Small cell lung cancer	+	−	+	+	+	−	−	+	+	+
Lymphoma	−	−	−	−	−	+	−	−	−	−
Melanoma	−/(+)	−	−	−	+	−	+	−	−	−
Ewing's sarcoma	−/(+)	−	−	−	+	−	-	+	−/(+)	−/(+)

MCC is negative for leukocyte common antigen (CD45, in contrast to lymphoma), S-100 (in contrast to melanoma), CK7 and TTF-1 (in contrast to small cell lung cancer) [12,13].

MCC diagnosis is based on histological features combined with CK20 expression and TTF-1 negativity. A subgroup of MCC does not stain accordingly, as 12.6% of MCC are CK20 negative, while 7% express TTF-1 [12]. CM2B4, which targets the MCV-large T antigen, may also be useful in diagnostics of virus-positive MCC (sensitivity 88.2%; specificity 94.3%) [6].

Proposed immunohistochemical panel: CK(AE1/AE3), CK20, CK7, TTF-1, CD45, S-100, CD-99 and neuroendocrine markers, such as synaptophysin, chromogranin A and CD56. In case of a nonconclusive result of the immunohistochemical panel, it may be necessary to do supplementary staining according to differential diagnosis.

Sentinel node (SN) protocol: A formalin-fixed SN will be split through the hilum and optionally further cut into parallel slices. The paraffin-embedded tissue slices will be cut in six serial steps with 50 μm intervals with a HE and CK(AE1/AE3) stained section at each level [14]. Isolated tumor cells within a lymphatic channel in the parenchyma of a lymph node, or its capsule, are classified as a metastasis.

Pathology report content should include:

Histopathological parameters: Macroscopic tumor diameter (microscopic in case of no clinical estimation, horizontal, in mm); Margin status (distance from lateral and deep resection margins, macroscopic and/or microscopic estimation, in mm); Tumor thickness (from stratum granulare, or highest tumor cell in case of ulceration, to deepest tumor cell, in mm with no decimals); Ulceration (total loss of epidermis with vital reaction, present/absent); Lymphovascular invasion (LVI) and perineural invasion (present/absent); Level of invasion (extracutaneous extension to muscle/fascia/cartilage/bone); Collision tumors (description if present).

Immunohistochemical analyses: Ki67 (% of tumor cells, estimate in 10% intervals if image counting tools are not available (more reproducible than hotspot counting)); Tumor infiltrating lymphocytes (number of CD8 positive lymphocytes per HPF); Viral status (MCPyV is positive when >1% of tumor cells stain independently of intensity when using the immunohistochemical clone CM2B4).

Sentinel lymph node status: Maximum diameter of the metastasis is reported, as well as number of involved/removed lymph nodes and extracapsular extension (presence of nodal metastasis extending through the lymph node capsule and into adjacent tissue) [15]; In-transit metastasis (present/absent): A discontinuous tumor distinct from the primary lesion and located between the primary lesion and the draining regional lymph nodes or distal to the primary lesion (Figure S1).

3. Work-Up & Staging

3.1. Work-Up

Patients should undergo full clinical skin and lymph node investigation with/without ultrasound (US) of the regional lymph node basin (Figure 2). Baseline F18-FDG whole-body positron emission tomography-computed tomography (PET-CT) scan may exclude differential diagnosis (mainly small cell lung cancer), lead to upstaging in 16% of patients and change management in up to 37% (Table 2) [16,17]. Until more data are produced, baseline PET-CT may be considered for all MCC patients.

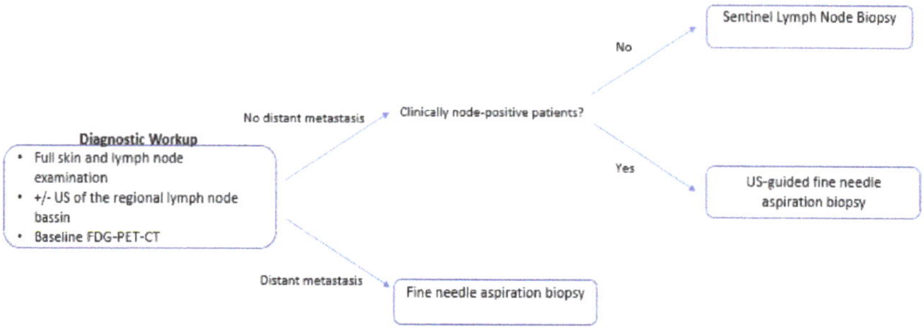

Figure 2. Diagnostic workup for MCC patients.

Table 2. Impact of staging with PET-CT adapted from Hawryluk et al. [16].

Stage	Patients Upstaged by PET/CT	Patients Upstaged by PET/CT (%)
IA	0/12	0%
IB	0/5	0%
IIA	0/3	0%
IIB	0/5	0%
IIIA	1/22	4.5%
IIIB	4/8	50%
IV	5/5	100%

Patients without clinically involved lymph nodes are recommended sentinel lymph node biopsy (SLNB), as 24–32% of these patients harbor clinically occult, microscopic regional metastasis [18–20]. Risk increases with tumor diameter, but patients with tumors down to 0.5 cm have been shown to be sentinel lymph node (SLN) positive in 14% [21]. SLNB should be carried out at the same time as wide excision to avoid inaccuracy and minimize the risk of false-negative results if performed after wide excision (15–17%) [18,22]. Patients with clinically involved lymph nodes should be offered US-guided fine needle aspiration biopsy (FNAB) to confirm the diagnosis.

3.2. Staging

The current staging is based on the eighth edition of the tumor, node, metastasis (TNM) staging system, recommend by the Union for International Cancer Control (UICC) and the American Joint Committee on Cancer (Figure S1). The expected 5-year overall survival (OS) for each stage is illustrated in Table 3 [8].

Table 3. Five-year overall survival with confidence interval (CI) [8].

Stage	Clinical Staging (95% CI)	Pathological Staging (95% CI)
I	45.0% (41.9–48.1%)	62.8% (59.6–65.8%)
IIA	30.9% (27.0–34.9%)	54.6% (49.3–59.7%)
IIB	27.3% (16.0–39.9%)	34.8% (25.6–44.1%)
IIIA	Data lacking	40.3% (37.5–43.0%)
IIIB	Data lacking	26.8% (23.4–30.4%)
IV	Data lacking	13.5% (11.0–16.3%)

The eighth edition of the TNM staging system is based on 9387 MCC cases with follow-up and staging data from 1998 to 2012 [8]. Staging can be clinical or pathological; the latter is more precise. Patients with local disease and negative SLNs have better prognosis (76% at 5 years) than those with clinically negative nodes (59% at 5 years) as the latter may harbor occult metastases that are not detected at time of diagnosis if SLNB is not performed [23]. Patients with nodal disease and unknown primary tumor are now given similar stage as patients with any tumor and microscopic nodal involvement (pathological stage IIIA), as patients with clinically detected nodal disease and unknown primary tumor show improved prognosis over cases with concurrent known primary tumor (OS 42% vs. 27%, respectively) [8].

4. Treatment

Due to the rarity of MCC, prospective clinical trials are rarely conducted. Therefore, current treatment recommendations are mostly based on retrospective studies with few patients (Figure S2).

5. Management of the Primary Tumor

5.1. Surgery

Traditionally, the recommended excisional margin for primary tumors has been 2–3 cm [24,25]. Recent studies comparing different excision margins show that patients ($n = 47$) treated with 1, 2 or 3 cm margins did not have a statistically significant difference in disease-free survival and OS [26]. Similarly, the largest single-institution study to date ($n = 240$) did not demonstrate a significant difference in local recurrence or disease-specific survival between patients treated with 1, 1.1–1.9 or >2 cm excisions [27]. Surgery-only ($n = 104$) with an excisional width of 1–2 cm to the tumor bed (tumor diameter < 2 cm) has demonstrated local recurrence rates down to 1.9% [19]. However, these studies were not randomized clinical trials so confounding by indication may be prevalent; larger excision margins may have been used for larger tumors. Regular randomized trials testing different resection margins are warranted but difficult to complete due to the small number of patients. A positive surgical margin is associated with reduced OS and should lead to re-excision [28,29]. Based on the above studies, an excisional margin of 1–2 cm is recommended.

5.2. Adjuvant Radiotherapy

Primary tumor: Radiotherapy (RT) is recommended following surgical excision [30]. In 4843 MCC cases, the largest cohort to date, it was shown that localized MCC (stage I and II) treated with primary surgery and adjuvant RT was associated with improved OS, compared to surgery alone (stage I: HR = 0.71, 95% CI = 0.64 to 0.80, $p < 0.001$; stage II: HR 0.77, 95 % CI = 0.66 to 0.89, $p < 0.001$) [28].

Recommended dose is 50–60 Gy at 2 Gy/d, 5 fractions per week (F/W) [31–33]. Adjuvant radiotherapy (RT) to the primary site has been shown to improve local control, and data from three pooled prospective trials, which included 88 high-risk MCC patients, showed that pre-radiation margin status (positive/negative) did not have an impact on time to loco-regional failure in patients receiving adjuvant RT [34]. As most MCCs are located in the head-and-neck area, a wide surgical margin is not always feasible and should not be pursued at all costs, but respect functionality and cosmesis, especially as adjuvant RT leads to a high degree of local control. Administration of RT should be carried out within 3 weeks after surgery to minimize disease progression prior to RT [35].

Adjuvant RT may be left out in patients with low-risk characteristics in their primary tumors (Figure S3). These include small primary tumors (≤1 cm diameter), negative margin status, no LVI, negative SLNB and no chronic immunosuppression (i.e., lymphoma/leukemia) [18,19,36]. In a small retrospective study on patients with low-risk head-and-neck primary tumors, adjuvant RT was associated with increased local control without a survival benefit [37]. Since all recurrences were salvaged by radiotherapy, adjuvant RT should not routinely be recommended for this patient subgroup but discussed per case.

Regional lymph nodes: Prophylactic regional RT is not recommended in SLNB-negative patients, as this has not shown to reduce the regional recurrence rate [38].

5.3. Definitive Radiotherapy—Nonresectable Disease

Definitive RT increases disease control but should be reserved for patients who are not candidates for complete, gross resection or refuse surgical intervention. A systematic review including 23 studies found that definitive RT to 136 primary tumor sites resulted in local recurrence rates of 7.6% with a median follow-up time of 24 months. Definitive RT was more effective in managing local disease at the primary tumor site, compared with the regional site (7.6% vs. 16%, $p = 0.02$) [39]. In terms of survival, a study of 50 patients with local disease based on clinical examination and ultrasound treated with definitive RT or conventional treatment (surgery and adjuvant RT) indicated no statistically significant difference in overall ($p = 0.18$) or disease-free survival ($p = 0.32$) between the groups [40]. However, no randomized studies have evaluated the effect of primary surgery and adjuvant RT versus definitive RT. The recommended doses are 56–60 Gy at 2 Gy/d.

Management of the primary tumor summarized:

A 1–2 cm clinical excision margin resulting in negative margins.

Adjuvant RT for primary tumors >1 cm and/or absence of low-risk characteristics (negative surgical margin, negative SLNB, no LVI and no chronic immunosuppression). Adjuvant RT offers disease control, but potential benefit should always be carefully weighed against morbidity and frailty of the patient.

Recommended dose: 50–60 Gy at 2 Gy/d, 5 F/W with 1–2 cm margins.

Definitive RT may be offered to patients who are not candidates for or refuse surgery.

Recommended dose: 56–60 Gy at 2 Gy/d, 5 F/W with 1–2 cm margins.

Prophylactic regional RT in SLNB-negative patients is not recommended.

6. Management of the Regional Lymph Nodes

The evidence on management of patients with nodal disease is particularly scarce, as most studies are retrospective, have too few patients and/or short follow-up periods and often have selection bias in patients receiving adjuvant therapy. Recent large single-institution studies suggest no difference between SLNB-positive patients treated with radiotherapy or therapeutic lymph node dissection (TLND) [41,42]. Patients with nodal disease have poor prognosis and high recurrence rates, compared to patients with localized disease [8,43]. This may warrant an aggressive treatment approach, despite lack of evidence on optimal management of patients with nodal disease.

6.1. Lymph Node Dissection and Locoregional Radiotherapy

All patients with pathologically confirmed regional lymph node metastases should be recommended TLND, whether it be a positive SLN or a palpable/radiologically confirmed metastasis. Additionally, TLND provides prognostic information as SLNB-positive patients with 1 or more positive non-SLNs are associated with a significantly worse prognosis [42]. Adjuvant regional RT is associated with increased morbidity, while the effect of adjuvant RT on OS seems less convincing, as stage III-patients ($n = 2065$) treated with surgery and adjuvant RT were not shown to have a statistically significant improvement in OS, compared to patients treated with surgery alone ($p = 0.80$) [28,36,41]. Indications for post-operative regional RT may therefore be restricted to patients with extracapsular disease to achieve regional control. The recommended doses are 50–60 Gy at 2 Gy/d, 5 F/W (Figure S3) [33]. No randomized controlled trials have compared regional nodal surgery to regional RT. Based on the above considerations and uncertainties, patients with nodal involvement should be evaluated individually in multidisciplinary tumor board consultations.

6.2. Definitive Radiotherapy – Nonresectable Disease

Definitive RT offers clinically meaningful disease control and is indicated for patients, who are not candidates for complete, gross resection or refuse surgical intervention. A systematic review and analysis of 23 studies found that 127 regional nodal sites treated with definitive RT resulted in a 16% recurrence rate with a median follow-up time of 24 months [39]. A prospective trial examining patients treated with macroscopic ($n = 24$) and microscopic ($n = 26$) nodal disease did not show a statistically significant difference in disease-specific survival when patients were treated with definitive RT versus TLND with/without regional RT ($p = 0.9$ and $p = 0.7$, respectively) [44]. The recommended dose to the regional lymph nodes is 56–60 Gy at 2 Gy/d per day. Collectively, definitive RT may offer clinically acceptable primary site and regional disease control for patients with unresectable disease or patients with serious co-morbidity preventing surgical intervention.

6.3. Other Treatment Regimens

Adjuvant chemotherapy is not recommended for resected stage III patients due to lack of association with improved OS ($n = 2065$, $p = 0.71$) [28]. Clinical trials with immunotherapy may be considered, albeit more specific recommendations will await the results of clinical studies, e.g., the ADAM trial—a randomized phase 3 trial with adjuvant avelumab in 100 stage III MCC patients [45].

Management of the regional lymph nodes summarized:

Therapeutic radical lymph node dissection is recommended in MCC patients with regional nodal involvement.

Adjuvant regional RT should be considered in the presence of extracapsular disease.

Recommended dose: 50–60 Gy at 2 Gy/d, 5 F/W.

Definitive RT should be offered to patients, who are not candidates for or refuse surgery.

Recommended dose: 56–60 Gy at 2 Gy/d, 5 F/W.

Patients with nodal involvement should be evaluated in multidisciplinary tumor board conferences.

7. Management of Distant Metastatic Disease

Traditionally, patients with distant metastatic disease were treated with conventional chemotherapy mainly with only palliative effect. There are currently no randomized controlled trials comparing chemotherapy with immunotherapy, however promising clinical trials have resulted in the recommendation of immunotherapy in the first-line setting in several recent guidelines [46,47].

7.1. Immune Checkpoint Inhibitors

Treatment with immune checkpoint inhibitors in the first-line setting is associated with response rates of >50%; some durable responses and are well tolerated [48]. Qualitative interviews with patients ($n = 19$) treated with both first-line chemotherapy and second-line treatment with the immune checkpoint inhibitor avelumab indicate a better quality of life during avelumab treatment [49]. This may be considered in the clinical setting, as MCC patients have a high average age and often multiple comorbidities.

Immunotherapy has been investigated in nonresectable stage IIIB and stage IV patients. There are no known predictive factors, as investigations of specific biomarkers (tumor PD-L1 expression, infiltrating lymphocyte PD-L1 expression, viral status, intratumoral CD8+ infiltration) have been unable to predict clinical response [2,50]. Exclusion criteria include ECOG Performance Status ≥2, steroid use in doses >10 mg prednisone daily, continued need of other immune suppressing agents, organ transplant recipients (heart, lungs and liver) or autoimmune disease with risk of unmanageable flares.

7.2. Avelumab (PD-L1 Antibody)

First-line setting: Treatment with avelumab in patients ($n = 29$) with metastatic MCC, no prior systemic treatment and at least 3 months-follow-up resulted in an objective response rate (ORR) of 62.1% (18/29 patients) with 4 (13.8%) complete responses (CR) and 14 (48.3%) partial responses (PR) [51]. At time of analysis, 14/18 (77.8%) of responses were ongoing with a 6-month response duration in 83% of responding patients. Among 39 patients evaluable for safety analysis, 28/39 (71.8%) experienced grade 1–3 treatment-related adverse events (TRAE) and 6/39 (15.4%) patients discontinued treatment. There were no grade 4 TRAE or treatment-related deaths.

Second-line setting: The largest clinical trial with immune therapy in stage IV MCC patients ($n = 88$) who had progressed after receiving chemotherapy resulted in an ORR of 29/88 (33%), including 19 (21.6%) PR and 10 (11.4%) CR [52]. Among the responders, the majority (74%) had duration of response of ≥1-year, while the 1-year OS rate was 52%. TRAE occurred in 62/80 (70%) of patients, including 4 (5%) patients with grade 3 TRAE. There were no deaths related to treatment [2]. These findings resulted in avelumab becoming the first approved drug for metastatic MCC by the Food and Drug Administration (FDA) and the European Medicines Agency (EMA) [53].

7.3. Pembrolizumab (PD-1 Antibody)

First-line setting: 50 stage IIIB/stage IV patients were administered pembrolizumab with an ORR of 56% (28/50 patients) including 12 with CR and 16 with PR [54]. Of the 28 patients with a confirmed response, the median response duration was not reached (ranged from 5.9 months to 34.5+ months) with an estimated 24 months-progressive-free survival of 48.3%. TRAE occurred in 48 (96%) of patients, including 14 (28%) patients with grade 3–4 TRAE. There was one treatment-related death. Pembrolizumab is approved for MCC in the US but not in Europe.

7.4. Other Immune Checkpoint Inhibitors

Case studies and clinical trials with other immune checkpoint inhibitors such as nivolumab have shown efficacy in treatment of metastatic MCC, but no applications for approval by the FDA or EMA have yet been filed [55].

7.5. Chemotherapy

Chemotherapy used for metastatic MCC is associated with high response rates, but responses are short-lived and the risk of adverse events, such as hematological toxicity and treatment-related death is not negligible [48]. Furthermore, the impact on OS is unclear since there are no comparisons with best supportive care. However, historical data of chemotherapy across regimens and centers over time has not indicated a benefit [28,56].

First-line setting: Response rates range from 53–61% with a median progression-free survival of 3.1 months and duration of response <8 months [48]. Ninety-five percent of patients ($n = 62$) treated may eventually develop progressive disease [57]. The primary recommended treatment regimens include cisplatin or carboplatin in combination with etoposide, as platinum-containing regimens may result in higher rates of complete (21% vs. 17%) and partial response (29% vs. 17%) compared with non-platinum-containing regimens [48,58].

Second-line-setting: Response rates range from 23–45% with a median progression-free survival of 2 months and duration of response <8 months [48].

Later-lines-setting: Chemotherapy in second-line or higher shows response rates of 10–29% with no complete responses. The median duration of response is less than 2 months with a progression-free survival and OS ranging from 2–3 months and 4–5 months, respectively [59,60].

Management of distant metastatic disease summarized:

Immune checkpoint inhibitors targeting the PD-1/PD-L1 axis should be strongly considered as first-line treatment for fit patients with no contraindications.

Avelumab has been approved by the FDA and EMA, whereas pembrolizumab has only been approved by the FDA. Small studies show effect of other checkpoint inhibitors, including nivolumab.

Immune checkpoint inhibitors are associated with high response rates, durable responses and relatively few adverse events.

Chemotherapy is primarily recommended for fit patients with contraindications to immunotherapy or after progression on immunotherapy (second line).

Recommended regimens are cisplatin or carboplatin in combination with etoposide.

No standard treatment can be recommended for patients progressing after chemotherapy; however, fit patients may be candidates for clinical trials.

8. Follow-Up

The patient follow-up exam should include full skin inspection and lymph node examinations with/without US of the regional lymph nodes. As 90% of MCC reoccurrences are seen within 2 years, follow-up is recommended every 3 months during this period, followed by every 6 months for the following 3 years [61]. PET-CT is the imaging modality of choice. It may be conducted one to two times a year or as clinically indicated [16,58]. Follow-up may be individualized based on patient risk factors, as there is no solid evidence for the optimal follow-up strategy in MCC patients. Risk factors, such as virus-negative and/or immune-compromised patients, may warrant closer follow-up due to increased risk of disease progression and MCC-related death [6].

Follow-up summarized:

Five-year follow-up period: Every 3 months for the first 2 years. Every 6 months for the following 3 years.

PET-CT is performed one to two times a year or if clinically indicated.

9. Registration

MCC is a very rare skin cancer with a dismal prognosis compared even to melanoma, and therefore a prospective systematic registration of the patients' clinicopathological characteristics, surgical and oncological treatments and clinical outcome is paramount to improve the patient outcome. A Danish national database is under planning and all medical specialties involved in diagnosis and treatment of these patients will participate in this registration.

10. Methods

The current recommendations were developed by a national, multidisciplinary expert group involved in the management of Merkel cell carcinoma patients. A systematic literature review was performed using a broad search with the following key word "Merkel Cell Carcinoma" in PubMed. The literature search was performed including papers from 1998 to 2019 with exclusion of non-English papers. Additional papers were included if found in reference lists. Relevant websites with guidelines by other MCC groups and organizations such as the European Organisation for Research and Treatment of Cancer and the National Comprehensive Cancer Network were also included. Where no firm conclusions could be made based on the retrieved literature, expert consensus was obtained during discussions in the expert group. A formal evaluation of the evidence level in the retrieved references was not performed. The current work is the first attempt to agree on national guidelines, and future updates will include more formal evaluation of the literature.

11. Conclusions

This review constitutes the consensus management recommendations developed by the Danish MCC expert group and is based on a systematic literature search. MCC is rare, and there is a lack of randomized controlled trials. Patients with localized disease are generally recommended surgical excision with a 1–2 cm margin and adjuvant radiotherapy to the primary site. Clinically node-negative patients should be offered sentinel node biopsy. Patients with regional lymph node involvement are recommended complete node removal and, in case of extracapsular disease, adjuvant radiotherapy. Definitive radiotherapy is recommended for patients not amenable for surgery. Although available data on radiotherapy are conflicting, results from large datasets including a recent metanalysis point to more restricted dosing, ensuring sufficient margins and lack of survival benefit in patients with nodal involvement [28,30–32]. The latter is interpreted as a consequence of early subclinical metastatic spread. Immune checkpoint inhibitors targeting the PD-1/PD-L1 axis should be considered as first-line treatment for fit patients with nonresectable stage IIIB/IV disease.

Prospective comparative data on harmonized patient characteristics and treatment modalities are needed to determine optimal treatment sequence and modus of MCC patients with primary and regional nodal disease. Acknowledging the facts that MCC is highly immunogenic and immunotherapy is considered the new standard treatment in the advanced setting, the future role of immunotherapy in MCC patients with primary and regional disease is highly awaited [45,62].

Supplementary Materials: The following are available online at http://www.mdpi.com/2072-6694/12/3/554/s1, Figure S1: The 8th edition UICC staging system, Figure S2: Treatment algorithm for MCC, Figure S3: Radiotherapy indication and doses for patients with MCC.

Author Contributions: Writing, S.N.; writing, review and editing with focus on surgical sections, M.L.B.-B., L.R.H., A.V.; writing, review and editing with focus on oncology sections S.W.L., E.T., M.L., N.J.; writing, review and editing with focus on pathology sections, T.S., S.K., M.S.L.; writing, review, editing and project administration, A.H.C. All authors have read and agreed to the published version of the manuscript.

Funding: This research received no external funding

Acknowledgments: A special acknowledgement to the members of the Danish MCC interest group: Martin Heje, Pia Sjøgren, Dorte Gad, Helle Skyum, Anne Lene Hagen Wagenblast, Christian Jordening, Pernille Lassen, Klaus Kallenbach, Lene Dissing Sjö, Lars Bastholt, Lars Bjørn Stolle, Ursula Falkmer, Anders Krarup-Hansen, Niels Gyldenkerne, Lars Gorm, Vasudha Rajiv Deshpande. Photography credits to Tina Rasmussen, clinical photographer at Rigshospitalet, Copenhagen University Hospital.

Conflicts of Interest: Merck & Pfeizer sponsored the venue for the first and second Danish MCC Meeting and a 1-day MCC preceptorship at the Charité Hospital in Berlin, Germany. They had no role in the design of the study; in the collection, analyses, or interpretation of data; in the writing of the manuscript, or in the decision to publish the results.

References

1. Feng, H.; Shuda, M.; Chang, Y.; Moore, P.S. Clonal integration of a polyomavirus in human Merkel cell carcinoma. *Science* **2008**, *319*, 1096–1100. [CrossRef]
2. Kaufman, H.L.; Russell, J.; Hamid, O.; Bhatia, S.; Terheyden, P.; D'Angelo, S.P.; Shih, K.C.; Lebbe, C.; Linette, G.P.; Milella, M.; et al. Avelumab in patients with chemotherapy-refractory metastatic Merkel cell carcinoma: A multicentre, single-group, open-label, phase 2 trial. *Lancet Oncol.* **2016**, *17*, 1374–1385. [CrossRef]
3. Lyhne, D.; Lock-Andersen, J.; Dahlstrom, K.; Drzewiecki, K.T.; Balslev, E.; Muhic, A.; Krarup-Hansen, A. Rising incidence of Merkel cell carcinoma. *J. Plast. Surg. Hand Surg.* **2011**, *45*, 274–280. [CrossRef] [PubMed]
4. Ezaldein, H.H.; Ventura, A.; DeRuyter, N.P.; Yin, E.S.; Giunta, A. Understanding the influence of patient demographics on disease severity, treatment strategy, and survival outcomes in merkel cell carcinoma: A surveillance, epidemiology, and end-results study. *Oncoscience* **2017**, *4*, 106–114. [CrossRef] [PubMed]
5. Wong, S.Q.; Waldeck, K.; Vergara, I.A.; Schroder, J.; Madore, J.; Wilmott, J.S.; Colebatch, A.J.; De Paoli-Iseppi, R.; Li, J.; Lupat, R.; et al. UV-associated mutations underlie the etiology of MCV-negative merkel cell carcinomas. *Cancer Res.* **2015**, *75*, 5228–5234. [CrossRef] [PubMed]
6. Moshiri, A.S.; Doumani, R.; Yelistratova, L.; Blom, A.; Lachance, K.; Shinohara, M.M.; Delaney, M.; Chang, O.; McArdle, S.; Thomas, H.; et al. Polyomavirus-Negative merkel cell carcinoma: A more aggressive subtype based on analysis of 282 cases using multimodal tumor virus detection. *J. Investig. Dermatol.* **2017**, *137*, 819–827. [CrossRef]
7. Hasan, S.; Liu, L.; Triplet, J.; Li, Z.; Mansur, D. The role of postoperative radiation and chemoradiation in merkel cell carcinoma: A systematic review of the literature. *Front. Oncol.* **2013**, *3*, 276. [CrossRef]
8. Harms, K.L.; Healy, M.A.; Nghiem, P.; Sober, A.J.; Johnson, T.M.; Bichakjian, C.K.; Wong, S.L. Analysis of prognostic factors from 9387 merkel cell carcinoma cases forms the basis for the new 8th edition ajcc staging system. *Ann. Surg. Oncol.* **2016**, *23*, 3564–3571. [CrossRef]
9. Heath, M.; Jaimes, N.; Lemos, B.; Mostaghimi, A.; Wang, L.C.; Penas, P.F.; Nghiem, P. Clinical characteristics of Merkel cell carcinoma at diagnosis in 195 patients: The AEIOU features. *J. Am. Acad. Dermatol.* **2008**, *58*, 375–381. [CrossRef]
10. Andea, A.A.; Coit, D.G.; Amin, B.; Busam, K.J. Merkel cell carcinoma: Histologic features and prognosis. *Cancer* **2008**, *113*, 2549–2558. [CrossRef]
11. Llombart, B.; Requena, C.; Cruz, J. Update on merkel cell carcinoma: Epidemiology, etiopathogenesis, clinical features, diagnosis, and staging. *Actas Dermo Sifiliogr.* **2017**, *108*, 108–119. [CrossRef] [PubMed]
12. Tetzlaff, M.T.; Harms, P.W. Danger is only skin deep: Aggressive epidermal carcinomas. An overview of the diagnosis, demographics, molecular-genetics, staging, prognostic biomarkers, and therapeutic advances in Merkel cell carcinoma. *Mod. Pathol.* **2020**, *33*, 42–55. [CrossRef] [PubMed]
13. Lebbe, C.; Becker, J.C.; Grob, J.J.; Malvehy, J.; Del Marmol, V.; Pehamberger, H.; Peris, K.; Saiag, P.; Middleton, M.R.; Bastholt, L.; et al. Diagnosis and treatment of Merkel Cell Carcinoma. European consensus-based interdisciplinary guideline. *Eur. J. Cancer* **2015**, *51*, 2396–2403. [CrossRef] [PubMed]

14. van der Ploeg, A.P.; van Akkooi, A.C.; Schmitz, P.I.; Koljenovic, S.; Verhoef, C.; Eggermont, A.M. EORTC Melanoma Group sentinel node protocol identifies high rate of submicrometastases according to Rotterdam Criteria. *Eur. J. Cancer* **2010**, *46*, 2414–2421. [CrossRef]
15. Gershenwald, J.E.; Scolyer, R.A.; Hess, K.R.; Sondak, V.K.; Long, G.V.; Ross, M.I.; Lazar, A.J.; Faries, M.B.; Kirkwood, J.M.; McArthur, G.A.; et al. Melanoma staging: Evidence-based changes in the American Joint Committee on Cancer eighth edition cancer staging manual. *CA Cancer J. Clin.* **2017**, *67*, 472–492. [CrossRef]
16. Hawryluk, E.B.; O'Regan, K.N.; Sheehy, N.; Guo, Y.; Dorosario, A.; Sakellis, C.G.; Jacene, H.A.; Wang, L.C. Positron emission tomography/computed tomography imaging in Merkel cell carcinoma: A study of 270 scans in 97 patients at the Dana-Farber/Brigham and Women's Cancer Center. *J. Am. Acad. Dermatol.* **2013**, *68*, 592–599. [CrossRef]
17. Siva, S.; Byrne, K.; Seel, M.; Bressel, M.; Jacobs, D.; Callahan, J.; Laing, J.; Macmanus, M.P.; Hicks, R.J. 18F-FDG PET provides high-impact and powerful prognostic stratification in the staging of Merkel cell carcinoma: A 15-year institutional experience. *J. Nucl. Med.* **2013**, *54*, 1223–1229. [CrossRef]
18. Gunaratne, D.A.; Howle, J.R.; Veness, M.J. Sentinel lymph node biopsy in Merkel cell carcinoma: A 15-year institutional experience and statistical analysis of 721 reported cases. *Br. J. Dermatol.* **2016**, *174*, 273–281. [CrossRef]
19. Frohm, M.L.; Griffith, K.A.; Harms, K.L.; Hayman, J.A.; Fullen, D.R.; Nelson, C.C.; Wong, S.L.; Schwartz, J.L.; Bichakjian, C.K. Recurrence and Survival in patients with merkel cell carcinoma undergoing surgery without adjuvant radiation therapy to the primary site. *JAMA Dermatol.* **2016**, *152*, 1001–1007. [CrossRef]
20. Gupta, S.G.; Wang, L.C.; Penas, P.F.; Gellenthin, M.; Lee, S.J.; Nghiem, P. Sentinel lymph node biopsy for evaluation and treatment of patients with Merkel cell carcinoma: The Dana-Farber experience and meta-analysis of the literature. *Arch. Dermatol.* **2006**, *142*, 685–690. [CrossRef]
21. Iyer, J.G.; Storer, B.E.; Paulson, K.G.; Lemos, B.; Phillips, J.L.; Bichakjian, C.K.; Zeitouni, N.; Gershenwald, J.E.; Sondak, V.; Otley, C.C.; et al. Relationships among primary tumor size, number of involved nodes, and survival for 8044 cases of Merkel cell carcinoma. *J. Am. Acad. Dermatol.* **2014**, *70*, 637–643. [CrossRef] [PubMed]
22. Hoeller, U.; Mueller, T.; Schubert, T.; Budach, V.; Ghadjar, P.; Brenner, W.; Kiecker, F.; Schicke, B.; Haase, O. Regional nodal relapse in surgically staged Merkel cell carcinoma. *Strahlenther. Onkol. Organ Dtsch. Rontgenges.* **2015**, *191*, 51–58. [CrossRef] [PubMed]
23. Lemos, B.D.; Storer, B.E.; Iyer, J.G.; Phillips, J.L.; Bichakjian, C.K.; Fang, L.C.; Johnson, T.M.; Liegeois-Kwon, N.J.; Otley, C.C.; Paulson, K.G.; et al. Pathologic nodal evaluation improves prognostic accuracy in Merkel cell carcinoma: Analysis of 5823 cases as the basis of the first consensus staging system. *J. Am. Acad. Dermatol.* **2010**, *63*, 751–761. [CrossRef] [PubMed]
24. Yiengpruksawan, A.; Coit, D.G.; Thaler, H.T.; Urmacher, C.; Knapper, W.K. Merkel cell carcinoma. Prognosis and management. *Arch. Surg.* **1991**, *126*, 1514–1519. [CrossRef]
25. Ott, M.J.; Tanabe, K.K.; Gadd, M.A.; Stark, P.; Smith, B.L.; Finkelstein, D.M.; Souba, W.W. Multimodality management of Merkel cell carcinoma. *Arch. Surg.* **1999**, *134*, 388–392, discussion 392–383. [CrossRef]
26. Sattler, E.; Geimer, T.; Sick, I.; Flaig, M.J.; Ruzicka, T.; Berking, C.; Kunte, C. Sentinel lymph node in Merkel cell carcinoma: To biopsy or not to biopsy? *J. Dermatol.* **2013**, *40*, 374–379. [CrossRef]
27. Perez, M.C.; de Pinho, F.R.; Holstein, A.; Oliver, D.E.; Naqvi, S.M.H.; Kim, Y.; Messina, J.L.; Burke, E.; Gonzalez, R.J.; Sarnaik, A.A.; et al. Resection Margins in merkel cell carcinoma: Is a 1-cm margin wide enough? *Ann. Surg. Oncol.* **2018**, *25*, 3334–3340. [CrossRef]
28. Bhatia, S.; Storer, B.E.; Iyer, J.G.; Moshiri, A.; Parvathaneni, U.; Byrd, D.; Sober, A.J.; Sondak, V.K.; Gershenwald, J.E.; Nghiem, P. Adjuvant Radiation therapy and chemotherapy in merkel cell carcinoma: Survival analyses of 6908 cases from the national cancer data base. *J. Natl. Cancer Inst.* **2016**, *108*. [CrossRef]
29. Chen, M.M.; Roman, S.A.; Sosa, J.A.; Judson, B.L. The role of adjuvant therapy in the management of head and neck merkel cell carcinoma: An analysis of 4815 patients. *JAMA Otolaryngol. Head Neck Surg.* **2015**, *141*, 137–141. [CrossRef]
30. Petrelli, F.; Ghidini, A.; Torchio, M.; Prinzi, N.; Trevisan, F.; Dallera, P.; De Stefani, A.; Russo, A.; Vitali, E.; Bruschieri, L.; et al. Adjuvant radiotherapy for Merkel cell carcinoma: A systematic review and meta-analysis. *Radiother. Oncol.* **2019**, *134*, 211–219. [CrossRef]

31. Patel, S.A.; Qureshi, M.M.; Sahni, D.; Truong, M.T. Identifying an optimal adjuvant radiotherapy dose for extremity and trunk merkel cell carcinoma following resection: An analysis of the national cancer database. *JAMA Dermatol.* **2017**, *153*, 1007–1014. [CrossRef] [PubMed]
32. Patel, S.A.; Qureshi, M.M.; Mak, K.S.; Sahni, D.; Giacalone, N.J.; Ezzat, W.; Jalisi, S.; Truong, M.T. Impact of total radiotherapy dose on survival for head and neck Merkel cell carcinoma after resection. *Head neck* **2017**, *39*, 1371–1377. [CrossRef] [PubMed]
33. Yusuf, M.; Gaskins, J.; Wall, W.; Tennant, P.; Bumpous, J.; Dunlap, N. Optimal adjuvant radiotherapy dose for stage I, II or III Merkel cell carcinoma: An analysis of the National Cancer Database. *Jpn. J. Clin. Oncol.* **2019**, *50*, 175–184. [CrossRef] [PubMed]
34. Finnigan, R.; Hruby, G.; Wratten, C.; Keller, J.; Tripcony, L.; Dickie, G.; Rischin, D.; Poulsen, M. The impact of preradiation residual disease volume on time to locoregional failure in cutaneous Merkel cell carcinoma–a TROG substudy. *Int. J. Radiat. Oncol. Biol. Phys.* **2013**, *86*, 91–95. [CrossRef] [PubMed]
35. Tsang, G.; O'Brien, P.; Robertson, R.; Hamilton, C.; Wratten, C.; Denham, J. All delays before radiotherapy risk progression of Merkel cell carcinoma. *Australas. Radiol.* **2004**, *48*, 371–375. [CrossRef] [PubMed]
36. Fields, R.C.; Busam, K.J.; Chou, J.F.; Panageas, K.S.; Pulitzer, M.P.; Allen, P.J.; Kraus, D.H.; Brady, M.S.; Coit, D.G. Recurrence after complete resection and selective use of adjuvant therapy for stage I through III Merkel cell carcinoma. *Cancer* **2012**, *118*, 3311–3320. [CrossRef] [PubMed]
37. Takagishi, S.R.; Marx, T.E.; Lewis, C.; Tarabadkar, E.S.; Juhlin, I.D.; Blom, A.; Iyer, J.G.; Liao, J.J.; Tseng, Y.D.; Fu, T.; et al. Postoperative radiation therapy is associated with a reduced risk of local recurrence among low risk Merkel cell carcinomas of the head and neck. *Adv. Radiat. Oncol.* **2016**, *1*, 244–251. [CrossRef]
38. Grotz, T.E.; Joseph, R.W.; Pockaj, B.A.; Foote, R.L.; Otley, C.C.; Bagaria, S.P.; Weaver, A.L.; Jakub, J.W. Negative sentinel lymph node biopsy in merkel cell carcinoma is associated with a low risk of same-nodal-basin recurrences. *Ann. Surg. Oncol.* **2015**, *22*, 4060–4066. [CrossRef]
39. Gunaratne, D.A.; Howle, J.R.; Veness, M.J. Definitive radiotherapy for Merkel cell carcinoma confers clinically meaningful in-field locoregional control: A review and analysis of the literature. *J. Am. Acad. Dermatol.* **2017**, *77*, 142–148. [CrossRef]
40. Pape, E.; Rezvoy, N.; Penel, N.; Salleron, J.; Martinot, V.; Guerreschi, P.; Dziwniel, V.; Darras, S.; Mirabel, X.; Mortier, L. Radiotherapy alone for Merkel cell carcinoma: A comparative and retrospective study of 25 patients. *J. Am. Acad. Dermatol.* **2011**, *65*, 983–990. [CrossRef]
41. Perez, M.C.; Oliver, D.E.; Weitman, E.S.; Boulware, D.; Messina, J.L.; Torres-Roca, J.; Cruse, C.W.; Gonzalez, R.J.; Sarnaik, A.A.; Sondak, V.K.; et al. Management of sentinel lymph node metastasis in merkel cell carcinoma: Completion lymphadenectomy, radiation, or both? *Ann. Surg. Oncol.* **2019**, *26*, 379–385. [CrossRef] [PubMed]
42. Lee, J.S.; Durham, A.B.; Bichakjian, C.K.; Harms, P.W.; Hayman, J.A.; McLean, S.A.; Harms, K.L.; Burns, W.R. Completion lymph node dissection or radiation therapy for sentinel node metastasis in merkel cell carcinoma. *Ann. Surg. Oncol.* **2019**, *26*, 386–394. [CrossRef] [PubMed]
43. Mehrany, K.; Otley, C.C.; Weenig, R.H.; Phillips, P.K.; Roenigk, R.K.; Nguyen, T.H. A meta-analysis of the prognostic significance of sentinel lymph node status in Merkel cell carcinoma. *Dermatol. Surg. Off. Publ. Am. Soc. Dermatol. Surg.* **2002**, *28*, 113–117, discussion 117. [CrossRef]
44. Fang, L.C.; Lemos, B.; Douglas, J.; Iyer, J.; Nghiem, P. Radiation monotherapy as regional treatment for lymph node-positive Merkel cell carcinoma. *Cancer* **2010**, *116*, 1783–1790. [CrossRef] [PubMed]
45. Adjuvant Avelumab in MCC (ADAM). Available online: https://clinicaltrials.gov/ct2/show/NCT03271372 (accessed on 11 March 2019).
46. Cornejo, C.; Miller, C.J. Merkel cell carcinoma: Updates on staging and management. *Dermatol. Clin.* **2019**, *37*, 269–277. [CrossRef] [PubMed]
47. Robinson, C.G.; Tan, D.; Yu, S.S. Recent advances in Merkel cell carcinoma. *F1000Res* **2019**, *8*. [CrossRef]
48. Nghiem, P.; Kaufman, H.L.; Bharmal, M.; Mahnke, L.; Phatak, H.; Becker, J.C. Systematic literature review of efficacy, safety and tolerability outcomes of chemotherapy regimens in patients with metastatic Merkel cell carcinoma. *Future Oncol.* **2017**, *13*, 1263–1279. [CrossRef]
49. Kaufman, H.; Lambert, J.; Barbosa, C.D.; Guillemin, I.; Mahnke, L.; Bharmal, M. Patient experiences with avelumab vs. chemotherapy for treating MCC: Results from protocol-specified qualitative research. *J. Clin. Oncol.* **2017**, *2017*, e21065. [CrossRef]

50. Nghiem, P.T.; Bhatia, S.; Lipson, E.J.; Kudchadkar, R.R.; Miller, N.J.; Annamalai, L.; Berry, S.; Chartash, E.K.; Daud, A.; Fling, S.P.; et al. PD-1 blockade with pembrolizumab in advanced merkel-cell carcinoma. *N. Engl. J. Med.* **2016**, *374*, 2542–2552. [CrossRef]
51. D'Angelo, S.P.; Russell, J.; Lebbe, C.; Chmielowski, B.; Gambichler, T.; Grob, J.J.; Kiecker, F.; Rabinowits, G.; Terheyden, P.; Zwiener, I.; et al. Efficacy and safety of first-line avelumab treatment in patients with stage iv metastatic merkel cell carcinoma: A preplanned interim analysis of a clinical trial. *JAMA Oncol.* **2018**, *4*, e180077. [CrossRef]
52. Kaufman, H.L.; Russell, J.S.; Hamid, O.; Bhatia, S.; Terheyden, P.; D'Angelo, S.P.; Shih, K.C.; Lebbe, C.; Milella, M.; Brownell, I.; et al. Updated efficacy of avelumab in patients with previously treated metastatic Merkel cell carcinoma after >/=1 year of follow-up: JAVELIN Merkel 200, a phase 2 clinical trial. *J. Immunother. Cancer* **2018**, *6*, 7. [CrossRef] [PubMed]
53. U.S. Food and Drug Administration (FDA). Approval Of Avelumab. Available online: https://www.fda.gov/NewsEvents/Newsroom/PressAnnouncements/ucm548278.htm (accessed on 29 January 2019).
54. Nghiem, P.; Bhatia, S.; Lipson, E.J.; Sharfman, W.H.; Kudchadkar, R.R.; Brohl, A.S.; Friedlander, P.A.; Daud, A.; Kluger, H.M.; Reddy, S.A.; et al. Durable tumor regression and overall survival in patients with advanced merkel cell carcinoma receiving pembrolizumab as first-line therapy. *J. Clin. Oncol.* **2019**, *37*, 693–702. [CrossRef] [PubMed]
55. Walocko, F.M.; Scheier, B.Y.; Harms, P.W.; Fecher, L.A.; Lao, C.D. Metastatic Merkel cell carcinoma response to nivolumab. *J. Immunother. Cancer* **2016**, *4*, 79. [CrossRef] [PubMed]
56. Sabol, A.J.; Markovic, S.; Otley, C. Benefit of chemotherapy in patients with Merkel cell carcinoma with distant metastases and visceral organ involvement. *J. Clin. Oncol.* **2016**. [CrossRef]
57. Iyer, J.G.; Blom, A.; Doumani, R.; Lewis, C.; Tarabadkar, E.S.; Anderson, A.; Ma, C.; Bestick, A.; Parvathaneni, U.; Bhatia, S.; et al. Response rates and durability of chemotherapy among 62 patients with metastatic Merkel cell carcinoma. *Cancer Med.* **2016**, *5*, 2294–2301. [CrossRef] [PubMed]
58. National Comprehensive Cancer Network (NCCN). Guidelines-Merkel Cell Carcinoma. Available online: https://www.nccn.org/professionals/physician_gls/pdf/mcc.pdf (accessed on 16 September 2019).
59. Becker, J.C.; Lorenz, E.; Ugurel, S.; Eigentler, T.K.; Kiecker, F.; Pfohler, C.; Kellner, I.; Meier, F.; Kahler, K.; Mohr, P.; et al. Evaluation of real-world treatment outcomes in patients with distant metastatic Merkel cell carcinoma following second-line chemotherapy in Europe. *Oncotarget* **2017**, *8*, 79731–79741. [CrossRef]
60. Cowey, C.L.; Mahnke, L.; Espirito, J.; Helwig, C.; Oksen, D.; Bharmal, M. Real-world treatment outcomes in patients with metastatic Merkel cell carcinoma treated with chemotherapy in the USA. *Future Oncol.* **2017**, *13*, 1699–1710. [CrossRef]
61. Allen, P.J.; Bowne, W.B.; Jaques, D.P.; Brennan, M.F.; Busam, K.; Coit, D.G. Merkel cell carcinoma: Prognosis and treatment of patients from a single institution. *J. Clin. Oncol. Off. J. Am. Soc. Clin. Oncol.* **2005**, *23*, 2300–2309. [CrossRef]
62. Nivolumab Treatment for Virus-Associated Cancers (CheckMate 358). Available online: https://clinicaltrials.gov/ct2/show/NCT02488759 (accessed on 18 February 2020).

MDPI
St. Alban-Anlage 66
4052 Basel
Switzerland
Tel. +41 61 683 77 34
Fax +41 61 302 89 18
www.mdpi.com

Cancers Editorial Office
E-mail: cancers@mdpi.com
www.mdpi.com/journal/cancers

www.ingramcontent.com/pod-product-compliance
Lightning Source LLC
Chambersburg PA
CBHW061519210926
51850CB00008B/1971